Continuous Probability
Lecture Slide Notes

Ralph E. Morganstern
Santa Clara University

ISBN-13: 978-1490337005
ISBN-10: 1490337008

Table of Contents

Table of Contents

Table of Contents

Table of Contents

Table of Contents

Preface

These Lecture Slide Notes have been used over the past several years for a two-quarter graduate level sequence in probability for engineers. Most textbooks delay presentation of some key probability concepts until later chapters where they are covered together in their discrete and continuous forms. Here we include these concepts in Part 1 on Discrete Probability so that students taking only a single quarter/semester walk away with a complete picture. Besides providing a degree of completeness for those students, this allows the transition to Part 2 on Continuous Probability to go forward with essentially no new concepts! In this manner, the increased level of mathematical sophistication encountered in the continuous domain, is not compounded by the introduction of unfamiliar concepts. Part 2 presupposes a working knowledge of the discrete probability concepts covered in Part 1 but is otherwise self-contained. The differential probability in an interval dx is determined by a continuous probability density function (PDF) which integrates to yield the cumulative distribution function (CDF). The concepts of joint, conditional, and marginal densities, expected values, and independence are easily transitioned to the continuous domain by emulating their discrete counterparts. The transformation between continuous probability densities is given a unique representation in terms of a composite 3-dimensional plot showing the before and after probability densities as well as the coordinate transformation curve. Both the Jacobian determinant and CDF transformation methods are covered with careful consideration of the integration and differentiation procedures involved. The CDF method for RV data simulation is motivated by a 3-dimensional plot using a "sample and hold" analog to digital coordinate transformation to generate a discrete (sampled) representation of a continuous distribution. Moment generating functions, RV sums, convolution, and "order statistics", are covered in the continuous domain, again with reference to their discrete counterparts. The distinction between counting the number events and the time between their arrivals are discussed as two complementary aspects of random processes. Continuous distributions and their relationship to limiting forms of discrete distributions are illustrated with a number of transition charts as well as a comparison of common discrete and continuous distributions. The central limit theorem, bounds for unknown distributions, and approximation methods relating sums of discrete RVs to Poisson, Gaussian, and r-Erlang estimates are also discussed. The Bivariate Gaussian distribution, its ellipses of concentration, eigenvalues, eigenvectors, and its interpretation in terms of a Bayesian measurement update for the conditional mean lead directly to the Gauss-Markov Theorem; the extension to a multivariate Gaussian distribution yields a powerful tool for multiple measurement updates in a Gaussian arena. As in Part 1 all concepts are illustrated with many examples, where the emphasis is on multiple visualizations using trees, graphical illustrations, Venn diagrams, comparison tables, coordinate axis transformations as well as integration strip illustrations, special 3-dimensional plots, and matrix partitioning graphics.

This "Lecture Slide Notes" format is convenient for self-study because it covers the subject matter in a concise and easily accessible manner by employing multiple visualization techniques in slide format together with focused explanatory notes. Each slide stands alone as a "one-page synopsis" that encapsulates a complete concept, algorithm, or theorem using a combination of equations, graphs, diagrams, and comparison tables. The explanatory notes are placed directly below each slide in order to reinforce and/or give additional insight into the particular technique or concept illustrated in that slide. A Table of Contents serves to organize the slides in terms of the main probability topics covered and gives a list of all slide titles with their page numbers. An index is also provided to link related aspects of topics and cross-reference key concepts, specific applications, and the various visualization aids. Although no problem sets have been included in these notes, a good number of examples are worked out in detail. References to a number of standard text books are given, but there has been no attempt to make an exhaustive bibliography.

1 Transition to Continuous Random Variables

Transition to Continuous Random Variables

1.1 Continuous, Discrete, and Mixed RVs

Continuous, Discrete, and Mixed RVs

Probability Density Function (PDF)

$Event:$ $\mathcal{E} = \{x : a \le x \le b\}$

$$\Pr[x \in \mathcal{E}] = \int_{\mathcal{E}} f_X(x)dx = \int_a^b f_X(x)dx$$

$$\Pr[x = 2.0] = \int_{x=2.0}^{2.0} f_X(x)dx = 0$$ Prob at a point = 0

Except for δ-fcn at a point

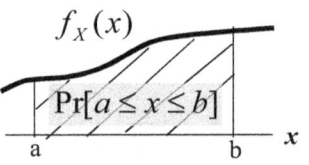

Mixed Continuous & Discrete Outcomes – Dirac δ-fcn

$$f_X(x) = \alpha\delta(x - x_0) + \frac{\beta}{(b-a)}$$

$$\int_a^b \alpha\delta(x - x_0)dx = \int_{x_0-\varepsilon}^{x_0+\varepsilon} \alpha\delta(x - x_0)dx = \alpha$$

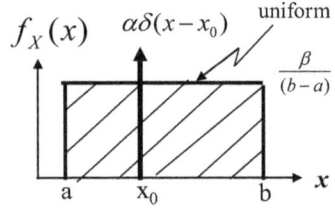

Sampled Continuous Fcn g(x)

$$f_X(x) = \sum_{k=0}^n \alpha_k\delta(x - x_k)$$

$$\alpha_k = \int_a^b g(x)\delta(x - x_k) = g(x_k)$$

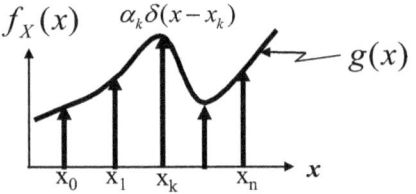

In discrete probability a random variable (RV) is characterized by its *probability mass function* (PMF) $p_X(x)$ which specifies the amount of probability associated with each point in the discrete sample space. Continuous probability generalizes this concept to a *probability density function* (PDF) $f_X(x)$ at each point in the continuous sample space. Probability has no meaning at a single point; rather a *differential probability* is defined over an interval dx to be $dP = f_X(x)dx$. Just as the sum of $p_X(x)$ over the whole sample space must be unity, now the integral of the density $f_X(x)$ over the whole sample space must also be unity. An event E is defined by a sum or integral over a portion of the sample space as shown by the shaded area in the upper figure between x=a, and x=b.

The middle panel gives an example of a mixed distribution containing both a continuous uniform distribution $\beta/(b-a)$ over the interval [a, b] and a Dirac δ-function, $\alpha\cdot\delta(x-x_0)$ corresponding to a discrete contribution at the point x_0. The uniform distribution is shown as a continuous horizontal line at "height" $\beta/(b-a)$ between a and b, and the Dirac δ-function is shown as an arrow corresponding to a probability mass "α" accumulated at a single point $x=x_0$. The integral over the continuous part gives $(b-a)\cdot\beta/(b-a) = \beta$ and the integral of the Dirac δ-function $\alpha * \delta(x-x_0)$ over *any interval containing* x_0 yields α. Thus, for this to be a valid probability density function, we require the sum of the two contributions be unity or $\alpha + \beta = 1$. Consider the continuous density given by the curve $f_X(x) = g(x)$ in the bottom panel and take the sum of products $f_X(x) = \sum\alpha_k\cdot\delta(x-x_k)$ corresponding to the δ-function arrows at arbitrary points on the curve g(x). Is this a valid discrete "PMF"? In order for this to be so, the sum of the contributions α_k must be unity. This makes it a valid density, but it does not necessarily represent a "sampled" version of the original function g(x). To do this, we need to develop a "sampling" transformation $Y_k=Y_k(X)$ for k=0,1,2,...,n so as to transform the original continuous $f_X(x)$ to a discrete $f_Y(y_k)$ (See Slide#2-14).

Transition to Continuous Random Variables

1.1.1 Dirac δ-function Representations

Dirac δ-function Representations

$$\delta(x) = \lim_{\varepsilon \to 0} \frac{1}{\pi} \cdot \frac{\varepsilon}{\varepsilon^2 + x^2} \qquad (1)$$

$$\delta(x) = \lim_{\varepsilon \to 0} \frac{1}{\sqrt{\pi}} \cdot \frac{e^{-x^2/\varepsilon^2}}{\varepsilon} \qquad (2)$$

$$\delta(x) = \lim_{N \to \infty} \frac{1}{\pi} \cdot \frac{\sin Nx}{x} \qquad (3)$$

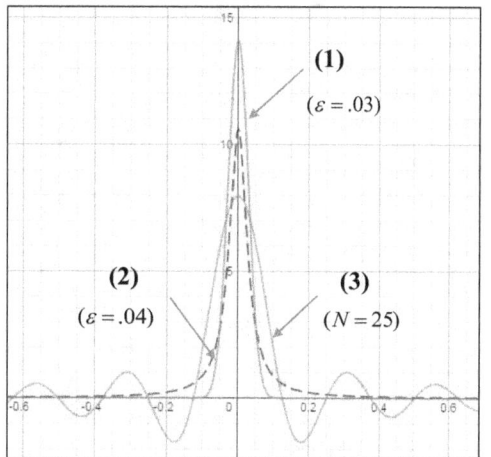

(1) ($\varepsilon = .03$)

(2) ($\varepsilon = .04$) (3) ($N = 25$)

$$(1) \quad \int_{-\infty}^{+\infty} \delta(x)\,dx = \frac{1}{\pi} \, L_{\varepsilon \to 0} \int_{-\infty}^{+\infty} \frac{\varepsilon}{\varepsilon^2 + x^2} \cdot dx$$

Substitute: $x = \varepsilon \tan\theta \quad ; \quad dx = \varepsilon \sec^2\theta \, d\theta$

$$= \frac{1}{\pi} \, L_{\varepsilon \to 0} \int_{-\pi/2}^{+\pi/2} \frac{\varepsilon \cdot \varepsilon \sec^2\theta \, d\theta}{\varepsilon^2 (1 + \tan^2\theta)} = \frac{1}{\pi}\left(\frac{\pi}{2} - \frac{-\pi}{2}\right) = 1$$

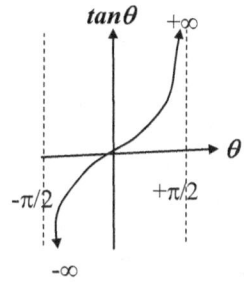

The Dirac δ-function $\delta(x-x_0)$ integrates to unity, while the product $f(x) \cdot \delta(x-x_0)$ evaluates the function at the point x_0 and thus integrates to $f(x_0)$ whenever the integration range *surrounds the point* x_0. Because integration of the product of any function $f(x)$ with a delta function $\delta(x-x_0)$ simply evaluates that function at $x = x_0$, the δ-function is often called a "sifting" function. The Kronecker delta δ_{ij} has a similar effect in sums as it sifts out the single term for which $i=j$.

Three of the many approximate representations of the Dirac δ-function are given in this slide as limiting forms; they all have the limiting characteristics of the δ-function, namely, (i) $\delta(x)$ is zero everywhere except at $x=x_0$ at which point it is infinite, and (ii) $\delta(x)$ integrates to unity for all intervals surrounding the point x_0.

Representation (1) $\pi^{-1} \varepsilon /(\varepsilon^2+x^2)$ integrates to unity by making a simple trigonometric substitution and taking the limit as $\varepsilon \to 0$. Also note that in the limit $\varepsilon \to 0$ this function is zero everywhere except at $x = 0$ where it is infinite, thus satisfying criterion (i).

Representation (2) is the limit of a Gaussian with zero mean and standard deviation $\sigma = \varepsilon /2^{1/2}$ and as $\varepsilon \to 0$ it is seen to approach zero everywhere, except at $x=0$ where it approaches infinity. Making the substitution $\varepsilon = 2^{1/2}\sigma$ into (2) yields a standard Gaussian density $\exp(-x^2/2\sigma^2) / \sigma(2\pi)^{1/2}$, which integrates to unity for any σ. Note that in the limit $\sigma \to 0$ the Gaussian becomes "infinitesimally thin" having an infinite value at $x=0$ and zero elsewhere, *i.e.,* it is a δ- function!

The proof of representation (3) is found in standard texts on Fourier analysis and digital processing.

1.1.2 Relationship Between the Heaviside "Step" Function and Dirac δ-function

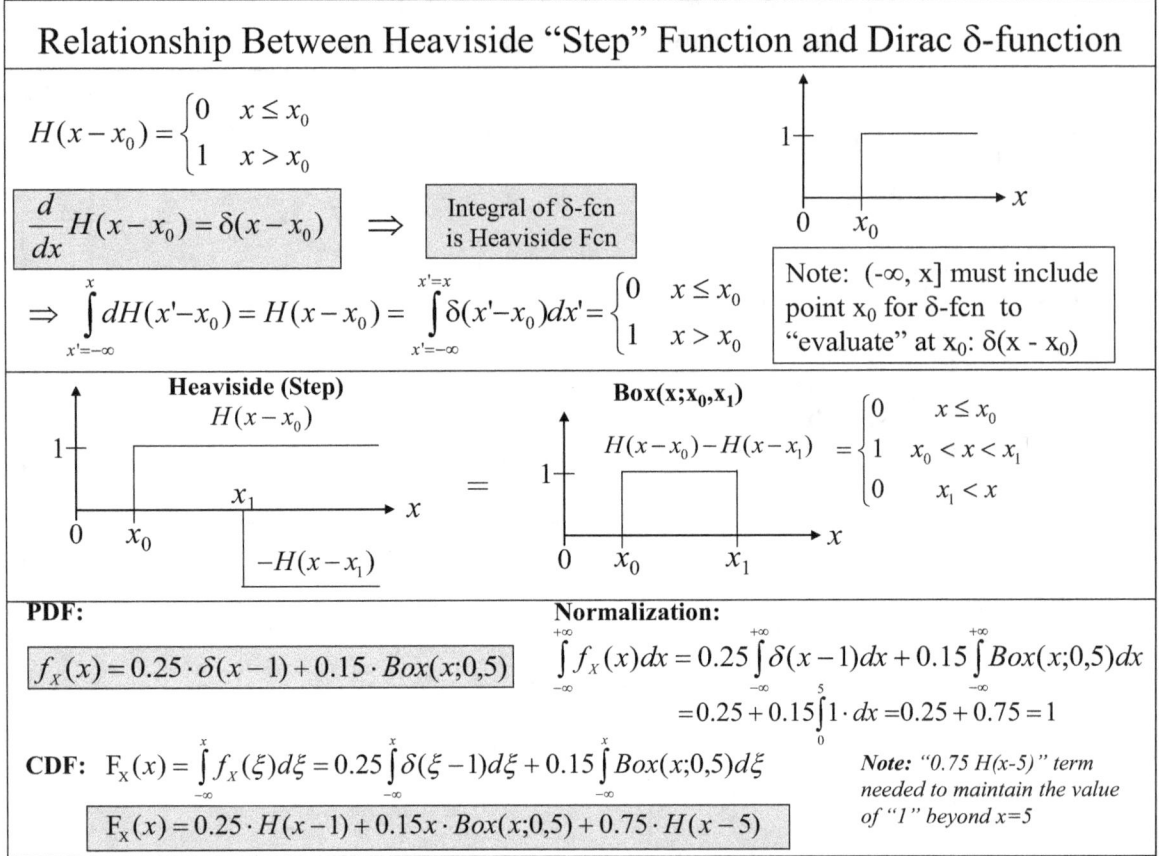

Relationship Between Heaviside "Step" Function and Dirac δ-function

$$H(x-x_0) = \begin{cases} 0 & x \le x_0 \\ 1 & x > x_0 \end{cases}$$

$$\frac{d}{dx}H(x-x_0) = \delta(x-x_0) \quad \Rightarrow \quad \boxed{\text{Integral of } \delta\text{-fcn is Heaviside Fcn}}$$

$$\Rightarrow \int_{x'=-\infty}^{x} dH(x'-x_0) = H(x-x_0) = \int_{x'=-\infty}^{x'=x} \delta(x'-x_0)dx' = \begin{cases} 0 & x \le x_0 \\ 1 & x > x_0 \end{cases}$$

Note: $(-\infty, x]$ must include point x_0 for δ-fcn to "evaluate" at x_0: $\delta(x - x_0)$

Heaviside (Step) $H(x-x_0)$

$$H(x-x_0) - H(x-x_1) = \begin{cases} 0 & x \le x_0 \\ 1 & x_0 < x < x_1 \\ 0 & x_1 < x \end{cases}$$

PDF:

$$\boxed{f_X(x) = 0.25 \cdot \delta(x-1) + 0.15 \cdot Box(x;0,5)}$$

Normalization:

$$\int_{-\infty}^{+\infty} f_X(x)dx = 0.25\int_{-\infty}^{+\infty}\delta(x-1)dx + 0.15\int_{-\infty}^{+\infty} Box(x;0,5)dx$$
$$= 0.25 + 0.15\int_0^5 1 \cdot dx = 0.25 + 0.75 = 1$$

CDF: $F_X(x) = \int_{-\infty}^{x} f_X(\xi)d\xi = 0.25\int_{-\infty}^{x}\delta(\xi-1)d\xi + 0.15\int_{-\infty}^{x} Box(x;0,5)d\xi$

Note: "0.75 H(x-5)" term needed to maintain the value of "1" beyond x=5

$$\boxed{F_X(x) = 0.25 \cdot H(x-1) + 0.15x \cdot Box(x;0,5) + 0.75 \cdot H(x-5)}$$

Another important mathematical function is the Unit Step or Heaviside function $H(x-x_0)$ shown in the top panel. This function is zero for x less than or equal to x_0 and thereafter it is unity (creating the step). The derivative of the Heaviside function $dH(x-x_0)/dx$ is the Dirac delta function $\delta(x-x_0)$. This is easily verified in two steps (i) integrate the LHS or the equality using the fundamental theorem of calculus to yield $H(x- x_0)$ and (ii) integrate the RHS using the properties of the integrated delta function which requires the integration interval to "surround the point x_0"; thus it integrates to zero for $x \le x_0$ and to one for $x > x_0$, which is precisely the definition of the Heaviside function. (Q.E.D.).

The middle panel illustrates how the subtraction of two Heaviside functions $H(x-x_0) - H(x-x_1)$ yields the box-shaped pulse between x_0 and x_1 (or $Box(x; x_0,x_1)$) by a simple graphical superposition of the two Heaviside functions one above the axis and one below the axis. Obviously changing the order of the two terms flips the box below the axis (not shown).

A typical example is given in the bottom panel. Note in the CDF "$0.25 \cdot H(x-1)$" starts the step at x=1 and continues out to $+\infty$ while the term "$0.15 \cdot Box(x; 0,5)$" ramps from 0.0 to 0.75 over the interval [0,5] bringing the sum of the first two terms to "1". However, beyond x=5 the "$0.15 \cdot Box(x; 0,5)$" is zero, so we must add the last term "$0.75 \cdot H(x-5)$ with maintains the CDF function at the value of "1" beyond x=5 out to $+\infty$.

1.1.3 Properties of PDF: Continuous and Discrete

Properties of PDF: Continuous and Discrete

	Continuous	Discrete
	Prob Density Fcn (PDF)	**Prob Mass Fcn (PMF)**
Normalization	$\int\limits_{-\infty}^{+\infty} f_X(x)dx = 1$	$\sum\limits_{x} p_X(x) = 1$
Positive	$f_X(x) \geq 0 \quad \text{all } x$	$p_X(x) \geq 0 \quad \text{all } x$
Prob. Interpr	$f_X(x)dx = \Pr[x \leq X \leq x+dx]$	$p_X(x) = \Pr[X = x]$

The table summarizes the normalization, positivity, and probability interpretation of continuous PDF and discrete PMF densities; the notation *pneumonic* is "p" for points as in $p_X(x)$ and "f" for function as in $f_X(x)$. We note that the sum of a discrete PMF over the entire sample space is unity translates to the integral of a continuous PDF over the entire sample space is unity; also both the densities must be ***non-negative*** in order to have a probability interpretation. *Thus, any calculation that leads to a **negative density** is incorrect – no exceptions!*

Finally, a discrete probability distribution evaluates the "mass" at a given point X=x,
$$p_X(x)= \Pr[X = x]$$
while a continuous probability density must be multiplied by a differential dx and has the interpretation of the probability "over an interval" between x and x+dx
$$f_X(x)dx = \Pr[x \leq X \leq x+dx]$$
Thus for a continuous distribution, the probability at a point is meaningless; it does not even have the correct dimensionless units because probability density is probability per length and must therefore be multiplied by a length to be dimensionless!

A discrete distribution can be represented as the sum of products $\sum \alpha_k \cdot \delta(x-x_k)$ where the α_k are the PMF values at the points x_k, and the properties of the δ-function inside an integral allow us to treat each term in the sum as if it were a continuous function. Conversely, a continuous distribution can always be "sampled" and re-written as an appropriate weighted sum $\sum \alpha_k \cdot \delta(x-x_k)$. However, in this case the coefficients α_k. are not *ad hoc*, but rather are obtained by integrating the original continuous distribution over the sample segments to obtain local concentrations at the sampling points as shown explicitly in Slide# 2-14.

1.2 *Cumulative Distribution Function (CDF)*

Cumulative Distribution Function (CDF)

$$F_X(x) = \Pr[X \le x] = \int_{x'=-\infty}^{x} f_X(x')dx'$$

Bdy Values: $F_X(-\infty) = 0$; $F_X(+\infty) = 1$

Monotone Non-decr.: $F_X(b) \ge F_X(a)$; if $b \ge a$

Prob Interpretation: $\Pr[a \le x \le b] = F_X(b) - F_X(a)$

Density PDF: $f_X(x) = \frac{d}{dx} F_X(x)$

or, $dF_X(x) = F_X(x+dx) - F_X(x) = f_X(x)dx$

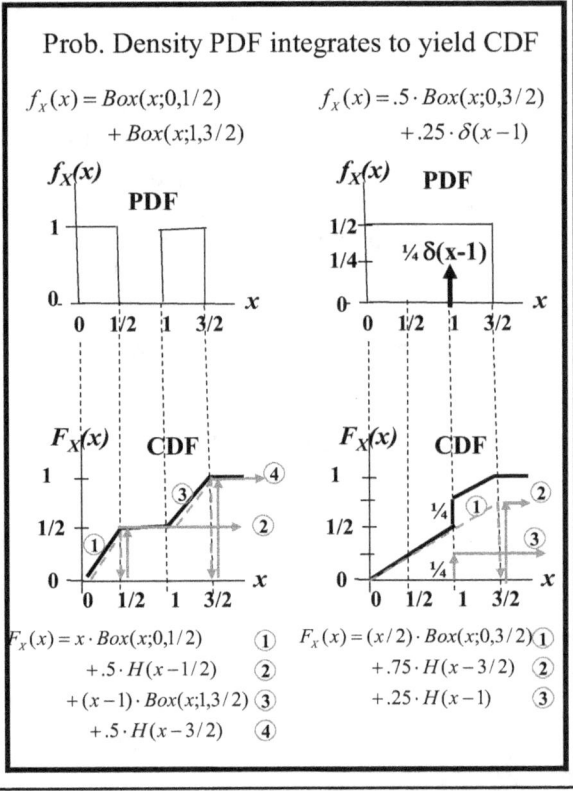

Prob. Density PDF integrates to yield CDF

$f_X(x) = Box(x;0,1/2)$
$\qquad + Box(x;1,3/2)$

$f_X(x) = .5 \cdot Box(x;0,3/2)$
$\qquad + .25 \cdot \delta(x-1)$

$F_X(x) = x \cdot Box(x;0,1/2)$ ①
$\qquad + .5 \cdot H(x-1/2)$ ②
$\qquad + (x-1) \cdot Box(x;1,3/2)$ ③
$\qquad + .5 \cdot H(x-3/2)$ ④

$F_X(x) = (x/2) \cdot Box(x;0,3/2)$ ①
$\qquad + .75 \cdot H(x-3/2)$ ②
$\qquad + .25 \cdot H(x-1)$ ③

The cumulative distribution function (CDF) for a continuous probability density function $f_X(x)$ is defined in a manner similar to that for discrete distributions $p_X(x)$ except that the cumulative sum over a discrete set is replaced by an integral over all X less than or equal to a value x. This integral yields the cumulative probability function of "x," $F_X(x) = \Pr[X \le x]$ which has the following important properties
(i) $F_X(x)$ always starts at 0 and ends at 1
(ii) $F_X(x)$ is continuous,
(iii) $F_X(x)$ is non-decreasing,
(iv) $F_X(x)$ is invertible; *i.e.*, $F_X^{-1}(x)$ exists, and
(v) The density $f_X(x) = d/dx\{F_X(x)\}$ (since exact differential $dF_X(x) = F_X(x+dx) - F_X(x) = f_X(x)dx$)
It is important to note all five properties of $F_X(x)$ as they have important consequences.
The figure shows the relationship between the density $f_X(x)$ and the cumulative distribution $F_X(x)$ for two cases
(i) two regions of constant density (two "boxes") and (ii) one region of constant density plus a delta function (one "box" and an arrow "spike") .
In case (i) $F_X(x)$ ramps linearly from a value of 0 to ½ in the interval [0, ½] because of the 1st constant density box, then remains constant at the value ½ over the interval [½ , 1], and finally ramps from ½ to 1 because of the 2nd constant density box. Note that the slopes of the two ramps are both "1" in this case and the total area under the density curves is unity: $1 \cdot [1/2-0] + 1 \cdot [3/2-1] = 1$.
In case (ii) $F_X(x)$ ramps from a value of 0 to ½ in the interval [0, 1] by virtue of the constant "½" density box, then jumps by "1/4" because of the delta function, and finally continues its ramp from the value ¾ to 1. Note that this is simply the superposition of a constant density of " ½" plus a delta function ¼* δ(x-1), and the total area under the density curves is unity: $(1/2) \cdot [3/2 - 0] + 1/4 = 1$. Note that the additional Heaviside functions are to ensure "persistence" of the last accumulated values because the box functions drop to zero as shown.

1.3 *Joint Probability Density Functions*

Joint Probability Density Functions

Joint PDF : $f_{XY}(x, y)$

$$Pr[E] = \iint_E f_{XY}(x, y)dxdy$$

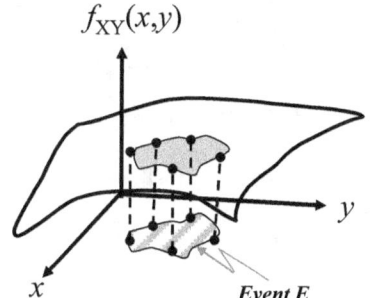

$f_{XY}(x,y)$

Marginal PDFs :

Pr[E] = Volume formed by curve and dotted lines to $f_{XY}(x,y)$ surface

$$f_X(x) = \int_{-\infty}^{+\infty} f_{XY}(x, y)dy$$

$$f_Y(y) = \int_{-\infty}^{+\infty} f_{XY}(x, y)dx$$

Joint PDF Properties

Positivity : $f_{XY}(x, y) \geq 0$ all (x,y)

Normalization : $\int_{-\infty}^{\infty} dx \int_{-\infty}^{\infty} dy\, f_{XY}(x, y) = 1$

Probability Interpretation :

$$Pr[x \leq X \leq x + dx, y \leq Y \leq y + dy] = f_{XY}(x, y)dxdy$$

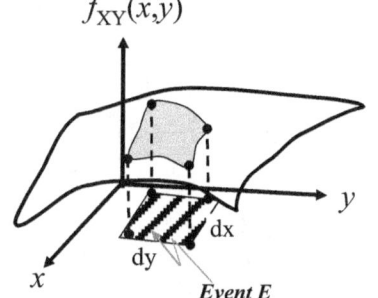

$f_{XY}(x,y)$

The joint PDF $f_{XY}(x,y)$ is a surface above the x-y plane whose z-values represent the probability density at each point in the x-y plane. An event E is a closed region in the x-y plane and the volume under the projection of the region E onto the probability density surface is simply the integral of differential probability $dP = f_{XY}(x,y)\, dxdy$ over the region E in the x-y plane; the integral yields the probability of the event E. The joint PDF has the following properties

(i) $f_{XY}(x,y)$ is non-negative for all (x,y),

(ii) integration over all (x,y) yields unity, and

(iii) the differential probability dP in a "box region" between x and x+dx and y and y+dy is simply the probability density multiplied by the differential area, *viz.*, $f_{XY}(x,y)\, dxdy$.

Marginal PDFs are defined as "sum projections" onto either the x- or y-axes in a manner analogous to that for discrete joint distributions where now the sums become integrals. Note that, as with discrete distributions, the marginal densities $f_X(x)$ and $f_Y(y)$ are new probability densities of a single variable and therefore must integrate to unity.

Transition to Continuous Random Variables

1.3.1 Joint PDF Example

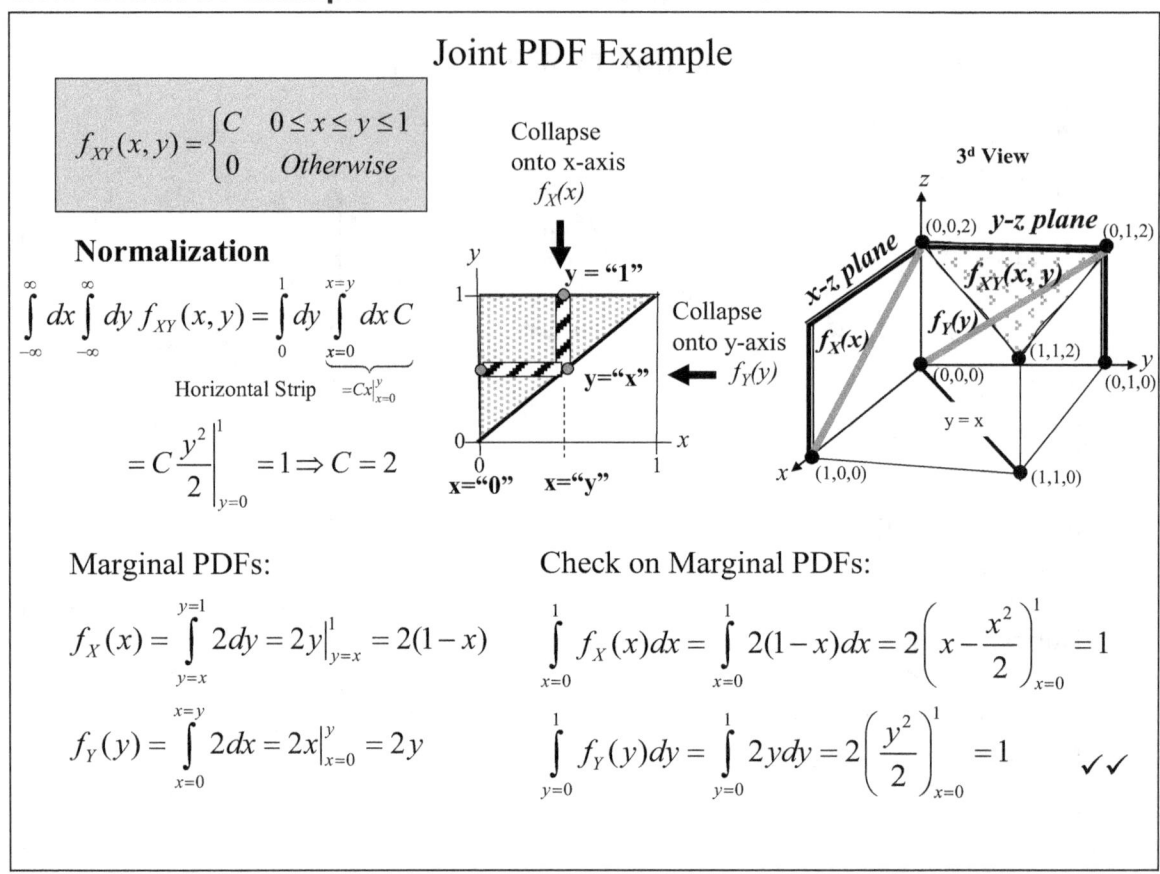

In this example the joint PDF is a positive constant C>0 over the upper part of the "unit square" above the line y=x. In order for this to be a valid density function it must be normalized, *i.e.*, it must integrate to unity. Since the density is a constant the normalization integration just gives half the area of the unit square times C: ½ $(1^2) \cdot C = 1$ or C=2. However, we have formally set up the integral using horizontal strips from x = "0" to x = "y" and equivalently with vertical strips from y ="x" to y ="1" because they are useful in the more general cases in which the PDF is not constant and varies with x and/or with y; moreover, they are precisely the set-up needed for computing the marginal distributions below.

The marginal $f_X(x)$ uses the vertical strips to collapse ("aggregate") all the probability contributions $f_{XY}(x,y)dy$ onto the x-axis by integrating between y ="x" to y ="1" to yield $f_X(x) = 2(1-x)$ for x∈[0,1] and 0 otherwise.

Similarly, the marginal $f_Y(y)$ uses the vertical strips to collapse ("aggregate") all the probability contributions $f_{XY}(x,y)dx$ onto the y-axis by integrating between x = "0" to x = "y" to yield $f_Y(y) = 2y$ for y∈[0,1] and 0 otherwise. The 3[d] plot shows the original joint uniform density as the shaded triangular region above the x-y plane as well as the two linear marginal densities in the x-z and y-z planes, the former $f_X(x)$ ramping down from a peak value of 2 and the latter $f_Y(y)$ ramping up to a peak of 2.

Note that it is the *integration limits* which give structure to the marginal distributions and make them functions of x and y respectively even though we started with the constant density of 2. If the density had been a constant over the "whole" unit square instead of just the upper half, then C would be 1 and the two marginals would also be constant distributions with constant density of 1.

We can check to see that the marginals $f_X(x)$ and $f_Y(y)$ are properly normalized to unity by integrating them over all x and y respectively.

1.3.2 Joint PDF Example – Probability of Specific Event E

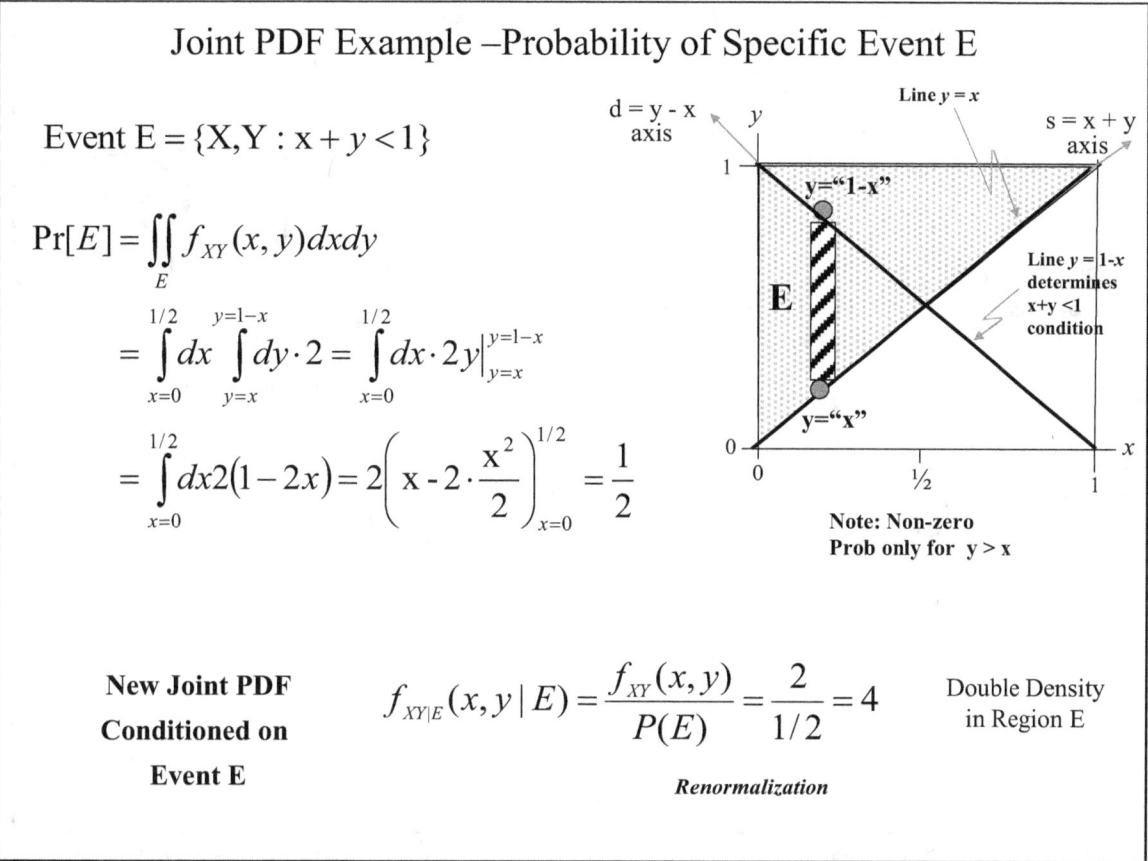

Joint PDF Example –Probability of Specific Event E

Event $E = \{X,Y : x+y < 1\}$

$$Pr[E] = \iint_E f_{XY}(x,y)\,dx\,dy$$

$$= \int_{x=0}^{1/2} dx \int_{y=x}^{y=1-x} dy \cdot 2 = \int_{x=0}^{1/2} dx \cdot 2y\Big|_{y=x}^{y=1-x}$$

$$= \int_{x=0}^{1/2} dx\,2(1-2x) = 2\left(x - 2\cdot\frac{x^2}{2}\right)_{x=0}^{1/2} = \frac{1}{2}$$

Note: Non-zero Prob only for $y > x$

New Joint PDF Conditioned on Event E

$$f_{XY|E}(x,y\,|\,E) = \frac{f_{XY}(x,y)}{P(E)} = \frac{2}{1/2} = 4$$

Double Density in Region E

Renormalization

Consider the same PDF as last slide and define the event E to be all points for which X+Y<1; geometrically this is determined by points with non-zero density that are below the anti-diagonal line y=1-x (negative slope) as shown in the figure. Inasmuch as the probability density is constant over the entire region, and the Event E takes up exactly ½ of the total geometrical area, it is obvious that the probability of the event E must be Pr[E] = ½. Using the definition of conditional joint probability on the next slide, we easily obtain the new joint PDF by renormalization as shown at the bottom of this slide.

Alternately, we can directly verify this result by performing the double integral of the constant density "2" over the region defined by the event E taking care to put in the appropriate integration limits. We choose a *vertical strip* and integrate over y first to obtain 2·y evaluated between y = "x" and y ="1-x" . This yields a function of x only "2(1-2x)" which is to be integrated over the limits x = "0" to x = "½" and evaluates to Pr[E] = ½ yielding the same value as the geometrical argument.

Note that although geometrical arguments are useful when the density is constant, they can no longer be used for more general density functions that depend upon the x and y coordinates.

Also note that sometimes the x and y variations are separable which allows independent integrations over horizontal and vertical strips, respectively; in general, however, we have to perform the two-dimensional integral with great care using appropriate coordinate limits involving the second variable.

Finally, note that for marginal distributions, we have a one-dimensional integral and **always use vertical or horizontal coordinate strips** to "aggregate" the joint density function contributions on either the x- or y-axes as appropriate.

1.4 Conditional Joint PDF

Conditional Joint PDF

Joint Probability Density Conditioned on Event A:

$$f_{XY|A}(x,y\,|\,A) = \begin{cases} \dfrac{f_{XY}(x,y)}{P(A)} & x,y \in A \\ 0 & otherwise \end{cases}$$

Recall Defn of Cond. Prob for Events

$$P(B\,|\,A) = \frac{P(A,B)}{P(A)}$$

Define "differential" Events A & B as:

$$A = Y \in [y, y+dy]$$
$$B = X \in [x, x+dx]$$

Substitution yields relation between joint and conditional densities:

$$f_{X|Y}(x\,|\,y)dx = \frac{f_{XY}(x,y)dxdy}{f_Y(y)dy} = \left[\frac{f_{XY}(x,y)}{f_Y(y)}\right]dx$$

$$f_{X|Y}(x\,|\,y) = \frac{f_{XY}(x,y)}{f_Y(y)} \qquad \Leftrightarrow \qquad p_{X|Y}(x\,|\,y) = \frac{p_{XY}(x,y)}{p_Y(y)}$$

| Continuous | Discrete |

The **Conditional Joint Density PDF** in the continuous case is defined in a manner analogous to the discrete conditional joint PMF where we simply renormalized the density by the probability of the conditioning event; thus, given a conditioning event A whose probability is P(A), we define the joint conditional density $f_{XY|A}(x,y|A)$ as the renormalized density $f_{XY}(x,y)$ / P(A) for all (x,y)ε A and zero otherwise. It is easy to check that this definition is properly normalized since the integral of $f_{XY}(x,y)$ over all (x,y) that are in the event A yields P(A); subsequent division by P(A) obviously yields unity.

The conditional probability density obtained by conditioning one variable, say x, on the second variable y, *i.e.,* $f_{X|Y}(x|y)$ is related to the underlying joint density $f_{XY}(x,y)$ by simply dividing by the *conditioning marginal* $f_Y(y)$ as given in the shaded boxed equation (along with its discrete counterpart.)

Note that this expression relating densities has a simple interpretation in terms of actual probabilities as follows:

(i) define events:{**A**:Y ε [y, y+dy]} , {**B**:X ε [x, x+dx]} and {**AB**:Y ε [y, y+dy]; X ε [x, x+dx]}
(ii) differential probabilities: P[A] = $f_Y(y)dy$, P[B] = $f_X(x)dx$, P[AB] = $f_{XY}(x,y)dxdy$
(iii) conditional probability definition: P[B|A] = P[A,B] / P[A]
(iv) substitution of (ii) into (iii) yields: $f_{X|Y}(x|y)$ dx = $f_{XY}(x,y)dxdy$ / $f_Y(y)dy$
(v) cancelling differentials yields: $f_{X|Y}(x|y) = f_{XY}(x,y)$ / $f_Y(y)$

1.4.1 Conditional Joint PDF Example: Density 4xy, Conditioning Event Y>X

Conditional Joint PDF Example: Density 4xy, Conditioning Event Y>X

Joint Prob Density:

$$f_{XY}(x,y) = \begin{cases} Cxy & x,y \in [0,1] \\ 0 & otherwise \end{cases}$$

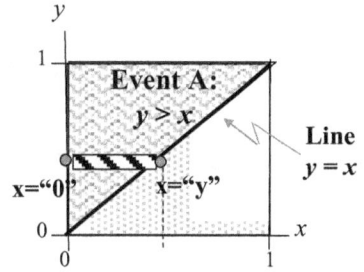

Normalization:

$$1 = \int_{x=0}^{x=1} dx \int_{y=0}^{y=1} dy \cdot Cxy = \int_{x=0}^{x=1} dx \cdot Cx \cdot \frac{y^2}{2}\Big|_{y=0}^{y=1}$$

$$= \int_{x=0}^{x=1} dx \cdot Cx \cdot \frac{1}{2} = \frac{C}{2}\frac{x^2}{2}\Big|_{x=0}^{1} = \frac{C}{4} \Rightarrow C = 4$$

Event A: $y > x$

$$f_{XY|A}(x,y \mid A) = \frac{f_{XY}(x,y)}{P(A)} = \frac{4xy}{\int_{y=0}^{y=1} dy \int_{x=0}^{x=y} dx \cdot 4xy} = \frac{4xy}{1/2} = 8xy$$

Joint Probability
Conditioned on Event A

Note: Joint density is "doubled" since Event A occupies only half the outcomes.

Marginals

$$f_{Y|A}(y \mid A) = \int_{x=0}^{x=y} f_{XY|A}(x,y \mid A)dx = 8y \cdot \frac{x^2}{2}\Big|_{x=0}^{x=y} = 4y^3$$

Marginal Prob. for Y
Conditioned on Event A

$$f_{X|A}(x|A) = \int_{y=x}^{y=1} f_{XY|A}(x,y|A)dy = 8x \cdot \frac{y^2}{2}\Big|_{y=x}^{y=1} = 4x(1-x^2)$$

Marginal Prob. for X
Conditioned on Event A

Here we are given a joint PDF $f_{XY}(x,y) = Cxy$ inside the unit square which depends upon both x and y and has "surfaces of constant density" that are hyperbolas (not shown) inside the unit square. To find the normalization constant C we integrate the density over the unit square by noting that the integral factors into a product of integrals over x and y separately; we easily find the value C=4.

Defining the event as the region above the line y=x {A: y>x }, we can immediately find the probability density conditioned on the event A from its definition

$$f_{XY|A}(x,y|A) = f_{XY}(x,y) / P(A) = 4xy / P(A) = 4xy / \tfrac{1}{2} = 8xy.$$

Note by symmetry of the density expression 8xy, the event A must contain ½ the total probability in the unit square. However, it is instructive to carry out the double integral, using the horizontal strips to integrate over x from x ="0" to x ="y"; note that the upper limit for the x-integral depends upon the value of "y" and hence the integration over y must be **done after the x-integration**. (Note that this dependence occurs because of constraints imposed by the definition of event A; this is in sharp contrast to the unconstrained normalization integral over the whole unit square which factored into two *independent integrals*. Also note that integrating over a vertical strip instead (not shown) reverses the roles of x and y.)

The conditional joint probability density $f_{XY|A}(x,y|A)$ = "8xy" yields two conditional marginal densities, namely, $f_{X|A}(x|A)$ and $f_{Y|A}(y|A)$ obtained by integrating over vertical and horizontal strips respectively. For example, we show the horizontal strip from x="0" to x = "y" which when integrated over these limits yields $f_{Y|A}(y|A) = 4y^3$. To find the other conditional marginal $f_{X|A}(x|A)$ we would integrate over the vertical strips from y ="x" to y = "1" to find $f_{X|A}(x|A) = 4x(1-x^2)$. The reader should verify that each of the conditional marginals integrate to unity over the interval [0,1].

Transition to Continuous Random Variables

1.5 *Independence and Expected Values*

Independence and Expected Values

Independence	$f_{XY}(x,y) = f_X(x) \cdot f_Y(y)$ *all* x, y ("factors")
	$f_{X|Y}(x \mid y) = f_X(x)$ *all* x, y ("condition on y" does not matter)
Expected Value	$E[g(X,Y)] = \int_{-\infty}^{+\infty} dx \int_{-\infty}^{+\infty} dy\, g(x,y) f_{XY}(x,y)$
Conditional Expected Value	$E[g(X,Y) \mid A] = \int_{-\infty}^{+\infty} dx \int_{-\infty}^{+\infty} dy\, g(x,y) f_{XY|A}(x,y \mid A)$

Example: previous slide $\quad f_{XY}(x,y) = 4xy \qquad\qquad$ ***X, Y are Independent***

$$f_X(x) = 2x \qquad E[X] = \int_0^1 x \cdot 2x\, dx = 2\frac{x^3}{3}\Big|_{x=0}^1 = \frac{2}{3}$$

$$f_Y(y) = 2y \qquad E[Y] = \int_0^1 y \cdot 2y\, dy = 2\frac{y^3}{3}\Big|_{y=0}^1 = \frac{2}{3}$$

Conditioned on Event A: Y>X $\quad f_{XY|A}(x,y|A) = 8xy \neq 4x(1-x^2)\cdot 4y^3 \qquad$ ***X, Y are Dependent (conditioned on A)***

$$f_{X|A}(x \mid A) = 4x(1-x^2) \qquad E[X \mid A] = \int_0^1 x \cdot 4x(1-x^2)\,dx = 4\left(\frac{x^3}{3} - \frac{x^5}{5}\right)\Big|_{x=0}^1 = \frac{8}{15}$$

$$f_{Y|A}(y \mid A) = 4y^3 \qquad E[Y \mid A] = \int_0^1 y \cdot 4y^3\, dy = 4\frac{y^5}{5}\Big|_{y=0}^1 = \frac{4}{5} = \frac{12}{15}$$

Independence of the RVs X and Y means that for all (x,y) the joint probability density factors into the product of individual densities $f_{XY}(x,y) = f_X(x)\,f_Y(y)$ or equivalently that the conditional density becomes the marginal density $f_{X|Y}(x|y) = f_X(x)$. The joint density of previous slide $f_{XY}(x,y) = 4xy$ clearly results from the product of two independent densities $f_X(x) = 2x$ and $f_Y(y) = 2y$. The conditioning event A defined by{A: Y>X }created a conditional joint density $f_{XY|A}(x,y|A) = 8xy$ and "conditional" marginals $f_{Y|A}(y|A) = 4y^3$, (yϵ[0,1]) and $f_{X|A}(x|A) = 4x(1-x^2)$, (xϵ[0,1]) are no longer independent because of the constraint Y>X imposed by event A. This fact is easily verified by noting that the product

$$f_{Y|A}(y|A)\, f_{X|A}(x|A) = 4y^3 \cdot 4x(1-x^2) = 16y^3 \cdot x(1-x^2)$$

does not equal $f_{XY|A}(x,y|A) = 8xy$.

Expected value and conditional expected value have definitions analogous to the discrete case where summations are replaced by integrations. The middle panel computes the expected values of X and Y to be equal E[X]=E[Y]=2/3, while the bottom panel computes the expected values conditioned on A as E[X|A]= 8/15 and E[Y|A]=12/15. The latter result that E[Y|A] > E[X|A] is intuitive from the fact that conditioning event A constrains Y>X.

2 Mathematical Transformations of Continuous RVs

2.1 CDF and PDF Probability Density Transformation Methods

CDF and PDF Probability Density Transformation Methods	
Methods of Calculating New PDF under Transformation Y = g(X)	**Example**
1) CDF Method: **Step#1**) Find CDF $F_X(x)$ by integrating $f_X(x)$:	$f_X(x) = 2x/3$, $x \in [1,2]$ $F_x(x) = (\xi^2/3)\big\|_{\xi=1}^{\xi=x} = \boxed{(x^2-1)/3}$
Step#2 (*a*) Express CDF for Y in terms of X: $F_Y(y) = \Pr(Y \le y) = \Pr(g(X) \le y)$ (*y= g(x) may not be "one-to-one" so there may be multiple contributions*)	$y = 4x-5$ (single contrib.)
(*b*) Perform **explicit inversion** of y=g(x) to find $x = g^{-1}(y)$	$x = g^{-1}(y) = \boxed{(y+5)/4}$
(*c*) Substitute $x = g^{-1}(y)$ into $F_X(x)$: $\quad F_Y(y) = \Pr(Y \le y) = \Pr(g(X) \le y)$ $= \Pr(X \le g^{-1}(y)) = F_X(g^{-1}(y))$	$F_Y(y) = F_X((y+5)/4)$ $= [((y+5)/4)^2 - 1]/3$ $= (y+5)^2/48 - 1/3$
Step#3) Differentiate wrt *y*: $$f_Y(y) = \frac{d}{dy}F_Y(y) = \frac{d}{dy}\int_{y'=-\infty}^{y'=y} f_Y(y')dy'$$	$f_Y(y) = (y+5)/24$ $y \in [-1,3]$
2) PDF Method: $\quad f_Y(y)dy = f_X(x)dx$; $y = g(x)$ Transform PDF $f_Y(y)$ using derivatives $\qquad f_Y(y) = \dfrac{f_X(x)}{\|dy/dx\|}$	$dy/dx = d/dx(4x-5) = 4$ $f_Y(y) = (2/3) \cdot x /4$ but, $x = g^{-1}(y) = (y+5)/4$
Express everything in terms of variable y *Note absolute value* $\longrightarrow$ $\quad f_Y(y) = \dfrac{f_X(x = g^{-1}(y))}{\|g'(x = g^{-1}(y))\|}$	$f_Y(y) = (2/3) \cdot [(y+5)/4] /4$ $f_Y(y) = (y+5)/24$ $y \in [-1,3]$

It is very important to understand how probability densities change under a coordinate transformation y=g(x). In the discrete case, we have seen how the game of craps, which is based on the face value sum of the pair of dice, is best described by transforming from the individual dice coordinates (d_1, d_2) to the *sum and difference* coordinates (s, d) (rotated 45 degrees). The dice transformation to *minimum and maximum* coordinates (z, w) corresponding to corner shaped surfaces of constant minimum or maximum values seem odd for dice but turns out to be a good first step in understanding order statistics for the continuous case (see Slide#3-2 *ff.*).

The two basic methods for transforming densities are (i) the CDF-method which requires integration first and then differentiation, and (ii) the PDF method which only requires differentiation. Although both methods are useful for one-dimensional PDFs $f_X(x)$, the PDF method is best for transforming joint $f_{XY}(x,y)$ densities because the derivative $g'(x)$ generalizes to a Jacobian determinant which is easily computed from the transformation.

The **CDF method** involves three distinct steps as indicated on the slide, namely (i) compute CDF $F_X(x)$, (ii) Relate $F_Y(y) = \Pr[Y \le y]$ to $F_X(x)$; then invert the transformation $x = g^{-1}(y)$ and substitute to find $F_Y(y)$ with a *redefined y domain*, and (iii) differentiate with respect to "y" to obtain the transformed probability density $f_Y(y)$. Note that if the function is multi-valued and therefore not invertible, it must be broken up into intervals for which it is invertible and appropriate "fold-over" multiplicities must be accounted for.

The **PDF Method** uses derivatives of the transformation to transfer densities from the original set of RVs to the new set; the Jacobian accounts for linear, areal, and volume changes between the coordinates. In one dimension the Jacobian is simply a derivative and is obtained by transferring the probability in the interval x to x+dx: $f_X(x)dx$ to the probability in the interval y to y+dy: $f_Y(y)dy$. Equating the two expressions yields $f_Y(y)$ $=f_X(x) / |dy/dx| = f_X(g^{-1}(y)) / |dy/dx|$. Note that the absolute value is necessary since $f_Y(y)$ must always be greater than or equal to zero.

2.1.1 Transformation Y=X⁻¹ on Uniform Density - CDF Method

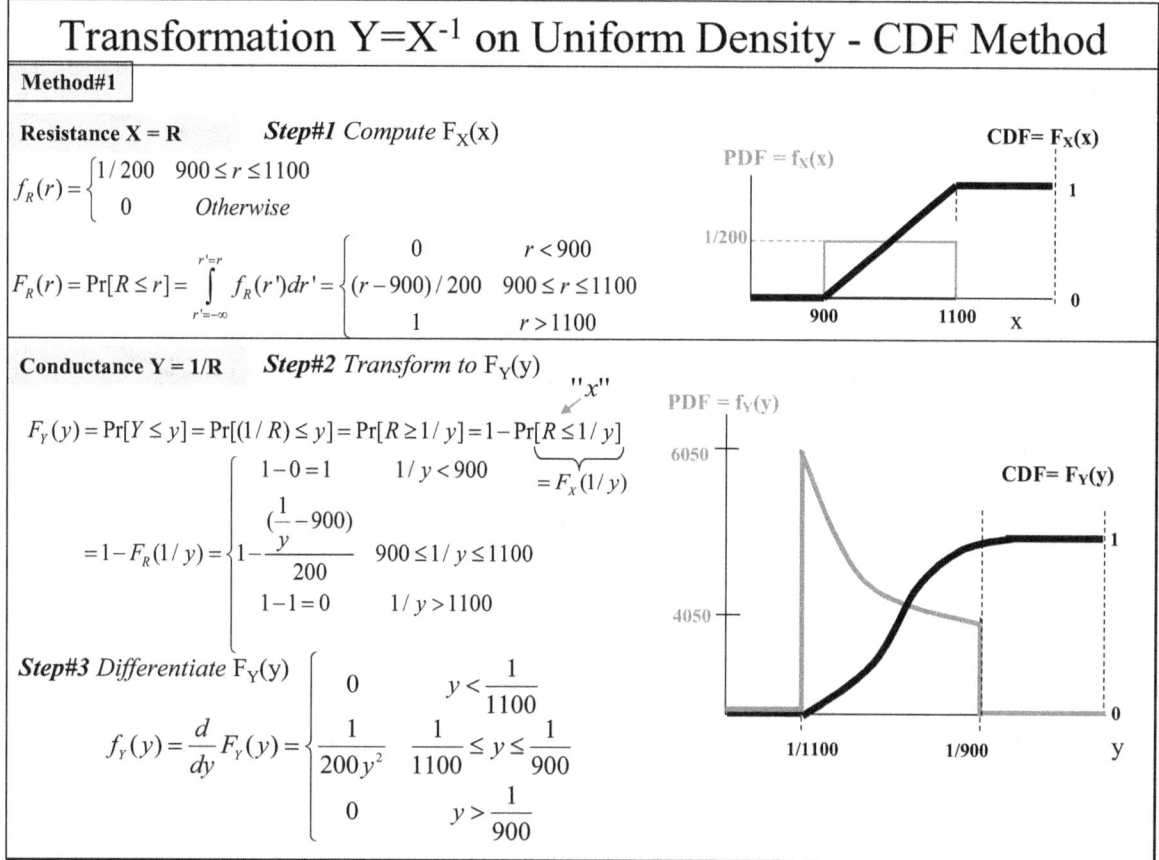

Transformation $Y=X^{-1}$ on Uniform Density - CDF Method

Method#1

Resistance X = R   ***Step#1*** *Compute* $F_X(x)$

$$f_R(r) = \begin{cases} 1/200 & 900 \le r \le 1100 \\ 0 & Otherwise \end{cases}$$

$$F_R(r) = \Pr[R \le r] = \int_{r'=-\infty}^{r'=r} f_R(r')dr' = \begin{cases} 0 & r < 900 \\ (r-900)/200 & 900 \le r \le 1100 \\ 1 & r > 1100 \end{cases}$$

PDF = $f_X(x)$ CDF= $F_X(x)$

Conductance Y = 1/R   ***Step#2*** *Transform to* $F_Y(y)$

$$F_Y(y) = \Pr[Y \le y] = \Pr[(1/R) \le y] = \Pr[R \ge 1/y] = 1 - \Pr[R \le 1/y]$$

$$= 1 - F_R(1/y) = \begin{cases} 1-0=1 & 1/y < 900 \\ 1 - \dfrac{(\frac{1}{y}-900)}{200} & 900 \le 1/y \le 1100 \\ 1-1=0 & 1/y > 1100 \end{cases}$$

$= F_X(1/y)$

PDF = $f_Y(y)$ CDF= $F_Y(y)$

Step#3 *Differentiate* $F_Y(y)$

$$f_Y(y) = \frac{d}{dy} F_Y(y) = \begin{cases} 0 & y < \dfrac{1}{1100} \\ \dfrac{1}{200 y^2} & \dfrac{1}{1100} \le y \le \dfrac{1}{900} \\ 0 & y > \dfrac{1}{900} \end{cases}$$

The resistance X=R of a circuit has a uniform probability density function $f_R(r)=1/200$ between 900 and 1100 ohms as shown in the top panel; the corresponding CDF $F_R(r)$ is the ramp function starting at "0" for R≤900 and reaching "1" at R=1100 and beyond as shown. The detailed analytic function is given in the slide and represents the result of Step#1 in the CDF-Method.

The problem is to find the PDF for the conductance Y=1/X = 1/R. We first write down the probability definition of $F_Y(y)$ for a given value Y=y and then re-express it as a function of R =1/Y

$$F_Y(y) = \Pr[Y \le y] = \Pr[R \ge (1/y)] = 1 - \Pr[R \le (1/y)]$$
$$= 1 - F_R(1/y)$$

The last expression is now evaluated (lower slide panel) by substituting r=1/y into the expression for $F_R(r)$ given in the upper panel. Note the resulting expression has been written down by direct substitution and the intervals have been left in terms of 1/y. (This constitutes step#2 of the method). Finally, differentiating $F_Y(y)$ with respect to "y" we find (step#3) the desired PDF $f_Y(y)$; we have also "flipped" the "1/y" interval specifications and reordered the resulting "y" intervals in the customary increasing order.

As seen in this example, the CDF method requires careful attention to the definition of the $F_Y(y)$ defined in terms of cumulative probability of the **variable Y**. Since Y=1/R, this leads to $F_Y(y) = 1 - F_R(1/y)$ and a reverse ordering of the inequalities for the intervals.

2.1.2 Transformation Y=X⁻¹ on Uniform Density - PDF Method

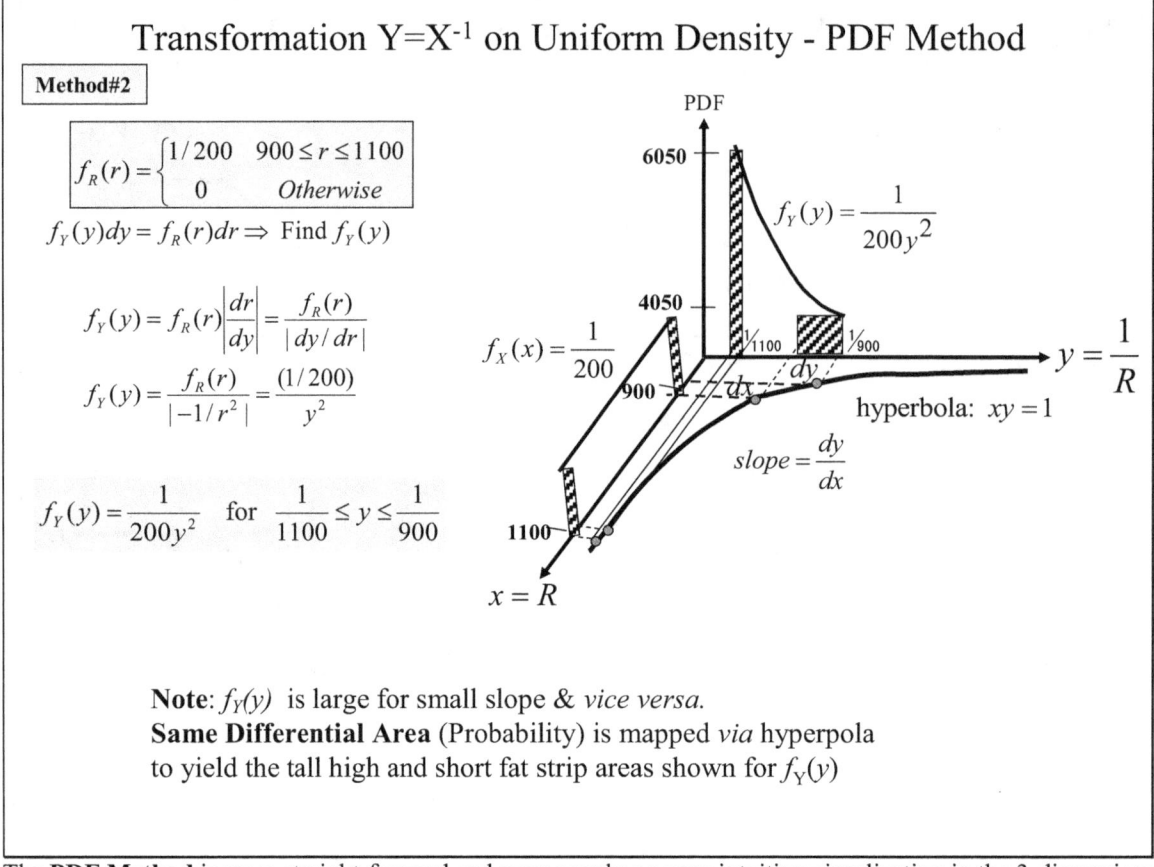

Transformation $Y=X^{-1}$ on Uniform Density - PDF Method

Method#2

$$f_R(r) = \begin{cases} 1/200 & 900 \le r \le 1100 \\ 0 & Otherwise \end{cases}$$

$$f_Y(y)dy = f_R(r)dr \Rightarrow \text{Find } f_Y(y)$$

$$f_Y(y) = f_R(r)\left|\frac{dr}{dy}\right| = \frac{f_R(r)}{|dy/dr|}$$

$$f_Y(y) = \frac{f_R(r)}{|-1/r^2|} = \frac{(1/200)}{y^2}$$

$$f_Y(y) = \frac{1}{200y^2} \quad \text{for} \quad \frac{1}{1100} \le y \le \frac{1}{900}$$

PDF

6050

4050

$$f_Y(y) = \frac{1}{200y^2}$$

$$f_X(x) = \frac{1}{200}$$

$$y = \frac{1}{R}$$

hyperbola: $xy = 1$

$$slope = \frac{dy}{dx}$$

900

1100

$x = R$

Note: $f_Y(y)$ is large for small slope & *vice versa*.
Same Differential Area (Probability) is mapped *via* hyperpola
to yield the tall high and short fat strip areas shown for $f_Y(y)$

The **PDF Method** is more straight-forward and moreover has a very intuitive visualization in the 3-dimensional plot shown on this slide. The uniform probability density function $f_R(r)=1/200$ between 900 and 1100 ohms is written explicitly in the first boxed equation. The Jacobian method just divides the constant density $f_R(r) = 1/200$ by $|dy/dr|=|-1/r^2| = y^2$ (magnitude of the derivative) to yield $f_Y(y)=1/(200y^2)$ for y ε [1/1100, 1/900]. The **3-dimensional plot** shows exactly what is going on:

i) The original uniform distribution $f_X(x)=1/200$ is displayed as a vertical rectangle in the x-z plane

ii) Sample strips at either end with width "dx" have the same small probability dP= $f_X(x)$dx as shown At R=900, the density $f_X(x)$ is divided by the large slope $|dy/dx|$ yielding a smaller magnitude for $f_Y(y)$ as illustrated, but this is compensated by a proportionately larger "dy" and thus transfers the same small probability dP= $f_Y(y)$dy.

iii) Conversely, the strip at R=1100 is divided by a small slope $|dy/dx|$ and yields a larger magnitude for $f_Y(y)$, which is compensated by a proportionately smaller "dy" again transferring the same dP.

iv) The end point values of the transformed density $f_Y(y)$ are illustrated in the figure. The strip width "dx" projects onto the transformation hyperbola curve at two red points which defines a small "dy" at x =1100 and a large "dy" at x = 900 as determined by the slope of the curve. The shape in between these end points is a result of the smoothly varying slope of the transformation hyperbola shown in the x-y plane.

Thus the slope of the transformation curve (hyperbola xy=constant in this case) in the x-y plane determines how each "dx" strip of the uniform distribution $f_X(x)=1/200$ in the x-z plane transfers to the new density $f_Y(y)$ shown in the y-z plane. This 3-dimensional representation de-mystifies the nature of the transformation of probability densities and makes it quite natural and intuitive for 1-dimensional density functions. It is easily extended to two-dimensional joint distributions.

2.1.3 Transformation $Y=X^3$ on Uniform Distribution - Both Methods

---

Transformation $Y=X^3$ on Uniform Distribution - Both Methods

PDF Method: Jacobian PDF Density Transformation

$$f_X(x) = \begin{cases} 1/2 & -1 \leq x \leq +1 \\ 0 & \text{Otherwise} \end{cases} \qquad f_Y(y) = \frac{f_X(x)}{|dy/dx|} = \frac{1/2}{|3x^2|} = \frac{1}{6y^{2/3}} = \begin{cases} \frac{1}{6}y^{-2/3} & -1 \leq y \leq +1 \\ 0 & \text{Otherwise} \end{cases}$$

Find PDF for $Y = X^3$, i.e., find $f_Y(y)$ $\quad y \in \left[(-1)^3, (+1)^3\right]$

CDF Method: i) $F_X(x)$; ii) $F_Y(y)$; iii) $d/dy[F_Y(y)]$

Step#1 Compute $F_X(x)$

$$F_X(x) = \Pr[X \leq x] = \int_{x'=-\infty}^{x'=x} f_X(x')dx' = \begin{cases} = 0 & x < -1 \\ \int_{x'=-1}^{x'=x}(1/2)dx' = (x+1)/2 & -1 \leq x \leq +1 \\ = 1 & x > +1 \end{cases}$$

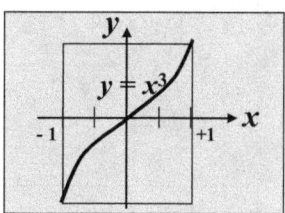

Step#2 Transform to $F_Y(y)$ $\quad Y = X^3$

$$F_Y(y) = \Pr[Y \leq y] = \Pr[X^3 \leq y] = \Pr[X \leq y^{1/3}] = F_X(y^{1/3})$$

$$= F_X(y^{1/3}) = \begin{cases} 0 & y^{1/3} < -1 \\ (y^{1/3}+1)/2 & -1 \leq y^{1/3} \leq +1 \\ 1 & y^{1/3} > +1 \end{cases}$$

Step#3 Differentiate $F_Y(y)$

$$f_Y(y) = \frac{d}{dy}F_Y(y) = \frac{d}{dy}\frac{(y^{1/3}+1)}{2} = \frac{1}{6}y^{-2/3}$$

$$F_Y(y) = F_X(y^{1/3}) = \begin{cases} 0 & y < -1 \\ (y^{1/3}+1)/2 & -1 \leq y \leq +1 \\ 1 & y > +1 \end{cases}$$

$$f_Y(y) = \begin{cases} \frac{1}{6}y^{-2/3} & -1 \leq x \leq +1 \\ 0 & \text{Otherwise} \end{cases}$$

Given a uniform distribution $f_X(x) = \frac{1}{2}$ on $[-1, +1]$, find $f_Y(y)$ for the transformation $Y=X^3$
PDF Method is straight forward and yields
$$f_Y(y) = \frac{1}{2} / (3x^2) = (1/6)\cdot 1/(y^{1/3})^2 = (1/6)y^{-2/3} \text{ on } [-1,1]$$

CDF Method computes $f_Y(y)$ by taking the derivative of $F_Y(y)$. The three steps displayed on the slide go through this process, *viz.*,
    (i) integrate and find $F_X(x) = (x+1)/2$ on $[-1,+1]$;
    (ii) formulate CDF(y) $F_Y(y) = \Pr[Y \leq y] = \Pr[X^3 \leq y] = \Pr[X \leq y^{1/3}] = F_X(x=y^{1/3}) = ([y^{1/3}]+1)/2$;
    (iii) $f_Y(y) = d/dy((y^{1/3}+1)/2) = (1/6) y^{-2/3}$ on $[-1,+1]$.
Note of caution: be sure to properly establish the domain of the variable y *via* the transformation; it will not always be the case that the domain of y is the same as that of x. For example, if the transformation is changed to $y=8x^3$, the PDF method gives instead
$$f_Y(y) = \frac{1}{2} / (24x^2) = (1/48)\cdot(y/8)^{-2/3} = (1/12) y^{-2/3} \text{ on } [-8,8] \textbf{ (not on } [-1,+1])$$
This is another reason why it is prudent to always check the normalization of your density; if the density does not integrate to unity over your domain, then either your integration or your domain is incorrect.

2.1.4 Transformation $Y=X^2$ on Gaussian RV – Multiplicity Factor α

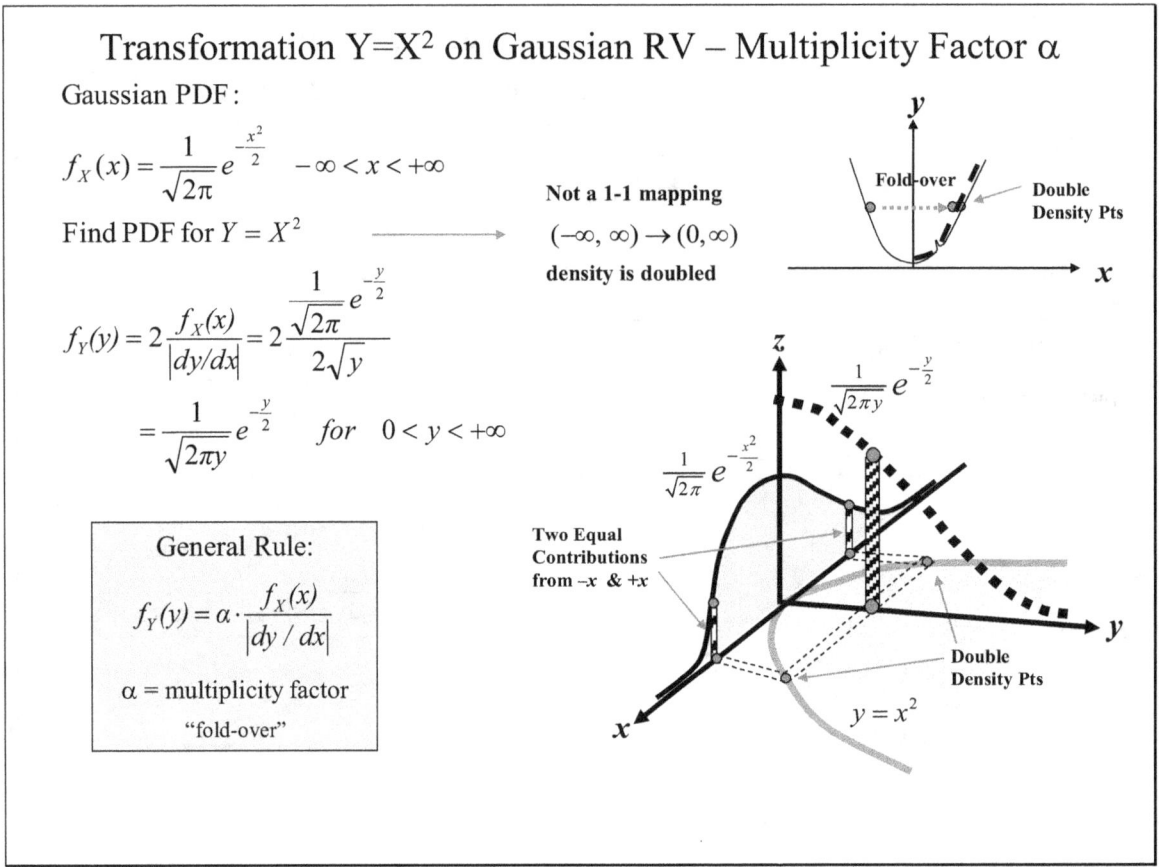

Transformation $Y=X^2$ on Gaussian RV – Multiplicity Factor α

Gaussian PDF:

$$f_X(x) = \frac{1}{\sqrt{2\pi}} e^{-\frac{x^2}{2}} \qquad -\infty < x < +\infty$$

Find PDF for $Y = X^2$

Not a 1-1 mapping

$(-\infty, \infty) \to (0, \infty)$

density is doubled

Fold-over

Double Density Pts

$$f_Y(y) = 2\frac{f_X(x)}{|dy/dx|} = 2\frac{\frac{1}{\sqrt{2\pi}} e^{-\frac{y}{2}}}{2\sqrt{y}}$$

$$= \frac{1}{\sqrt{2\pi y}} e^{-\frac{y}{2}} \qquad for \quad 0 < y < +\infty$$

General Rule:

$$f_Y(y) = \alpha \cdot \frac{f_X(x)}{|dy/dx|}$$

α = multiplicity factor

"fold-over"

$$\frac{1}{\sqrt{2\pi y}} e^{-\frac{y}{2}}$$

$$\frac{1}{\sqrt{2\pi}} e^{-\frac{x^2}{2}}$$

Two Equal Contributions from $-x$ **&** $+x$

Double Density Pts

$$y = x^2$$

A Gaussian PDF under the transformation $Y=X^2$ is easily computed using the PDF method provided one incorporates a multiplicity factor α as shown in the boxed density equation . The multiplicity factor arises because there are two contributions to the same y-value, one from $-x$ and the other from $+x$ as illustrated in the upper figure; thus folding the parabola across the x=0 symmetry line yields twice the density on positive x and this corresponds to a multiplicity factor $\alpha=2$ in the boxed density transformation equation.

The 3[d] plot shows the original Gaussian density function (grey) in the vertical x-z plane, the transformation $y=x^2$ in the horizontal x-y plane, and the resulting distribution shown as a dashed curve in the vertical y-z plane. The two thin vertical slices at $-x$ and $+x$ are mapped to the same y-value and hence doubles the density contribution to $f_Y(y)$ as shown.

Once again it is prudent to **check the normalization** of your density; if the density does not integrate to unity over your domain, then either your **integration** or your **domain** or your **multiplicity factor** is incorrect.

2.1.5 Transformation Y=$cos\pi$X on Uniform RV - Multiplicity Factor α

Transformation Y=$\cos\pi$X on Uniform RV - Multiplicity Factor α

PDF:

$$f_X(x) = \begin{cases} 1/4 & x \in [-2,2] \\ 0 & Otherwise \end{cases} \quad -\infty < x < +\infty$$

Find PDF for: $Y = \cos\pi X$

$$f_Y(y) = \alpha \frac{f_X(x)}{|dy/dx|} = 4 \frac{1/4}{|-\pi\sin\pi x|}$$

$$= \frac{1}{\pi\sin\pi(\pi^{-1}\cos^{-1}y)} = \frac{1}{\pi}\frac{1}{\sqrt{1-y^2}}$$

Trigonometry Stuff

$$\cos = \frac{y}{1} \Leftarrow \theta = \cos^{-1}y$$

$$\Rightarrow \sin\theta = \sin(\cos^{-1}y) = \frac{\sqrt{1-y^2}}{1}$$

Normalization

$$\int_{y=-1}^{+1} f_Y(y)dy = \frac{1}{\pi}\int_{y=-1}^{+1}\frac{1}{\sqrt{1-y^2}}dy \underset{y=\cos\theta}{=} \frac{1}{\pi}\int_{\theta=+\pi}^{0}\frac{-\sin\theta}{\sqrt{1-\cos^2\theta}}d\theta = -\frac{1}{\pi}\int_{\theta=+\pi}^{0}d\theta = 1 \quad ✓✓$$

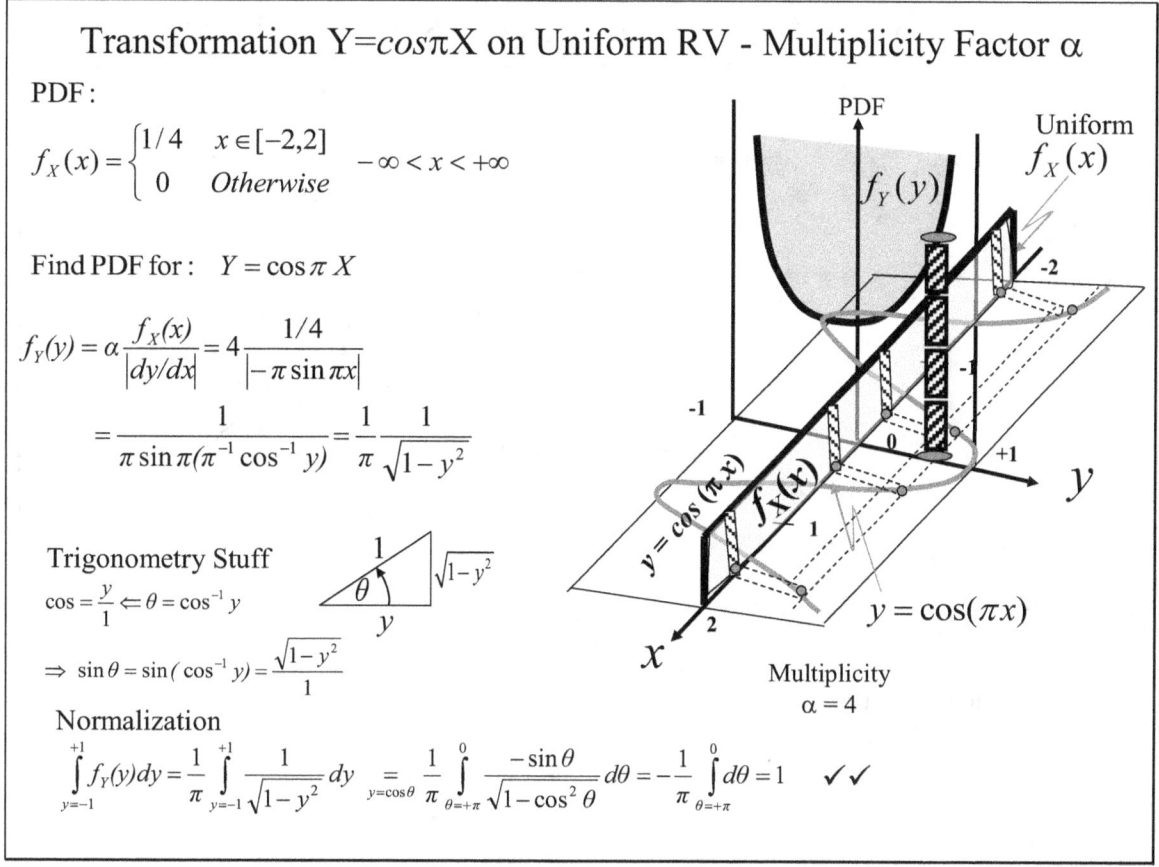

A uniform PDF $f_X(x)$ under the transformation Y=$\cos\pi$X is easily computed using the PDF method again with a multiplicity factor which in this case turns out to be α=4. As illustrated in the 3[d] plot the x-domain [-2,2] covers two full periods of the cosine transformation and thus makes 4 contributions (four solid dots in the x-y plane) to each fixed y value. The figure also shows the original Uniform density function (yellow) in the vertical x-z plane, the (grey) transformation y=$\cos\pi$x in the horizontal x-y plane, and the resulting distribution shown as a solid curve in the vertical y-z plane. The four thin vertical slices are mapped to the same y-value and hence quadruple the density contribution to every $f_Y(y)$ as shown.

The normalization of the resulting distribution is verified by integrating $f_Y(y)$ over [-1, 1]. Note that if we forgot to include the multiplicity factor α=4, the resulting normalization integral would yield ¼ instead of 1 and we would realize that there is a missing factor of 4 in the density.

2.2 *Transformation Y=3X-1 on Uniform $f_Y(y)$ – Slope Density Multiplier*

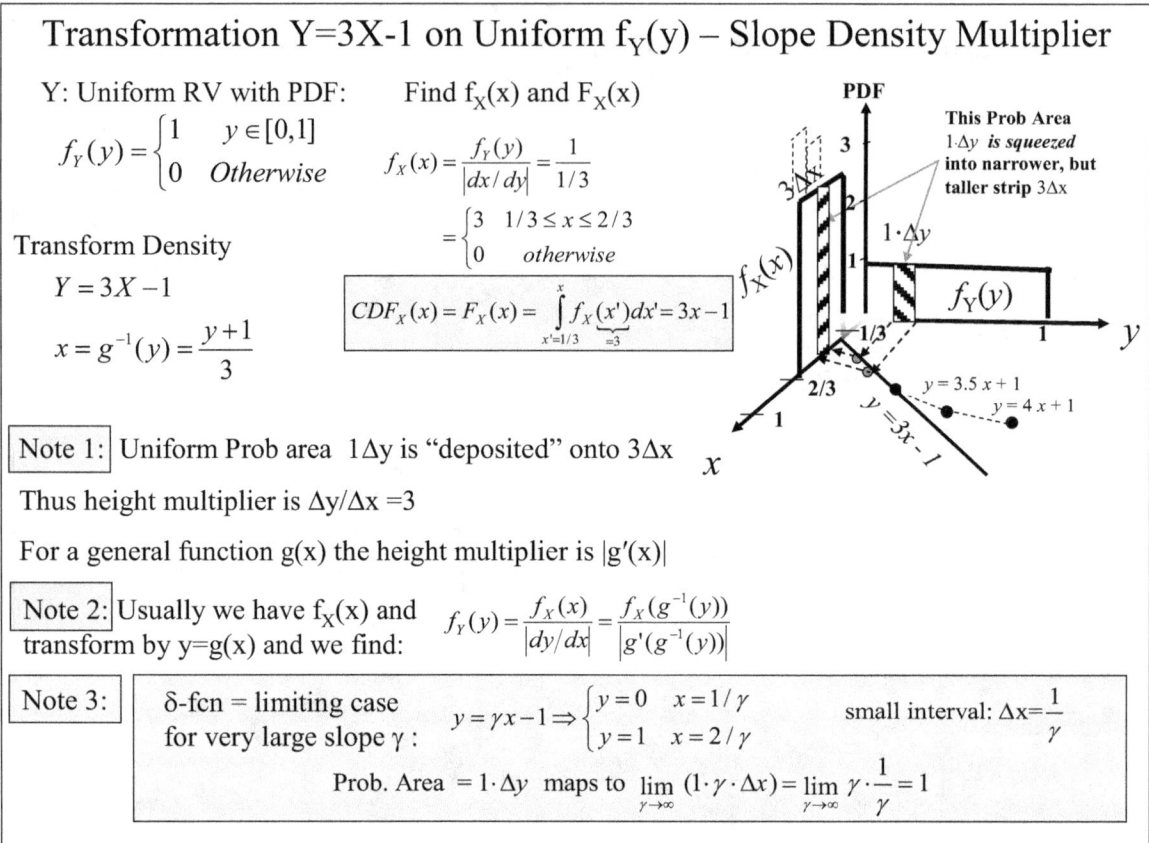

Transformation Y=3X-1 on Uniform $f_Y(y)$ – Slope Density Multiplier

Y: Uniform RV with PDF: Find $f_X(x)$ and $F_X(x)$

$$f_Y(y) = \begin{cases} 1 & y \in [0,1] \\ 0 & Otherwise \end{cases}$$

$$f_X(x) = \frac{f_Y(y)}{|dx/dy|} = \frac{1}{1/3}$$

$$= \begin{cases} 3 & 1/3 \le x \le 2/3 \\ 0 & otherwise \end{cases}$$

Transform Density

$$Y = 3X - 1$$

$$x = g^{-1}(y) = \frac{y+1}{3}$$

$$CDF_X(x) = F_X(x) = \int_{x'=1/3}^{x} \underbrace{f_X(x')}_{=3} dx' = 3x - 1$$

This Prob Area $1 \cdot \Delta y$ is *squeezed* into narrower, but taller strip $3\Delta x$

$y = 3.5 x + 1$
$y = 4 x + 1$

| Note 1: | Uniform Prob area $1\Delta y$ is "deposited" onto $3\Delta x$ |

Thus height multiplier is $\Delta y/\Delta x = 3$

For a general function g(x) the height multiplier is $|g'(x)|$

| Note 2: | Usually we have $f_X(x)$ and transform by y=g(x) and we find: |

$$f_Y(y) = \frac{f_X(x)}{|dy/dx|} = \frac{f_X(g^{-1}(y))}{|g'(g^{-1}(y))|}$$

| Note 3: | δ-fcn = limiting case for very large slope γ : |

$$y = \gamma x - 1 \Rightarrow \begin{cases} y = 0 & x = 1/\gamma \\ y = 1 & x = 2/\gamma \end{cases}$$

small interval: $\Delta x = \dfrac{1}{\gamma}$

Prob. Area $= 1 \cdot \Delta y$ maps to $\displaystyle\lim_{\gamma \to \infty} (1 \cdot \gamma \cdot \Delta x) = \lim_{\gamma \to \infty} \gamma \cdot \frac{1}{\gamma} = 1$

This example is given in order to introduce the key concepts needed to actually generate a RV having a given density function $f_X(x)$ from a **uniform RV Y**: $f_Y(y)=1$ in [0,1]. If we make a simple linear transformation y=g(x)=3x-1 and apply the Jacobian PDF Method we find $f_X(x) = f_Y(y)/ |dx/dy|=1/(1/3) =3$ over the x-interval [1/3, 2/3]; we also find by direct integration the cumulative distribution $F_X(x) = 3x-1$ which ramps from "0" at x=1/3 to "1" at x=2/3. The 3^d figure illustrates the transformation process in which each $1 \cdot \Delta y$ strip in the original uniform distribution $f_Y(y)=1$ in the y-z plane is "deposited" by the linear transformation to a strip $3 \cdot \Delta x$ in the x-z plane. In this process, probability normalization is preserved because all the probability in the y-interval [0, 1] is transferred to a 1/3 smaller x-interval [1/3,2/3] with a height multiplier of '3". Note that the CDF is $F_X(x) = 3x-1 = g(x)$ and that the density (height) multiplier g'(x)=3 is the derivative of the CDF.

Alternately, a "segmented line" is "dashed in" and yields three different height multipliers and a transformed density with three different heights as shown. Extending this approach, to a general transformation, setting g(x) = $F_X(x)$ for the "desired CDF", the slope g'(x) =d/dx[$F_X(x)$] gives a point-by-point height multiplier that is exactly the desired probability density, *viz.*, $f_X(x) =d/dx(F_X(x)) = g'(x)$! The next slide gives a formal proof of this CDF generation technique for a RV with that distribution.

Note 3 briefly discusses how the limiting form of the linear transformation y= γ x -1 yields a Dirac delta function δ(x) as the slope $\gamma \to \infty$. Under this transformation, the **uniform RV Y**: $f_Y(y)=1$ in y-interval [0,1] is transferred to the x-interval [1/γ , 2/γ] with a multiplier γ so in the limit as $\gamma \to \infty$ the interval shrinks to the point x=0 and the density $f_X(0) \to \infty$ while the area $f_X(x) \cdot \Delta x = \gamma \cdot (1/ \gamma)=1$ as required for a Dirac delta function. Thus the total probability under the uniform distribution $f_Y(y)=1$ is mapped to a single point x=0 in the y-z plane and is represented by a Dirac delta function arrow at x=0. We shall see examples of this when we talk about the Half-wave Rectifier (Slide#2-13) and discrete sampled versions of probability distributions (Slide#2-14) .

Mathematical Transformations of Continuous RVs

2.2.1 Generating a Random Variable with a Desired CDF

Generating a Random Variable with a Desired CDF
(Inverse CDF Transformation Technique)

Y is Uniform RV *with PDF*

$$f_Y(y) = \begin{cases} 1 & y \in [0,1] \\ 0 & Otherwise \end{cases}$$

Transform Density using the "desired" CDF: $F_X(x)$

$$y = g(x) \equiv F_X(x)$$

$$x = g^{-1}(y)$$

$$f_X(x)dx = f_Y(y)dy$$

$$f_X(x) = f_Y(y) \cdot \underbrace{\frac{dy}{dx}}_{=1} = 1 \cdot g'(x) \; ; \quad g^{-1}(y=0) \le x \le g^{-1}(y=1)$$

$$CDF_X(x) = F_X(x) = \int_{\xi=-\infty}^{\xi=x} f_X(\xi)d\xi = \int_{\xi=-\infty}^{\xi=x} g'(\xi)d\xi = g(x) - \underbrace{g(-\infty)}_{=0} \quad \Rightarrow \quad CDF_X(x) = g(x)$$

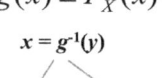

$F_X(x) = g(x)$

"Desired" CDF(X)   Monotone "Non-Decreasing"

Properties of CDF(x) means that
$$F_X^{-1}(y) = g^{-1}(y) \text{ Exists!}$$

Theorem for Generating RVs:
Given: $Y = g(X)$ is uniform RV
Then: $CDF_X = F_X(x) = g(x)$

Converse of Theorem :
Given : $CDF_X = F_X(x) = g(x)$
Then : $Y = g(X)$ is uniform RV

$$\text{Proof}: f_Y(y) = \frac{f_X(x)}{|dy/dx|} = \frac{\frac{d}{dx}F_X(x)}{g'(x)} = \frac{g'(x)}{g'(x)} = 1 \text{ on } [0,1]$$

$$\therefore Y \text{ is uniform RV}$$

Motivated by discussions of the previous slide, we now give a mathematical proof of the CDF Method for generating realizations of a RV having a desired distribution $f_X(x)$ from a uniform distribution $f_Y(y)$ on [0,1]. The proof hinges on the fact that the CDF $F_X(x) = g(x)$ is a **monotone non-decreasing function of x** and therefore has a unique **inverse $g^{-1}(y)$**. The theorem and its converse are given in the boxed statements at the bottom of the slide and basically state that the uniform RV Y on [0,1] yields the desired RV X under the transformation $y=g(x)$, *if and only if* the transformation $g(x) = F_X(x)$ is in fact the desired CDF.

Theorem: *Given $Y = g(X)$ is a uniform RV on [0,1], then the required transformation is $g(x) = F_X(x)$.*
The proof follows by simply equating the differential probabilities
$$dP = f_X(x)dx = f_Y(y)dy = 1 \cdot dy$$
which yields
$$f_X(x) = 1 \cdot dy/dx = |g'(x)| \quad \text{for all} \quad x \, \varepsilon \, [g^{-1}(y=0), \, g^{-1}(y=1)],$$
Note that the domain evaluation above required the inverse $g^{-1}(y)$ to exist and the absolute value is taken to insure the new probability density is never negative. The cumulative distribution $F_X(x)$ is the integral of the density $f_X(x)$ between $-\infty$ and x, so
$$F_X(x) = \int_{-\infty}^{x} f_X(x)dx = \int_{-\infty}^{x} g'(x)dx = g(x) - g(-\infty) = g(x) \quad \text{(Q.E.D.)}$$
where the property $g(-\infty) = 0$ is used in the last equality.

Converse: *Given $g(x) = F_X(x)$ (desired CDF) for all $x \, \varepsilon \, (-\infty, +\infty)$, then Y is uniform on [0,1].*
The Y density is obtained by the PDF (Jacobian) method as,
$$f_Y(y) = f_X(x) / |dy/dx| .$$
Recognizing that $d/dx \, [F_X(x)] = g'(x)$ and $dy/dx = g'(x)$ the above the density relationship yields
$$f_Y(y) == g'(x) / g'(x) = 1 \text{ for all } y \, \varepsilon \, [g(x = -\infty), g(x = +\infty)] = [0,1];$$
i.e., Y is uniform on [0,1]. (Q.E.D.)

2.2.2 Generating a Random Variable with a Desired CDF – 3ᵈ Visualization

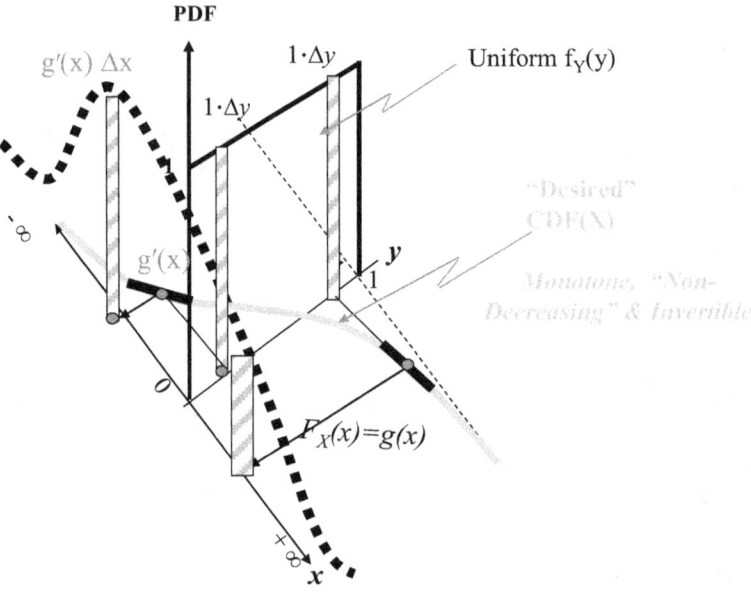

Generating a Random Variable with a Desired CDF -3ᵈ Visualization

Uniform Probability strip area 1Δy is "deposited" onto |g'(x)| Δx = f_X(x) Δx

The "height multiplier" is the slope |g'(x)| =d/dx{F_X(x)}=f_X(x)

Hence the transformed density is that of the desired distribution

This slide gives an intuitive illustration of how the slopes $g'(x)$ of the desired CDF $F_X(x) = g(x)$ act as a density transfer function to generate the desired PDF $f_X(x)$ on $(-\infty, +\infty)$ from a uniform distribution $f_Y(y)$ on [0,1]. The figure illustrates two vertical strip areas on the uniform distribution $f_Y(y)$ in the vertical y-z plane which obviously contain the same differential probability $dP = 1 \cdot \Delta y$. These two strips project onto different parts of the $y=g(x)$ transformation curve and therefore have different density multipliers (slopes) $g'(x)$ and yet must transfer the *same amount of probability* $dP = g'(x) \cdot 1 \cdot \Delta x$ to every point; this can only be the case if the strip widths Δx change inversely with the height multiplier $g'(x)$. This variation is illustrated in the vertical x-z plane of the figure by the foreground strip which is wider in order to compensate for the lower height and correspondingly for the strip behind it which is thinner because of its greater height. It is precisely this compensation generated by the slope $g'(x)$ that allows the uniform density to generate the desired PDF which is in general non-uniform as shown. Since the effective probability density (height) is the derivative of the desired CDF, *viz.*, $f_X(x) = g'(x) = d/dx[F_X(x)]$, we have thus generated a PDF that has the correct relationship to the desired CDF.

We note that the inverse $x = g^{-1}(y)$ must exist in order to map back from y to x and this is guaranteed by the unique defining properties of the cumulative distribution function, *i.e.*, it is continuous and monotonic non-decreasing on $(-\infty, \infty)$ and is therefore single-valued and invertible.

2.2.3 Simulating an Exponential RV

Simulating an Exponential RV

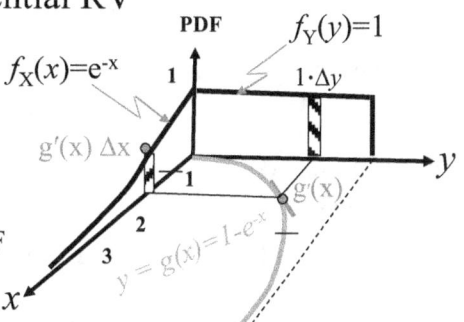

Thus to **Generate RV** with "desired" $CDF_X(x) = g(x)$

1) *Generate a Uniform RV:* $Y = \{y_1, y_2, ..., y_{1000}\}$
(Frequency of occurrence histogram for y's ~Uniform)

2) *Transform* $x_k = g^{-1}(y_k) = \{x_1, x_2, ..., x_{1000}\}$
Frequency of occurrence histogram for x's ~ desired PDF
Cumulative sum will approximate $CDF_X(x) = g(x)$

Prob Area $1 \cdot \Delta y$ has slope multiplier $g'(x)$ and yields strip $g'(x) \cdot \Delta x$

Example: "Exponential" CDF: $CDF_X = F_X(x) = g(x) = 1 - e^{-x} = y$

Inverse of g(x): $x = g^{-1}(y) = -\ln(1-y)$

If $Y = g(X)$ is uniform RV, then "X is exponential"

Note: $f_X(x) = e^{-x}$
integrates to yield

$$F_X(x) = \int_{\xi=0}^{x} e^{-\xi} d\xi$$

$$= -e^{-\xi}\Big|_{\xi=0}^{x} = 1 - e^{-x}$$

See program on Slide#8-7

> **Note:** This process explicitly uses the continuous, one-to-one nature of the cumulative distribution function CDF=g(x).
>
> This insures that the inverse always exists so as to make the transformation to the x_k

---

This slide gives an explicit algorithm to generate realizations $X = \{x_1, x_2, ..., x_{1000}\}$ of an exponential RV from a standard uniform RV $Y = \{y_1, y_2, ..., y_{1000}\}$ on [0,1]. The frequency-of-occurrence histogram for the set of y's generated by the computer random number generator will approximate a uniform distribution. We wish to generate a new set of x's from the set of y's in such a manner that their frequency-of-occurrence histogram will approximate an exponential distribution. According to the CDF theorem the required transformation is $x_k = g^{-1}(y_k) = \{x_1, x_2, ..., x_{1000}\}$.

For the exponential CDF we have $F_X(x) = g(x) = 1 - e^{-x} = y$ and solving for x yields the inverse function $x = g^{-1}(y) = -\ln(1-y)$. Thus applying this function to each of the 1000 y's we generate the desired x's and are assured that the histogram will approximate the density $f_X(x) = d/dx[1 - e^{-x}] = e^{-x}$.

The figure illustrates the uniform distribution $f_Y(y)$ in the vertical y-z plane, the desired CDF $F_X(x) = g(x) = 1 - e^{-x}$ in the horizontal x-y plane, and the resulting (desired) density $f_X(x) = e^{-x}$ in the vertical x-z plane and shows the transfer of one vertical strip from the uniform to the desired density planes.

A simple MatLab® script is easy enough to write and histograms of the input and output distributions verify this simple algorithm. A sample script is given on Slide# 8-7 where it is used to verify the Central Limit Theorem for sums of exponential RVs.

2.3 *Half Wave Rectifier Transformation yields Box plus δ-Function*

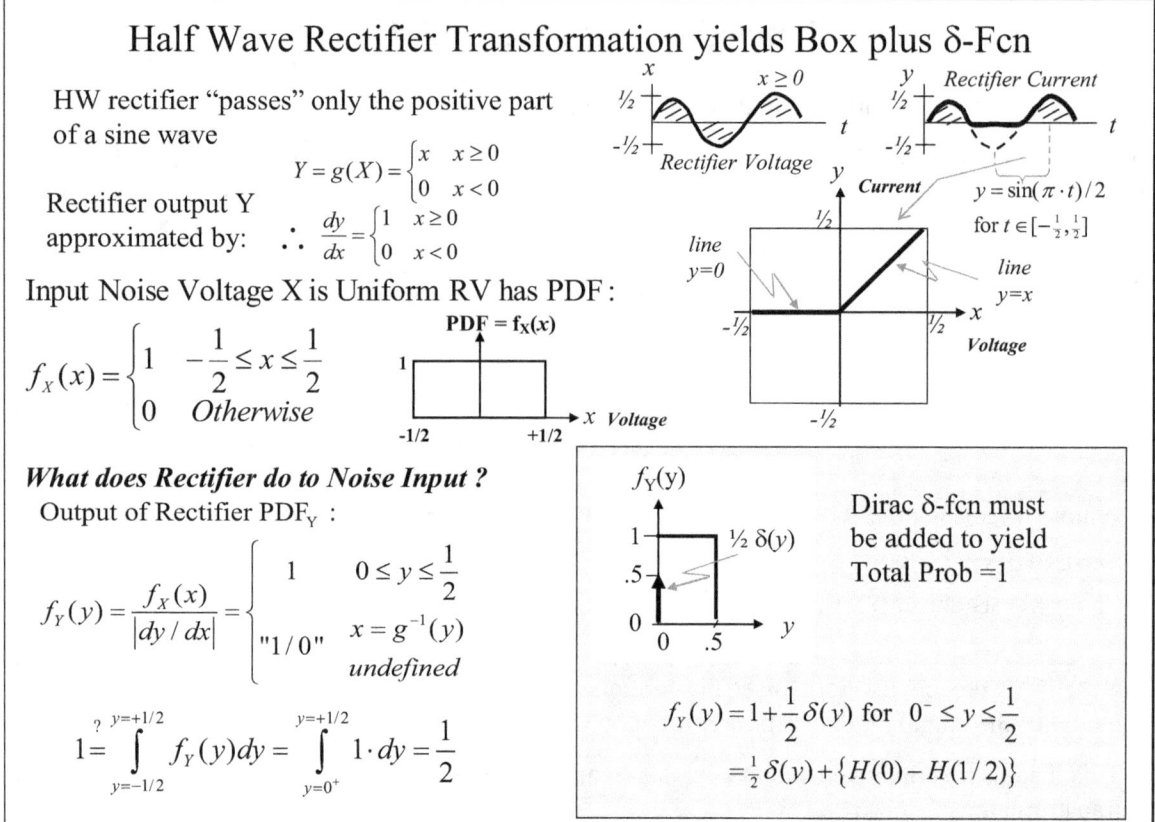

Half Wave Rectifier Transformation yields Box plus δ-Fcn

HW rectifier "passes" only the positive part of a sine wave

$$Y = g(X) = \begin{cases} x & x \geq 0 \\ 0 & x < 0 \end{cases}$$

Rectifier output Y approximated by:

$$\therefore \quad \frac{dy}{dx} = \begin{cases} 1 & x \geq 0 \\ 0 & x < 0 \end{cases}$$

Input Noise Voltage X is Uniform RV has PDF :

$$f_X(x) = \begin{cases} 1 & -\frac{1}{2} \leq x \leq \frac{1}{2} \\ 0 & Otherwise \end{cases}$$

PDF = $f_X(x)$

$y = \sin(\pi \cdot t)/2$ for $t \in [-\frac{1}{2}, \frac{1}{2}]$

What does Rectifier do to Noise Input ?

Output of Rectifier PDF_Y :

$$f_Y(y) = \frac{f_X(x)}{|dy/dx|} = \begin{cases} 1 & 0 \leq y \leq \frac{1}{2} \\ "1/0" & x = g^{-1}(y) \\ & undefined \end{cases}$$

$$1 = \int_{y=-1/2}^{\overset{?}{y=+1/2}} f_Y(y)dy = \int_{y=0^+}^{y=+1/2} 1 \cdot dy = \frac{1}{2}$$

$f_Y(y)$

½ δ(y)

Dirac δ-fcn must be added to yield Total Prob =1

$$f_Y(y) = 1 + \frac{1}{2}\delta(y) \text{ for } 0^- \leq y \leq \frac{1}{2}$$
$$= \frac{1}{2}\delta(y) + \{H(0) - H(1/2)\}$$

A half-wave rectifier is an electronic device that conducts electricity only when the voltage is positive; thus for the sinusoidal voltage waveform shown at the top, current flows only when the voltage swings positive as shown in the second plot at the top for the rectifier current. The action of a half-wave rectifier on one cycle of the sine curve is described by the coordinate transformation y = g(x) = x for x>=0 and 0 for x<0. A plot of this function, which is zero for negative x and rises linearly for positive x, is shown for a single cycle of the rectifier current indicated by the (red) horizontal brace.

Suppose that on top of the sinusoidal input signal we add some uniform noise $f_X(x) = 1$ on [-½, ½]. Once the (signal + noise) has passed through the rectifier, the input noise distribution (and signal) will be transformed by y=g(x). In order to separate signal from noise we need to characterize the noise density on the output of the rectifier; that is, we need to determine $f_Y(y)$ on [0, ½]. The PDF (Jacobian) method calculation requires we evaluate $f_Y(y) = f_X(g^{-1}(y)) / |dy/dx|$, which for x ε [0, ½] yields $f_Y(y)=1/1=1$ on [0, ½]. There are two problems with this result, namely (i) it ignores half the input noise probability in the interval x ε [-½, 0) for which the derivative g'(x)=0 and x = $g^{-1}(y)$ is undefined and (ii) the normalization integral of $f_Y(y)=1$ over [0, ½] only yields ½. Obviously, the other ½ in probability did not just disappear, rather it has piled up as a Dirac delta function at y=0, *i.e.*, we need to add a term +½ δ(y-0). On Slide#2-8 we gave a justification for this "piling up" at zero by taking the limiting case of a linear transformation where the **slope becomes zero**; that is precisely what we have here for x ε [-½, 0).

The boxed result gives the probability density function $f_Y(y)$ in terms of the Heaviside and delta functions as follows: $f_Y(y)$ = ½ δ(y) + {H(0) – H(1/2)}. This mixed density is displayed in the inset figure as a dark arrow at y=0 with magnitude ½ for the delta function and a box with height "1" extending from y="0" to y = " ½ ". Thus, half the noise power occurs at the origin (y=0) where it does no damage and the other half is uniformly added to the output signal. The next slide gives the 3[d] plot of the Half-wave rectifier transformation and a more detailed description of the ½ δ(y) contribution.

2.3.1 3^{d} View Half-Wave Rectifier Transformation of Probability Density

3^{d} View Half-Wave Rectifier Transformation of Probability Density

Region A) x ε [0, ½]: Probability "strips" transferred uniformly by g(x) in 1:1 manner from $f_X(x)$ on x ε [0, ½] to $f_Y(y)$ on y ε [0, ½]

Region B) x ε [-½, 0]: All Probability "strips" along "x" map to a "single point" at origin y=0 yielding δ-fcn = 1/2 δ(y-0) (Red arrow at origin)

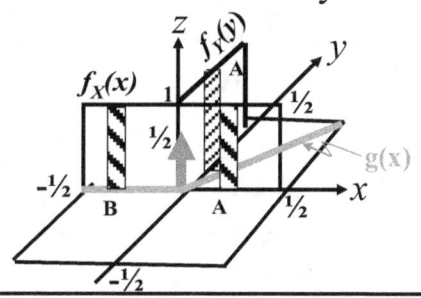

Formation of δ-fcn

Consider : $Y = g_\varepsilon(X) = \begin{cases} x & x \geq 0 \\ \varepsilon x - \varepsilon/2 & x < 0 \end{cases}$

For x ≥ 0 nothing changes

but for x<0 we have: $f_Y(y) = \dfrac{f_X(x)}{|dy/dx|} = \begin{cases} 1 & 0 \leq y \leq \frac{1}{2} \\ 1/\varepsilon & -\varepsilon \leq y < -\varepsilon/2 \end{cases}$

For $\varepsilon = .01$, $f_Y(y) = 100$ in strip $-\varepsilon \leq y \leq -\dfrac{\varepsilon}{2}$ $\ or\ -.010 \leq y \leq -.005$

As $\varepsilon \to 0$, magnificat ion factor $\approx \dfrac{1}{\varepsilon} = \dfrac{1}{.01} = 100$

Squeezed into region of width $\dfrac{\varepsilon}{2}$

CDF: $P[Y \leq 0] = F_Y(y=0) = \displaystyle\int_{y'=-1/2}^{0^+} f_Y(y')dy' = \lim_{\varepsilon \to 0} \dfrac{1}{\varepsilon} \cdot \dfrac{\varepsilon}{2} = \dfrac{1}{2}$

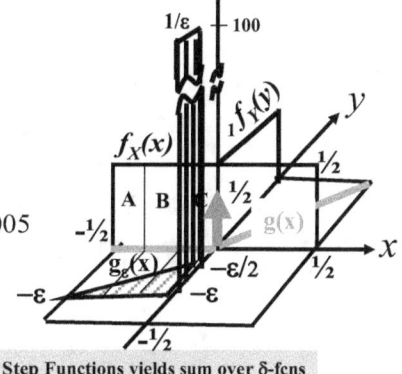

Series of Step Functions yields sum over δ-fcns

$$f_Y(y) = \sum_k \alpha_k \cdot \delta(y - y_k)$$

The upper panel and figure illustrate the uniform input distribution $f_X(x)$ as a rectangle extending from x=- ½ to x = +½ in the x-z plane, the half-wave rectifier transformation function y = g(x) in the x-y plane which split into two halves "A" for x ε [0,1/2], and "B" for x ε [-1/2,0) , and the resulting density $f_Y(y) = ½ δ(y) + \{H(0) - H(1/2)\}$ in the y-z plane shown by the thick red arrow at the origin y=0 plus a rectangle extending from y=0 to y = ½ just as in the previous slide. Note that in the region A (on the right) we have an invertible $x=g^{-1}(y)$ and each vertical strip in A is transferred in a 1:1 manner from $f_X(x)$ to $f_Y(y)$ over the whole interval [0, ½]. On the other hand, vertical strips in the left hand region B all pile up at the origin y=0 to give ½ δ(y) (thick red arrow at y=0). Details of this process are given in the lower panel.

The lower panel and figure illustrate the formation of the δ(y) contribution by introducing a transformation that reduces to the half-wave rectifier as the parameter ε→0. This introduces two changes, namely (i) the inverse can now be computed for x < 0 by the linear transformation x =$g_\varepsilon^{-1}(y)$ = (y+ ε/2)/ε, and (ii) there are now density contributions $f_Y(y)$ = 1/ε at negative values of y in the interval [-ε, -ε /2]. Upon taking the limit ε→0, these contributions all pile up at a single point, y=0 to give the delta function contribution corresponding to the left half of the uniform input distribution. The figure illustrates exactly how this happens. In the lower figure nothing changes for x >0, but for x<0 we have the new transformation y = $g_\varepsilon(x)$ = ε x - ε /2 in left half of the x-y plane; "y= $g_\varepsilon(x)$" ranges from y =- ε at x=-1/2 to y= -ε /2 at x=0 as illustrated. Further, we have broken up the left half of the uniform input distribution $f_X(x)$ into three strips A, B, C extending from x=- ½ to x = 0 in the x-z plane and show how they transform into the three tall strips in the y-z plane with amplitude $f_Y(y)$ = 1/ε. The constant density $f_Y(y)$ =1/ε extends from -ε to -ε /2, so Δy = ε /2 and the total probability contribution is $f_Y(y) \cdot Δy$ = (1/ε)*(ε /2) = ½; hence the limit ε →0 yields a δ(y) with amplitude ½. Generalization from just one to a series of "zero-slope" flat steps yields the boxed formula denoting a weighted sum of δ-functions; this will be useful in discussing sampled distributions on the next slide.

2.4 Analog to Digital (A/D) Converter – Sampled Distributions

Analog to Digital (A/D) Converter – Sampled Distributions

Continuous Representation of Discrete "sampled" Distributions

A/D converter

Mapping Fcn $Y = g(X) = k+1 \; ; \; k < x \le k+1$

Mapped Density $f_Y(y) = \sum_k \alpha_k \cdot \delta(y - y_k)$

a) Exponential	b) Gaussian	b) Uniform
$PDF_X = f_X(x) = \begin{cases} ae^{-ax} & x \ge 0 \\ 0 & x < 0 \end{cases}$	$PDF_X = f_X(x) = \dfrac{1}{\sqrt{2\pi}} e^{-x^2/2}$ $-\infty < x < \infty$	$PDF_X = f_X(x) = \begin{cases} 1 & 0 \le x \le 10 \\ 0 & \text{otherwise} \end{cases}$
$\alpha_k = \int_{x=k-1}^{k} f_X(x)dx = \begin{cases} \int_{x=k-1}^{k} ae^{-ax}dx = -e^{-ax}\big|_{x=k-1}^{k} & x \ge 0 \\ 0 & x < 0 \end{cases}$ $= \begin{cases} e^{-ak}(e^a - 1) & x \ge 0 \\ 0 & x < 0 \end{cases} ; k = 1,2,...$	$\alpha_k = \int_{x=k-1}^{k} \dfrac{1}{\sqrt{2\pi}} e^{-x^2/2} dx = \varphi(k) - \varphi(k-1)$ $\varphi(k) \equiv \int_{x=-\infty}^{x=k} \dfrac{1}{\sqrt{2\pi}} e^{-x^2/2} dx ; k \in (-\infty, \infty)$	$\alpha_k = \int_{x=k-1}^{k} \dfrac{1}{10} dx = \dfrac{k-(k-1)}{10}$ $= \dfrac{1}{10} ; k = 1,2,\cdots,10$
$f_Y(y) = \sum_{k=1}^{\infty} e^{-ak}(e^a - 1) \cdot \delta(y-k)$	$f_Y(y) = \sum_k \alpha_k \cdot \delta(y - y_k)$	$f_Y(y) = \sum_{k=1}^{10} \dfrac{1}{10} \delta(y-k)$

let $a = 0.1$: $\alpha_k = e^{-(0.1)k}(e^{0.1} - 1) = .105 \cdot e^{-(0.1)k}$

bin centers at $y_k = 0.1 \cdot (k-.5)$

See Slide#8-7

k	α_k
1	0.095
2	0.086
3	0.078
11	0.035

In discussing the half-wave rectifier on the last slide we found that the effect of a "zero" slope transformation function was to pile up all the probability in the x-interval into a single δ-function at the constant y="0" value associated with that part of the transformation. Here we extend that concept to a "sample and hold" type mapping function typical of an Analog to Digital (A/D) converter. The specific mapping function y=g(x) = k+1 for k < x ≤ k+1 is illustrated in the grey box as a series of horizontal steps over the entire range of x [-3, 3]; the y-values for these steps range from y=-2 to y=+3. Each horizontal (zero-slope) line accumulates the integral of $f_X(x)$ from x=k to k+1 onto its associated y-value shown as a red circle with the tip of a δ-function arrow pointing up out of the page and having an amplitude given by the integral for that interval denoted by the symbol α_k.

The table shows several examples of a digitally sampled representation for a) Exponential, b) Gaussian, and c) Uniform distributions in the three columns. The rows of the table give the specific continuous densities for each, the computations for the amplitudes of the discrete digital samples α_k, the resulting sum of δ-functions, and finally a plot showing arrows of different lengths to represent the δ-functions of the sampled distributions.

It is of interest to look at the exponential density distribution case in more detail; typically the x values have some resolution scale (bin size), say Δx, so that x really takes on values k Δx instead of the integers k given on the slide. This means that x=k is replaced by x=k·Δx, which yields new coefficients α_k =exp(-a·k·Δx) [exp(a·Δx) -1]; fixing the scale or "bin size" to be Δx=0.1, yields the coefficients shown on the slide, *viz.*, α_k= exp(-0.1·a·k)· [exp(0.1·a)-1] (valid for any value of the exponential parameter "a"). On the slide we simply set a=0.1 effectively absorbing the scale into the exponential parameter "a" whereas it really belongs with the coordinate bin size as described here. (See MatLab® script on Slide#8-7)

2.4.1 Sampled Representation of Continuous Distributions - 3^d View

Sampled Representation of Continuous Distributions - 3^d View

A/D converter *Mapping Fcn:*	$Y = g(X) = k+1 \; ; \; k \leq x \leq k+1 \; ; \; x = -3,-2,-1,0,1,2$	
Exponential	***Gaussian***	***Uniform***

$$f_Y(y) = \sum_{k=1}^{3} e^{-ak}(e^a - 1) \cdot \delta(y-k)$$

k	α_k
1	0.095
2	0.086
3	0.078

$$f_Y(y) = \sum_{k=-2}^{+3} \alpha_k \cdot \delta(y-k)$$

$$\alpha_k = \int_{x=k-1}^{k} \frac{1}{\sqrt{2\pi}} e^{-x^2/2} dx = \varphi(k) - \varphi(k-1)$$

$$\varphi(k) \equiv \int_{x=-\infty}^{x=k} \frac{1}{\sqrt{2\pi}} e^{-x^2/2} dx \; ; \; k \in (-\infty, \infty)$$

$$f_Y(y) = \sum_{k=-2}^{+3} \frac{1}{6} \cdot \delta(y-k)$$

$$\alpha_k = \frac{1}{6}$$

This slide illustrates the three dimensional representation of the sampled exponential, Gaussian, and uniform distributions of the last slide for the case of 6 samples with k=-2,-1,0,1,2,3. The red δ-function arrows have magnitudes α_k, given in the grey boxes above the plots; the δ-functions are located at the y-values of the zero slope steps of the sampling function in accordance with the discussion on the last slide. Note that for the exponential distribution the first δ-function arrow appears at y =1; there are no δ-function contributions for y=-2,-1, or 0 because the PDF (grey curve) is zero for x<0, and hence there is no accumulation of probability density over the sampling intervals [-3, -2], [-2, -1], and [-1, 0].

Note that in all three cases the δ-function (red) arrows appear at *y-values* corresponding to the intersection of the "steps" with the y-axis.

2.5 *Transformation of Joint PDFs using Jacobian Density Multiplier*

Transformation of Joint PDFs using Jacobian Density Multiplier

Linear Density

$$f_Y(y)dy = f_X(x)dx \quad ; \quad y = g(x)$$

$$f_Y(y) = \frac{f_X(x)}{|dy/dx|} \qquad \text{Express everything in terms of variable "y"}$$

Areal Density

$$f_{Q_1Q_2}(q_1,q_2)dq_1dq_2 = f_{XY}(x,y)dxdy$$

$$q_1 = q_1(x,y) \quad ; \quad q_2 = q_2(x,y)$$

$$f_{Q_1Q_2}(q_1,q_2)dq_1dq_2 = \left[f_{XY}(x,y) \cdot J\begin{pmatrix} x & y \\ q_1 & q_2 \end{pmatrix} \right] dq_1dq_2$$

Jacobian Transformation

$$f_{Q_1Q_2}(q_1,q_2) = \frac{f_{XY}(x,y)}{\left| J\begin{pmatrix} q_1 & q_2 \\ x & y \end{pmatrix} \right|} \qquad J\begin{pmatrix} q_1 & q_2 \\ x & y \end{pmatrix} \equiv \det \begin{bmatrix} \dfrac{\partial q_1}{\partial x} & \dfrac{\partial q_1}{\partial y} \\ \dfrac{\partial q_2}{\partial x} & \dfrac{\partial q_2}{\partial y} \end{bmatrix}$$

Up to this point we have dealt exclusively with single variable probability densities and deposited differential probabilities dP from density plane to the other *via* a coordinate transformation y=g(x). The transformed density is obtained from the equality dP= $f_Y(y)$ dy = $f_X(x)$ dx according to $f_Y(y)$ = $f_X(x)$/ |dy/dx| and this process is visualized in terms a density multiplier given by the inverse 1/ |dy/dx| of the slope of the transformation |g′(x)|. This concept of a linear density multiplier (slope magnitude) generalizes to a Jacobian determinant representing an areal density multiplier in two dimensions; in higher dimensions the Jacobian determinant generalizes to a volume or hyper-volume density multiplier.

In two dimensions, the pair of coordinate transformations q_1=$q_1(x,y)$ and q_2=$q_2(x,y)$ yields the following identity between the joint densities:

$$f_{Q1,Q2}(q_1,q_2) \, dq_1 \, dq_2 = f_{XY}(x,y) \, dx \, dy.$$

The determinant of the Jacobian partials relates the coordinate areas according to

$$dq_1 \, dq_2 = J(q_1,q_2|x,y) \cdot dx \, dy$$

Upon substitution into the density identity, we find

$$f_{Q1,Q2}(q_1,q_2) \, J(q_1,q_2|x,y) \cdot dx \, dy = f_{XY}(x,y) \, dx \, dy$$

so that the transformed density is

$$f_{Q1,Q2}(q_1,q_2) = f_{XY}(x,y) \, / \, | \, J(q_1,q_2|x,y) \, |$$

as shown in the boxed equation. This expression can be generalized to as many dimensions as needed under general coordinate transformations of the form.

$$q_1 = q_1(x_1,x_2,...,x_n), \ q_2 = q_2(x_1,x_2,...,x_n), \ ... \ , \ q_n = q_n(x_1,x_2,...,x_n)$$

$$f_{Q1,Q2,...Qn}(q_1,q_2,...,q_n) = f_{X1,X2,...,Xn}(x_1,x_2,...,x_n) \, / \, | \, J(q_1,q_2,...,q_n|x_1,x_2,...,x_n) \, |$$

2.5.1 Cartesian to Polar Coordinate Transformation – Jacobian Multiplier

Cartesian to Polar Coordinate Transformation – Jacobian Multiplier

Transformation

$$\left. \begin{array}{c} r = \sqrt{x^2 + y^2} \\ \theta = \tan^{-1}(y/x) \end{array} \right\} \underbrace{}_{\text{New Variables}} \Rightarrow \underbrace{\left\{ \begin{array}{l} x = r\cos\theta \\ y = r\sin\theta \end{array} \right.}_{\text{Old Variables}}$$

$dx'dy' = dr\,(r\,d\theta)$

Rotation through θ makes Areas identical

Partials

$$\frac{\partial r}{\partial x} = \frac{1}{2}(x^2 + y^2)^{-1/2} \cdot 2x = \frac{x}{r}$$

$$\frac{\partial r}{\partial y} = \frac{1}{2}(x^2 + y^2)^{-1/2} \cdot 2y = \frac{y}{r}$$

$$\tan\theta = \frac{y}{x}: \quad \left\{ \begin{array}{l} \sec^2\theta \dfrac{\partial \theta}{\partial x} = \dfrac{-y}{x^2} \\[2mm] \sec^2\theta \dfrac{\partial \theta}{\partial y} = \dfrac{1}{x} \end{array} \right.$$

$$\Rightarrow \frac{\partial \theta}{\partial x} = \frac{-r\sin\theta}{(r\cos\theta)^2} \cdot \cos^2\theta = \frac{-\sin\theta}{r} = \frac{-y}{r^2}$$

$$\Rightarrow \frac{\partial \theta}{\partial y} = \frac{1}{r\cos\theta} \cdot \cos^2\theta = \frac{\cos\theta}{r} = \frac{x}{r^2}$$

Jacobian

$$J\begin{pmatrix} r & \theta \\ x & y \end{pmatrix} \equiv \det \begin{bmatrix} \dfrac{\partial r}{\partial x} & \dfrac{\partial r}{\partial y} \\[2mm] \dfrac{\partial \theta}{\partial x} & \dfrac{\partial \theta}{\partial y} \end{bmatrix} = \begin{bmatrix} \dfrac{x}{r} & \dfrac{y}{r} \\[2mm] \dfrac{-y}{r^2} & \dfrac{x}{r^2} \end{bmatrix} = \frac{x^2}{r^3} + \frac{y^2}{r^3} = \frac{1}{r}$$

$$f_{R\Theta}(r,\theta) = \frac{f_{XY}(r\cos\theta, r\sin\theta)}{J\begin{pmatrix} r & \theta \\ x & y \end{pmatrix}} = f_{XY}(r\cos\theta, r\sin\theta) \cdot r$$

The transformation from Cartesian (x,y) to polar coordinates (r, θ) and its inverse are defined by

$$x = r\cos\theta; \quad y = r\sin\theta \quad \text{and} \quad r = (x^2 + y^2)^{1/2}; \quad \theta = \arctan(y/x)$$

The grey differential area can be read directly off the figure in both Cartesian and polar coordinates

$$dA = dx\,dy = dr(r\,d\theta) = r \cdot dr\,d\theta\ .$$

Thus, directly from the geometry, the Jacobian areal density multiplier is seen to be the factor "r".

More formally, the partials are computed in the slide and the Jacobian determinant J(r, θ | x,y) = 1/r, which allows us to write down the Jacobian density transformation formula explicitly as

$$f_{R\Theta}(r, \theta) = f_{XY}(x,y) / |\,J(r, \theta\,|\,x,y)\,| = f_{XY}(r\cos\theta, r\sin\theta) / (1/r) = r\,f_{XY}(r\cos\theta, r\sin\theta)$$

In the last equality, the radial distance "r" acts as the areal density multiplier and upon making the replacements x→rcosθ and y→rsinθ, in $f_{XY}(x,y)$ we create a *different function* of r and θ, which yields the desired probability density function $f_{R\Theta}(r, \theta)$.

Note that this density function $f_{R\Theta}(r, \theta)$ yields a differential probability by simply multiplying it by the product of coordinate differentials, *viz.*,

$$dP = f_{R\Theta}(r, \theta)\,dr\,d\theta$$

because the factor "r" has been absorbed into the density function itself. Further note that the product of coordinates does not have the units of area as it is the product of a length dr times a dimensionless quantity dθ (in radians); the probability density function itself carries the units of inverse length so that the differential probability dP remains dimensionless.

Thus, generalized coordinates such as r and θ, need not have the dimensions of length and the "coordinate" probability density $f_{R\Theta}(r, \theta)$ must have appropriate dimensions in order to make the probability dP dimensionless.

2.5.2 Rayleigh Distribution from Bivariate Gaussian PDF

Rayleigh Distribution from Bivariate Gaussian PDF

Consider 2^d Gaussian $f_{XY}(x,y)$ and transform to polar coordinates (r, θ):

$$f_{XY}(x,y) = \frac{1}{\sqrt{2\pi}}e^{-x^2/2} \cdot \frac{1}{\sqrt{2\pi}}e^{-y^2/2} = \frac{1}{2\pi}e^{-(x^2+y^2)/2}$$

$$x = r\cos\theta \qquad r = \sqrt{x^2+y^2}$$
$$y = r\sin\theta \qquad \theta = \tan^{-1}(y/x) \qquad \Rightarrow J\begin{pmatrix} r & \theta \\ x & y \end{pmatrix} = \frac{1}{r}$$

$$f_{R\Theta}(r,\theta) = \frac{f_{XY}(x,y)}{J\begin{pmatrix} r & \theta \\ x & y \end{pmatrix}} = \frac{\frac{1}{2\pi}\cdot e^{-r^2/2}}{1/r} = \underbrace{\frac{1}{2\pi}}_{=f_\Theta(\theta)}\cdot\underbrace{r\cdot e^{-r^2/2}}_{=f_R(r)} = f_\Theta(\theta)\cdot f_R(r)$$

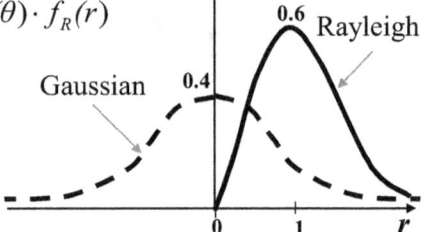

$f_R(r)$

Gaussian 0.4   0.6 Rayleigh

Uniform Distribution

$$f_\Theta(\theta) = \begin{cases} 1/(2\pi) & 0 \le \theta \le 2\pi \\ 0 & \text{else} \end{cases}$$

Rayleigh Distribution

$$f_R(r) = \begin{cases} r\cdot e^{-r^2/2} & r \ge 0 \\ 0 & \text{else} \end{cases}$$

Consider slightly different transformation as follows:

$$d = r^2 = x^2+y^2 \;\; ; \;\; x = d^{1/2}\cos\theta \qquad \Rightarrow J\begin{pmatrix} d & \theta \\ x & y \end{pmatrix} = 2$$
$$\theta = \tan^{-1}(y/x) \;\; ; \;\; y = d^{1/2}\sin\theta$$

or $\quad dA_{XY} = dxdy = rdrd\theta = d(r^2/2)d\theta = (1/2)d(d)d\theta$

$f_D(d)$

½   Exponential

$$f_{D\Theta}(d,\theta) = \frac{f_{XY}(x,y)}{J\begin{pmatrix} d & \theta \\ x & y \end{pmatrix}} = \frac{\frac{1}{2\pi}\cdot e^{-d/2}}{2} = \underbrace{\frac{1}{2\pi}}_{=f_\Theta(\theta)}\cdot\underbrace{e^{-d/2}\cdot\frac{1}{2}}_{f_D(d)}$$

Uniform Distribution

$$f_\Theta(\theta) = \begin{cases} 1/(2\pi) & 0 \le \theta \le 2\pi \\ 0 & \text{else} \end{cases}$$

Exponential Distribution

$$f_D(d) = \begin{cases} \frac{1}{2}\cdot e^{-d/2} & d \ge 0 \\ 0 & \text{else} \end{cases}$$

In the top panel, we consider a 2^d or Bivariate Gaussian $f_{XY}(x,y)$ given as the product of two independent 1^d Gaussians and make the transformation from Cartesian to polar coordinates. The new probability in polar coordinates results from substitution of $x=r\cos\theta$ and $y=r\sin\theta$ into the Cartesian Gaussian and use of the Pythagorean identity $x^2 + y^2 = r^2$ and yields an expression that only depends upon r. Upon dividing by the Jacobian $J(d,\theta|x,y) = 1/r$ we find the resulting $f_{R\Theta}(r, \theta)$ factors into two distributions, namely, a uniform distribution $f_\Theta(\theta) = 1/2\pi$ in θ and a Rayleigh distribution $f_R(r)= r\cdot\exp(-r^2)$ in r. The figure compares a 1^d Gaussian in the variable "r" with the Rayleigh distribution; note that the Gaussian is symmetric about r=0 and extends from $-\infty$ to $+\infty$, while the Rayleigh density is non-zero for positive values of r only. Although the Rayleigh density $f_R(r)$ "looks" somewhat like a Gaussian it must be noted that the $\exp(-r^2)$ is multiplied by "r" and moreover, the Rayleigh r is a polar coordinate taking on only positive values, while r is unrestricted for the Gaussian. Moreover, by setting it's derivative $d/dr(f_R(r)) = \exp(-r^2/2)(1-r^2) = 0 \Rightarrow r=1$ we see that the Rayleigh peaks at r=1 with a value $f_R(r=1) = \exp(-1/2) = 0.6065$. The independent Gaussians peak at x=y=0 with a value of $1/\sqrt{2\pi} = .3989$.

The bottom panel performs a different transformation from Cartesian coordinates (x,y) to the coordinates (r^2, θ), denoted (d, θ) for convenience. The Jacobian $J(x,y|d, \theta) = 1/2$ is obtained directly by comparing the differentials as illustrated on the slide. Alternately, we can compute the Jacobian determinant $J(d,\theta|x,y) = 2$ directly from the coordinate transformations. Also note the determinant relationship $J(d,\theta|x,y)=1/J(x,y|d,\theta) = 1/(½) = 2$. Substituting $x=d^{1/2}\cos\theta$ and $y=d^{1/2}\sin\theta$ and using the identity $x^2 + y^2 = d$ yields an expression that only depends upon d. Upon dividing by the Jacobian $J(d,\theta|x,y) = 2$ we find the resulting $f_{D\Theta}(d, \theta)$ factors into two distributions: a uniform distribution $f_\Theta(\theta) = 1/2\pi$ in θ and an exponential distribution $f_D(d)= ½ \cdot\exp(-d/2)$ in d. The plot is shown in the lower figure and again only positive values of d contribute.

2.5.3 Polar Coordinate Transformation *via* Rotation

Polar Coordinate Transformation *via* Rotation

$$x = r\cos\theta \;\;;\;\; dx = \cos\theta\, dr - r\sin\theta\, d\theta$$
$$y = r\sin\theta \;\;\;\;\; dy = \sin\theta\, dr + r\cos\theta\, d\theta$$

Rotation of differentials through angle θ
so that: $y' \parallel \theta$ and $x' \parallel r$

$$\begin{bmatrix} dx \\ dy \end{bmatrix} = \begin{bmatrix} \cos\theta & -r\sin\theta \\ \sin\theta & r\cos\theta \end{bmatrix}\begin{bmatrix} dr \\ d\theta \end{bmatrix}$$

$$\begin{bmatrix} dx' \\ dy' \end{bmatrix} = \underbrace{\begin{bmatrix} \cos\theta & \sin\theta \\ -\sin\theta & \cos\theta \end{bmatrix}}_{\text{coord. rotation t hru } \theta} \cdot \underbrace{\begin{bmatrix} \cos\theta & -r\sin\theta \\ \sin\theta & r\cos\theta \end{bmatrix}\cdot\begin{bmatrix} dr \\ d\theta \end{bmatrix}}_{=\begin{bmatrix} dx \\ dy \end{bmatrix}}$$

$$\begin{bmatrix} dx' \\ dy' \end{bmatrix} = \begin{bmatrix} C_\theta{}^2 + S_\theta{}^2 & -rS_\theta C_\theta + rS_\theta C_\theta \\ -C_\theta S_\theta + C_\theta S_\theta & r(C_\theta{}^2 + S_\theta{}^2) \end{bmatrix} = \begin{bmatrix} 1 & 0 \\ 0 & r \end{bmatrix}\cdot\begin{bmatrix} dr \\ d\theta \end{bmatrix}$$

$$= \begin{bmatrix} dr \\ r\cdot d\theta \end{bmatrix}$$

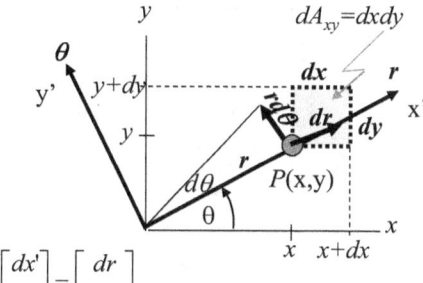

$$\begin{bmatrix} dx' \\ dy' \end{bmatrix} = \begin{bmatrix} dr \\ r\,d\theta \end{bmatrix}$$

$$dA_{xy} = dxdy = dx'dy' = dr\cdot rd\theta$$

$$f_{R\Theta}(r,\theta)drd\theta = f_{XY}(r\cos\theta, r\sin\theta)\cdot dx'dy'$$
$$= \underbrace{f_{XY}(r\cos\theta, r\sin\theta)r\cdot drd\theta}_{\equiv f_{R\theta}(r,\theta)}$$

Note: Computing "dxdy" before rotation yields:
(which is correct but useless!)

$$dxdy = (\cos\theta\, dr - r\sin\theta\, d\theta)(\sin\theta\, dr + r\cos\theta\, d\theta)$$
$$= C_\theta S_\theta dr^2 + r(C^2{}_\theta - S^2{}_\theta)drd\theta - r^2 S_\theta C_\theta d\theta^2$$

Since the product of two scalar lengths cannot be changed by rotation, *i.e.*, $dx'dy'=dxdy$,
we can use the canonical form $dx'dy'=dr\, rd\theta$

This slide gives an explicit proof of the intuitive idea that the grey Cartesian differential area $dA_{xy} = dx\, dy$ illustrated in the figure is equivalent to the polar box $dr\cdot(rd\theta)$ by simply rotating both dx and dy through an angle θ. Given the transformation $x=r\cos\theta$ and $y=r\sin\theta$ we easily find the differentials $dx = \cos\theta\, dr - r\sin\theta\, d\theta$ and $dy = \sin\theta\, dr + r\cos\theta\, d\theta$. These two equations may be re-expressed as the multiplication of a matrix times a column vector $[dr, d\theta]^T$ to obtain the new column vector $[dx, dy]^T$ as shown.

Now if we perform a coordinate rotation $R(\theta)$ through an angle θ on the Cartesian vector $[dx, dy]^T$ we obtain a new Cartesian vector $[dx', dy']^T$. Clearly this rotation leaves the differential area dA unchanged, so we must have $dxdy=dA = dA_{xy}(\theta) = dx'dy'$. The matrix multiplications performed on the slide show that rotated Cartesian vector $[dx', dy']^T$ can be expressed in terms of r and θ, simply as $[dx', dy']^T = [dr, rd\theta]^T$. This result allows us to formally equate the components: $dx' =dr$ and $dy'=rd\theta$. Substitution into of these components into the Cartesian expression $dA_{xy}(\theta) = dx'dy'$ finally yields $dA_{xy}(\theta) = dr(rd\theta)$, thus proving that the polar area $dr(rd\theta)$ is precisely the same as the rotated Cartesian area $dA_{xy}(\theta)$.

2.5.4 Jacobian Transformation of Joint PDFs - Examples

Jacobian Transformation of Joint PDFs - Examples

Continuous Sum and Difference	Non-orthogonal Coordinate Transform

Left panel:

$$\left.\begin{array}{l} q_1 : s = x + y \\ q_2 : d = y - x \end{array}\right\} \Rightarrow \left\{\begin{array}{l} x = (s-d)/2 \\ y = (s+d)/2 \end{array}\right.$$

$$J\begin{pmatrix} q_1 & q_2 \\ x & y \end{pmatrix} \equiv \left|\det\begin{bmatrix} \frac{\partial s}{\partial x} & \frac{\partial s}{\partial y} \\ \frac{\partial d}{\partial x} & \frac{\partial d}{\partial y} \end{bmatrix}\right| = \left|\det\begin{bmatrix} 1 & 1 \\ -1 & 1 \end{bmatrix}\right| = 2$$

$$f_{SD}(s,d) = \frac{f_{XY}\left(\frac{s-d}{2}, \frac{s+d}{2}\right)}{2}$$

Right panel:

$$\left.\begin{array}{l} q_1 = x - \frac{C_\theta}{S_\theta} y \\ q_2 = \frac{y}{S_\theta} \end{array}\right\} \Rightarrow \left\{\begin{array}{l} x = q_1 + q_2 C_\theta \\ y = q_2 S_\theta \end{array}\right.$$

$$J\begin{pmatrix} q_1 & q_2 \\ x & y \end{pmatrix} \equiv \left|\det\begin{bmatrix} \frac{\partial q_1}{\partial x} & \frac{\partial q_1}{\partial y} \\ \frac{\partial q_2}{\partial x} & \frac{\partial q_2}{\partial y} \end{bmatrix}\right| = \left|\det\begin{bmatrix} 1 & \frac{-C_\theta}{S_\theta} \\ 0 & \frac{-1}{S_\theta} \end{bmatrix}\right| = \frac{1}{|S_\theta|}$$

$$f_{Q_1 Q_2}(q_1, q_2) = \left[\frac{f_{XY}(q_1 + q_2 C_\theta, q_2 S_\theta)}{\left|\frac{1}{S_\theta}\right|}\right]$$

$$= f_{XY}(q_1 + q_2 C_\theta, q_2 S_\theta) \cdot |S_\theta|$$

Here are two examples of the Jacobian transformation of 2^d PDFs (PDF Method) for (i) the continuous version of the sum and difference coordinates (recall 6-sided pair of dice), and (ii) a non-orthogonal coordinate transformation which proves to be quite useful in certain circumstances.

The **left panel** defines the sum (q_1): s=x+y and difference (q_2): d=y-x and gives the resulting inverse transformations x=(s-d)/2 and y=(s+d)/2. The Jacobian is easily shown to be $J(q_1,q_2|x,y) = J(s,d|x,y)=2$ and the resulting density is easily evaluated as $f_{SD}(s,d)= f_{XY}((s-d)/2, (s+d)/2) / 2$.

The **right panel** defines the 1st variable q_1=x–y[cosθ /sinθ] and the 2nd variable q_2=y/sinθ and gives the resulting inverse transformations x=q_1 + q_2 cosθ and y= q_2 sinθ. The Jacobian is easily shown to be $J(q_1,q_2| x,y)$ = 1/ |sinθ| and the resulting density is easily evaluated as

$$f_{Q1, Q2}(q_1, q_2) = f_{XY}(q_1 + q_2 \cos\theta, q_2 \sin\theta) \cdot |\sin\theta|.$$

This general procedure is quite straight forward, but its application to specific problems may seem a bit novel. For example, we have a pair of independent RVs with two different distributions and wish to find the PDF for their sum. We can solve this problem by first recognizing that the joint distribution for a pair of independent RVs is simply the product of the two individual distributions, i.e., $f_{XY}(x,y) = f_X(x) \cdot f_Y(y)$. Then we proceed by transforming from the (x, y) coordinates to the sum-difference (s, d) coordinates to obtain $f_{SD}(s,d)$ which is given in its general form in the left panel of this slide. We next integrate over the difference coordinate d to obtain the marginal distribution $f_S(s)$ which is precisely the probability density that we seek. Introducing a second coordinate ("d" in this case) allows us to make use of the Jacobian method to obtain the **joint distribution** on two variables (s, d) (one of which we don't need); subsequent integration over the "unneeded" variable yields the single variable or marginal distribution to solve the problem. This represents a general method which at times may require a non-orthogonal transformation similar to the one given in the right hand panel. An explicit example is discussed on the next two slides and others will pop up from time to time.

2.6 Sum of Two Independent Exponential RVs - Jacobian Method

Sum of Two Independent Exponential RVs - Jacobian Method

Ex. 5: X & Y Indep Exponential RVs with PDFs   Find PDF for sum variable $f_S(s)$

$$f_X(x) = \begin{cases} \alpha e^{-\alpha x} & x \geq 0 \\ 0 & x < 0 \end{cases} \quad ; \quad f_Y(y) = \begin{cases} \beta e^{-\beta y} & y \geq 0 \\ 0 & y < 0 \end{cases} \qquad \begin{matrix} s = x + y \\ d = y - x \end{matrix} \Rightarrow \begin{cases} x = (s-d)/2 \\ y = (s+d)/2 \end{cases} \qquad J\begin{pmatrix} s & d \\ x & y \end{pmatrix} = 2$$

Must Integrate Joint Distribution $f_{SD}(s,d)$ over all d
to find PDF for sum variable $f_S(s)$

$$f_{SD}(s,d) = \frac{1}{2} f_{XY}\left(\frac{s-d}{2}, \frac{s+d}{2}\right) \underset{\text{X,Y Indep RVs}}{=} \frac{1}{2} f_X\left(\frac{s-d}{2}\right) \cdot f_Y\left(\frac{s+d}{2}\right)$$

$$= \frac{\alpha\beta}{2} e^{-\alpha\frac{s-d}{2}} e^{-\beta\frac{s+d}{2}} = \frac{\alpha\beta}{2} e^{-\frac{s}{2}(\alpha+\beta)} e^{-\frac{d}{2}(\beta-\alpha)}$$

Strips go from $d=-s$ to $d=+s$

$$f_S(s) = \int_{d=-s}^{d=+s} dd\, f_{SD}(s,d) = \frac{\alpha\beta}{2} e^{-\frac{s}{2}(\alpha+\beta)} \int_{d=-s}^{d=+s} dd\, e^{-\frac{d}{2}(\beta-\alpha)}$$

$$= \frac{\alpha\beta}{2} e^{-\frac{s}{2}(\alpha+\beta)} \left\{ \int_{d=-s}^{d=0} dd\, e^{-\frac{d}{2}(\beta-\alpha)} + \int_{d=0}^{d=+s} dd\, e^{-\frac{d}{2}(\beta-\alpha)} \right\}$$

$$= \frac{\alpha\beta}{2} e^{-\frac{s}{2}(\alpha+\beta)} \left\{ \frac{e^{-\frac{(\beta-\alpha)}{2}d}}{-\frac{(\beta-\alpha)}{2}} \Bigg|_{d=-s}^{0} + \frac{e^{-\frac{(\beta-\alpha)}{2}d}}{-\frac{(\beta-\alpha)}{2}} \Bigg|_{d=0}^{+s} \right\}$$

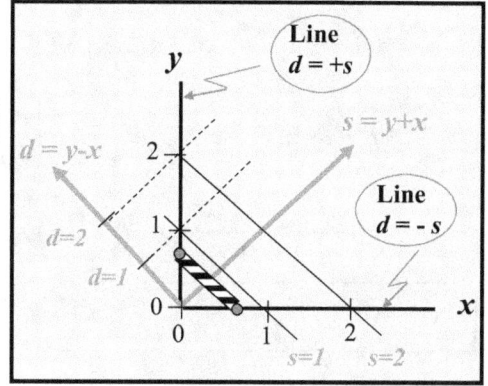

$$f_S(s) = \frac{\alpha\beta}{(\alpha-\beta)} e^{-\frac{s}{2}(\alpha+\beta)} \left\{ 1 - e^{-\frac{(\beta-\alpha)}{2}(-s)} + e^{\frac{(\beta-\alpha)}{2}s} - 1 \right\}$$

$$\boxed{f_S(s) = \frac{\alpha\beta}{(\alpha-\beta)} \left\{ e^{-\beta s} - e^{-\alpha s} \right\} \quad ; \quad s \geq 0}$$

We are given a pair of independent exponential RVs having parameters α and β respectively and wish to find the distribution of their sum. One approach is to note that because of independence the joint distribution $f_{XY}(x,y)$ is simply the product $f_X(x) \cdot f_Y(y)$ and then transform from (x, y) to sum/difference coordinates (s, d) *via* $x=x(s,d) = (s-d)/2$ and $y=y(s,d) = (s+d)/2$. The Jacobian method with $|J|=2$ applied to this product yields

$$f_{SD}(s,d) = f_X((s-d)/2) \cdot f_Y((s+d)/2)/2 = (\alpha\beta/2)\, e^{-s(\alpha+\beta)/2} e^{-d(\beta-\alpha)/2}$$

The idea is to find marginal distribution $f_S(s)$ by integrating over the difference coordinate d; this is straight-forward, but care must be taken to set up the integral properly. The figure shows the sum "s" axis is at $+45°$ degrees and the difference "d" axis is at $+135°$ and labels the x-axis as the *line* d=-s and the y-axis as the *line* d=+s. First note that the probability density is non-zero only in the first quadrant where both x and y are positive. We need to "collapse" the probability onto the s-axis, and accordingly, choose the integration strip along the d-coordinate as shown in the figure with "d" ranging from the x-axis d="-s" to the y-axis d="+s".

The term involving the variable "s" is brought outside of the integral an   d the integral over the variable "d" is split into two parts: one from d="-s" to d="0" and the other from d="0" to d="+s" . (Note these two parts are not equal because of the different parameters α and β along the x- and y-axes.) The integration of the exponential function is straight forward and after a little algebra, yields

$$f_S(s) = [\alpha\beta/(\alpha-\beta)] \{e^{-\beta \cdot s} - e^{-\alpha \cdot s}\} \quad \text{for } s \geq 0 \text{ and } \alpha \neq \beta$$

For the special case $\alpha=\beta$, the integrand is "1" and the integral over "d" reduces to 2s to give

$$f_S(s) = \alpha^2 s\, e^{-\alpha \cdot s} \quad \text{for } s \geq 0 \text{ and } \alpha=\beta$$

Alternately, let $\alpha-\beta=\varepsilon$ and take the limit as $\varepsilon \to 0$ (*i.e.*, $\alpha=\beta$) to yield

$$L_{\varepsilon \to 0} \frac{\alpha\beta}{\varepsilon} \cdot e^{-\beta \cdot s} \cdot \{1 - e^{-\varepsilon \cdot s}\} = L_{\varepsilon \to 0} \frac{\alpha^2}{\varepsilon} \cdot e^{-\beta \cdot s} \cdot \{1 - (1 - \varepsilon \cdot s)\} = \alpha^2 \cdot s \cdot e^{-\alpha \cdot s}$$

2.6.1 Sum of Two Independent Exponential RVs - CDF Method

Sum of Two Independent Exponential RVs-CDF Method

Double Integral for $CDF_S(s)$

$$F_S(s) = \int_{y=0}^{y=s} dy \int_{x=0}^{x=s-y} dx\, f_{XY}(x,y) = \int_{y=0}^{y=s} dy\, \beta e^{-\beta y}\left\{\int_{x=0}^{x=s-y} dx\, \alpha e^{-\alpha x}\right\}$$

$$= \alpha\beta \int_{y=0}^{y=s} dy\, e^{-\beta y}\left.\frac{e^{-\alpha x}}{-\alpha}\right|_{x=0}^{x=s-y} = \beta \int_{y=0}^{y=s} dy\, e^{-\beta y}\left\{1-e^{-\alpha(s-y)}\right\}$$

$$= \beta \int_{y=0}^{y=s} dy\left\{e^{-\beta y} - e^{-(\beta-\alpha)y}e^{-\alpha s}\right\} = \beta\left\{\left.\frac{e^{-\beta y}}{-\beta}\right|_{y=0}^{s} - \left.\frac{e^{-(\beta-\alpha)y}}{-(\beta-\alpha)}e^{-\alpha s}\right|_{y=0}^{s}\right\}$$

$$= \left\{(1-e^{-\beta s}) + \frac{\beta e^{-\alpha s}}{(\beta-\alpha)}(e^{-(\beta-\alpha)s}-1)\right\}$$

$$\boxed{F_S(s) = (1-e^{-\beta s}) + \frac{\beta}{(\beta-\alpha)}(e^{-\beta s}-e^{-\alpha s})}$$

Differentiate $F_S(s)$ to obtain $f_S(s)$

$$f_S(s) = \frac{\alpha\beta}{(\alpha-\beta)}\left\{e^{-\beta s} - e^{-\alpha s}\right\}\ ;\ s\geq o$$

(figure: y vs x plot with "surface" $s=x+y=const.$, line s, $y="s"$, $y="0"$, $x="0"$, $x="s-y"$)

|

Details

↓

$$f_S(s) = \frac{d}{ds}F_S(s) = \frac{d}{ds}\left\{(1-e^{-\beta s}) + \frac{\beta}{(\beta-\alpha)}(e^{-\beta s}-e^{-\alpha s})\right\}$$

$$= +\beta e^{-\beta s} + \frac{\beta}{\beta-\alpha}\left(-\beta e^{-\beta s} + \alpha e^{-\alpha s}\right) = \frac{\alpha\beta}{(\alpha-\beta)}\left\{e^{-\beta s} - e^{-\alpha s}\right\}$$

As an alternative to the Jacobian PDF Method of the last slide, the CDF Method may also be used to find the probability density $f_S(s)$ for the sum RV. To do this it is necessary to find the cumulative distribution function $F_S(s)$ and then differentiate with respect to s. In the case of a single variable X the procedure followed three steps, namely, first finding $F_X(x)$, then relating it to the CDF for the new RV Y : $F_Y(y)$, and finally performing the differentiation to find the density $f_Y(y) = d/dy[F_Y(y)]$. Clearly, with two RVs X and Y there is no way of directly relating the individual cumulative distributions $F_X(x)$ and $F_Y(y)$ to $F_S(s)$. Instead, it is necessary to go back to the basic probability definition of the cumulative distribution function, *viz.*, $F_S(s) = \Pr[S\leq s]$; this requires us to accumulate the differential probability $dP=f_{XY}(x,y)dxdy$ over the whole x-y region up to the line s=x+y which is the "surface of constant s". This is accomplished by choosing a horizontal integration strip (fixed value of "y") from x=0 to the x-value on the "surface (line)" *i.e.*, x= "s-y" as shown in the figure. This sets the limits for the integration over "x" which must be performed first (because it is evaluated at a fixed value of "y"). The subsequent integration from y= "0" to y="s" covers all horizontal strips within the given "s=const. surface".

The 1st integration over x yields an exponential to be evaluated at the end points 0 and s-y and this leaves the "circled" integral over y to be evaluated between the limits y= "0" and y="s" . This integral has been circled because there is no need to integrate it as it can be differentiated directly using Leibnitz's Rule as will be shown on the next slide. However, it is simple enough to perform the integration directly to find the cumulative distribution $F_S(s)$; subsequent differentiation yields the same result for $f_S(s)$ as found using the Jacobian method
$$f_S(s) = [\alpha\beta/(\alpha-\beta)]\ \{e^{-\beta\cdot s} - e^{-\alpha\cdot s}\}\ \text{for}\ s\geq 0\ \text{and}\ \alpha\neq\beta$$

2.7 Leibnitz's Rule for Differentiating Integrals

Leibnitz's Rule for Differentiating Integrals

$$I(t) = \int_{a(t)}^{b(t)} p(t,x)\,dx$$

$$\frac{dI(t)}{dt} = \lim_{\Delta t \to 0} \frac{I(t+\Delta t) - I(t)}{\Delta t}$$

$$\frac{dI(t)}{dt} = \frac{db}{dt}p(t,b) - \frac{da}{dt}p(t,a) + \int_{a(t)}^{b(t)} \frac{\partial p(t,x)}{\partial t}\,dx$$

Example: Instead of integrating to obtain explicit form of $F_S(s)$ and then $f_S(s) = d/ds\{F_S(s)$

Differentiate "integral form" $F_S(s)$ using Liebnitz's Rule to obtain $f_S(s)$

$$F_S(s) = \int_{y=0}^{y=s} dy\, \beta e^{-\beta y}\underbrace{\left\{1 - e^{-\alpha(s-y)}\right\}}_{= f_Y(s,y)}$$

$$f_S(s) = \frac{dF_S(s)}{ds} = \frac{ds}{ds}\cdot \beta e^{-\beta y}\underbrace{\left\{1 - e^{-\alpha(s-y)}\right\}}_{\text{Evaluates to 0}}\Bigg|_{y=s} + \int_{y=0}^{y=s} dy\, \frac{\partial}{\partial s}\underbrace{\left[\beta e^{-\beta y}\left\{1 - e^{-\alpha(s-y)}\right\}\right]}_{= -\beta e^{-(\beta-\alpha)y}\cdot e^{-\alpha s}(-\alpha)}$$

$$f_S(s) = \alpha\beta e^{-\alpha s}\cdot \int_{y=0}^{y=s} dy\, e^{-(\beta-\alpha)y} = \alpha\beta e^{-\alpha s}\cdot \frac{e^{-(\beta-\alpha)y}}{-(\beta-\alpha)}\Bigg|_{y=0}^{y=s} = \boxed{\frac{\alpha\beta}{(\alpha-\beta)}\left\{e^{-\beta s} - e^{-\alpha s}\right\}}$$

The integral of a function f(x,t) over x between the **constant limits** x=a and x=b is a function I(t). The derivative of this integral d/dt [I(t)] = lim$_{\Delta t \to 0}$ I(t+Δt)-I(t)]/Δt yields the integral of the partial derivative f_t(x,t), which is the last term in the Leibnitz's Rule boxed equation. In the more general situation in which the integration limits a(t) and b(t) are functions of the parameter "t", the two terms involving the time derivatives of da(t)/dt and db(t)/dt must be included.

In the previous slide we computed $F_S(s)$ and we circled the intermediate result after the 1st integration over x because it left an integral of a function f(y,s) over y and had limits y=0 to y=s. Transcribing the variables (y,s) to (x,t) yields exactly the form for which Leibnitz's Rule is applicable. For convenience, we have repeated the intermediate form for $F_S(s)$ given in the last slide and then apply Leibnitz's Rule directly to it to obtain correct result for $f_S(s)$ as shown.

2.7.1 Leibnitz's Rule Proof

Leibnitz's Rule Proof

$$I(t) = \int_{a(t)}^{b(t)} p(t,x)dx \quad ; \quad \frac{dI(t)}{dt} = \frac{db}{dt}p(t,b) - \frac{da}{dt}p(t,a) + \int_{a(t)}^{b(t)} \frac{\partial p(t,x)}{\partial t}dx$$

$$\frac{dI(t)}{dt} = \underset{\Delta t \to 0}{Lim}\left(\frac{I(t+\Delta t) - I(t)}{\Delta t}\right) = \underset{\Delta t \to 0}{Lim}\frac{1}{\Delta t}\left\{\int_{a(t+\Delta t)}^{b(t+\Delta t)} p(t+\Delta t,x)dx - \int_{a(t)}^{b(t)} p(t,x)dx\right\}$$

$$b(t+\Delta t) = b(t) + \Delta b = b(t) + b'(t)\cdot\Delta t \quad ; \quad a(t+\Delta t) = a(t) + \Delta a = a(t) + a'(t)\cdot\Delta t$$

$$p(t+\Delta t,x) = p(t,x) + \Delta p = p(t,x) + p_t(t,x)\cdot\Delta t$$

$$\frac{dI(t)}{dt} = \underset{\Delta t \to 0}{Lim}\frac{1}{\Delta t}\left\{\int_{a+\Delta a}^{b+\Delta b}(p+\Delta p)dx - \int_{a(t)}^{b(t)}pdx\right\} = \underset{\Delta t \to 0}{Lim}\frac{1}{\Delta t}\left\{\left[\int_{a+\Delta a}^{b+\Delta b} - \int_{a}^{b}\right]p(t,x)dx + \int_{a+\Delta a}^{b+\Delta b}p_t(t,x)\cdot\Delta t\cdot dx\right\}$$

$$= \underset{\Delta t \to 0}{Lim}\frac{1}{\Delta t}\left\{\left[\int_{a+\Delta a}^{b} + \int_{b}^{b+\Delta b} - \int_{a}^{a+\Delta a} - \int_{a+\Delta a}^{b}\right]p(t,x)dx + \left[\int_{a}^{b} - \int_{a}^{a+\Delta a} + \int_{b}^{b+\Delta b}\right]p_t(t,x)\cdot\Delta t\cdot dx\right\}$$

cancel

$$= \underset{\Delta t \to 0}{Lim}\frac{1}{\Delta t}\left\{\left[p(t,x)\cdot\Delta b - p(t,x)\cdot\Delta a\right] + \left[\Delta t\cdot\int_{a}^{b}p_t(t,x)dx - p_t(t,x)\cdot\underset{\sim\Delta t^2}{\underline{\Delta t\cdot\Delta a}} + p_t(t,x)\cdot\underset{\sim\Delta t^2}{\underline{\Delta t\cdot\Delta b}}\right]\right\}$$

$$\to 0 \qquad \to 0$$

$$\frac{dI(t)}{dt} = \frac{db}{dt}p(t,b) - \frac{da}{dt}p(t,a) + \int_{a(t)}^{b(t)} \frac{\partial p(t,x)}{\partial t}dx \qquad \text{QED}$$

The derivation of Leibnitz's rule for differentiation of an integral of $p(t,x)$ between the limits $x=a(t)$ and $x=b(t)$ is performed by applying the formal calculus definition of the derivative of an integral $I(t)$

$$d/dt\,[I(t)] = \lim_{\Delta t \to 0} I(t+\Delta t) - I(t)]/\Delta t$$

First we approximate all terms appearing in the expressions for $I(t+\Delta t)$ and $I(t)$ by performing Taylor expansions in t. Thus, the integral limits are expanded to 1st order as

$$a(t+\Delta t) = a(t) + \Delta a \quad ; \quad b(t+\Delta t) = b(t) + \Delta b \,;$$

and the integrand itself is approximated to 1st order as

$$p(t+\Delta t,\,x) = p(t,x) + \Delta p = p(t,x) + p_t(t,x)\cdot\Delta t$$

Now break out simpler integrals, some of which cancel as indicated on the slide; two of the remaining integrals are over the infinitesimal intervals $[a, a+\Delta a]$, $[b, b+\Delta b]$ and may be evaluated directly as single strips $p(t,a)\cdot\Delta a$ and $p(t,b)\cdot\Delta b$; upon dividing these two terms by Δt and taking limit $\Delta t \to 0$ gives the two derivative terms. A third integral contains terms of order $(\Delta t)^2$, which can be dropped because upon dividing by Δt and taking limit $\Delta t \to 0$ they vanish. Finally, there is a term containing the partial $p_t(t,x)\cdot\Delta t$ integrated over the x-interval $[a(t), b(t)]$; dividing this term by Δt and taking the limit $\Delta t \to 0$, yields the "integral of the partial derivative" term in Leibnitz's Rule. The boxed equation at the bottom gives the final result which is the Leibnitz rule.

2.8 *Generating Functions, RV Sums, and Convolution*

Generating Functions, RV Sums, and Convolution

Moment Generating Functions
$$\phi_X(t) \equiv E[e^{Xt}] = \int_{-\infty}^{+\infty} dx\, f_X(x) e^{Xt}$$

PDF_Z   Z=X+Y ==> Convolution
$$\phi_Z(t) \equiv E[e^{(X+Y)t}] = E[e^{Xt}] \cdot E[e^{Yt}] = \phi_X(t) \cdot \phi_Y(t)$$

$$Z = X + Y \quad X, Y \text{ are independent}$$

Method 1 - Direct Transformation of PDFs - Jacobian Sum-Difference Coordinates

$$Z = X + Y \; ; \; W = Y - X$$

$$f_{ZW}(z,w) = \frac{f_X(x) \cdot f_Y(y)}{J\left(\begin{smallmatrix} z & w \\ x & y \end{smallmatrix}\right)} = \frac{f_X(\frac{z-w}{2}) \cdot f_Y(\frac{z+w}{2})}{2}$$

$$f_Z(z) = \int_{w=-\infty}^{+\infty} dw \frac{f_X(\frac{z-w}{2}) \cdot f_Y(\frac{z+w}{2})}{2}$$

Substitute: $\xi = \frac{z+w}{2} \Rightarrow d\xi = \frac{dw}{2} \; ; \quad z - \xi = \frac{z-w}{2}$

$$f_Z(z) = \int_{\xi=-\infty}^{+\infty} d\xi\, f_X(z-\xi) \cdot f_Y(\xi) = \underbrace{f_X * f_Y[z]}_{Z=X+Y}$$

Convolution

Sum Variable Z = X + Y

Prob Densities: Sum ==> Convolution $f_Z(z) = f_X * f_Y[z]$

Generating Fcns: Sum ==> Prod $\phi_Z(t) = \phi_X(t) \cdot \phi_Y(t)$

Recall that the moment generating function of a RV X is defined to be the expectation of the exponential of its product with a parameter "t" *viz.*, $\phi_X(t) = E[e^{X \cdot t}]$. In the discrete case, the expectation is a sum of $e^{X \cdot t}$ over all values of the RV X and leads to a function of the parameter "t" $\phi_X(t)$ called the moment generating function. For continuous RVs the sum becomes an integral as given in the first equation. The definition of the generating function puts the RVs in the exponent so the sum of two variables in the exponent becomes the product of terms. This leads directly to the result that the generating function for the sum of two independent RVs is simply the product of the two generators, *viz.*, $\phi_Z(t) = \phi_X(t)\,\phi_Y(t)$. Both the PDF $f_X(x)$ and the generating function $\phi_X(t)$ give complete information about the random variable X and they may be used *interchangeably*. In the discrete case, we used the equation Z=X+Y and polynomial multiplication to show that the PDF of the sum variable $f_Z(z)$ is equal to the convolutional sum of $f_X(x)$ and $f_Y(y)$, expressed in common "*" notation as $f_Z(z) = (f_X * f_Y)[z]$.

Similarly, for continuous variables, $f_Z(z)$ is equal to the convolutional integral of $f_X(x)$ and $f_Y(y)$ given in the 1st boxed equation. This may be proved using either the CDF method applied to the sum variable Z or the Jacobian method applied to the pair (Z, W) of sum and difference variables. The joint density $f_{ZW}(z,w)$ is obtained by computing the Jacobian of the transformation J(z,w; x,y)=2 and substituting the inverse transformation x(z,w) and y(z,w) into $f_X(x)$ and $f_Y(y)$ respectively. Integration of this joint density over "w" is facilitated by the substitutions $\xi = (z+w)/2$, $d\xi = dw/2$, and $z - \xi = (z-w)/2$ to yield the desired result. The CDF method is developed on the next slide.

The second boxed equation summarizes results for the sum of two random variables Z=X+Y in terms of the two equivalent RV representations; it is seen that the PDF for the sum function requires a convolutional integral, while the generating function only requires a multiplication. We shall see that the generating function facilitates many otherwise tedious probability computations.

2.8.1 Convolution Integral for Sum RV – CDF Method

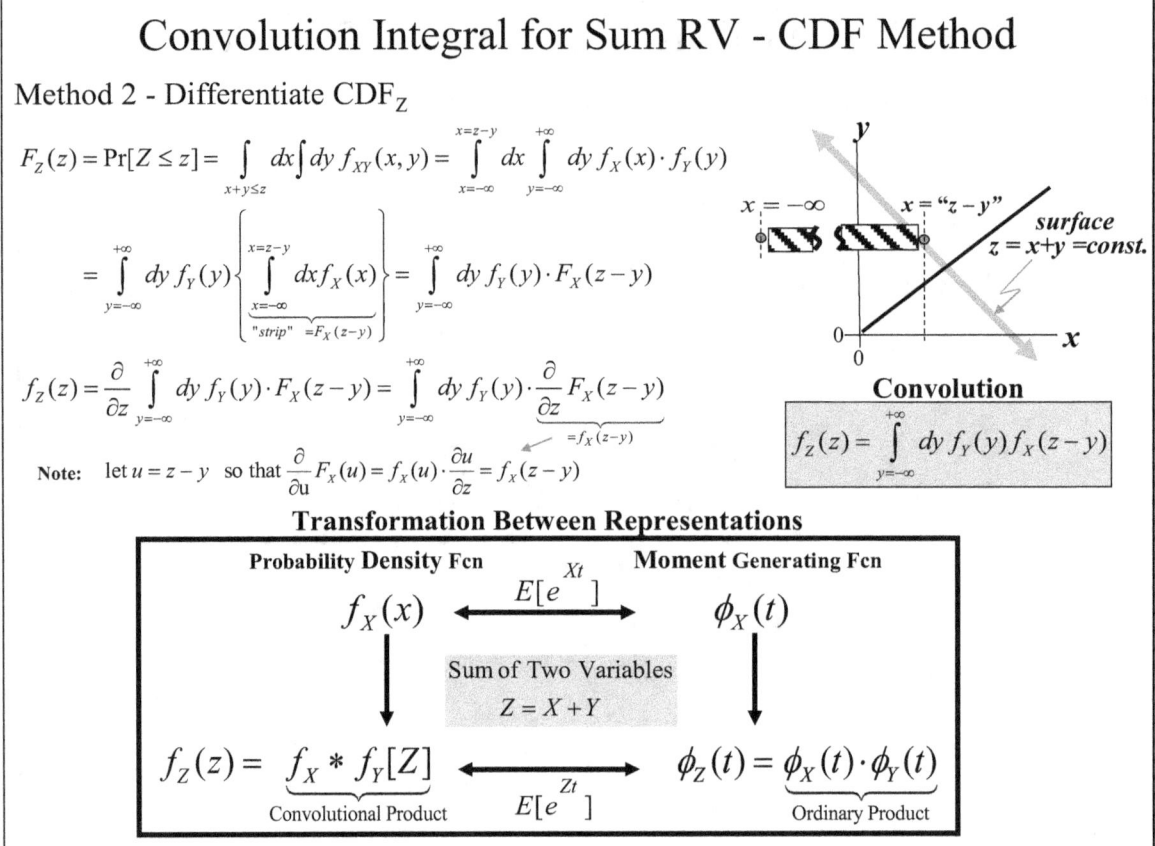

Convolution Integral for Sum RV - CDF Method

Method 2 - Differentiate CDF_Z

$$F_Z(z) = \Pr[Z \le z] = \int_{x+y \le z} dx \int dy \, f_{XY}(x,y) = \int_{x=-\infty}^{x=z-y} dx \int_{y=-\infty}^{+\infty} dy \, f_X(x) \cdot f_Y(y)$$

$$= \int_{y=-\infty}^{+\infty} dy \, f_Y(y) \underbrace{\left\{ \int_{x=-\infty}^{x=z-y} dx f_X(x) \right\}}_{\text{"strip"} \ = F_X(z-y)} = \int_{y=-\infty}^{+\infty} dy \, f_Y(y) \cdot F_X(z-y)$$

$$f_Z(z) = \frac{\partial}{\partial z} \int_{y=-\infty}^{+\infty} dy \, f_Y(y) \cdot F_X(z-y) = \int_{y=-\infty}^{+\infty} dy \, f_Y(y) \cdot \underbrace{\frac{\partial}{\partial z} F_X(z-y)}_{=f_X(z-y)}$$

Note: let $u = z - y$ so that $\frac{\partial}{\partial u} F_X(u) = f_X(u) \cdot \frac{\partial u}{\partial z} = f_X(z-y)$

Convolution

$$f_Z(z) = \int_{y=-\infty}^{+\infty} dy \, f_Y(y) f_X(z-y)$$

$x = -\infty$ $x = \text{"}z-y\text{"}$ *surface* $z = x+y = const.$

Transformation Between Representations

Probability Density Fcn   **Moment Generating Fcn**

$$f_X(x) \xleftrightarrow{\quad E[e^{Xt}] \quad} \phi_X(t)$$

Sum of Two Variables
$Z = X + Y$

$$f_Z(z) = \underbrace{f_X * f_Y[Z]}_{\text{Convolutional Product}} \xleftrightarrow{\quad E[e^{Zt}] \quad} \phi_Z(t) = \underbrace{\phi_X(t) \cdot \phi_Y(t)}_{\text{Ordinary Product}}$$

The proof of the convolutional integral using the CDF method requires that we first compute the cumulative distribution for the sum variable, $F_Z(z) = \Pr[Z \le z]$; the figure shows the surface of constant Z as the dark (double arrow) line at -45°. The double integral of $f_{XY}(x,y)$ is performed over **all x and y for which x+y=z** and the double arrow symbolizes the fact that the line extends to infinity in both directions. The integrand factors $f_{XY}(x,y) = f_X(x) f_Y(y)$ because of independence; the integration limits on "x" are established by considering a typical integration strip from $x = -\infty$ to the constant z surface x="z-y" and the limits on "y" are $(-\infty, +\infty)$ as is evident from the figure. It is observed that the integral over "x" is the probability $\Pr[X \le z-y] = F_X(z-y)$, *i.e.*, the cumulative sum for the RV X evaluated at X=z-y. This leaves a single integral $F_Z(z) = \int f_Y(y) F_X(z-y) \, dy$ for "y" over $(-\infty, +\infty)$. Using Leibnitz's Rule, we differentiate the integrand with respect to the "parameter" z: $f_Z(z) = d/dz \, [F_Z(z)] = \int f_Y(y) \cdot d/dz \{F_X(z-y)\} \, dy = \int f_Y(y) \cdot f_X(z-y) dy$, which is the desired convolutional integral for the sum variable Z.

The box at the bottom summarizes methods for obtaining the sum of two random variables Z=X+Y in terms of the two equivalent representations $f_X(x)$ and $\phi_X(t)$. On the top we indicate this back and forth equivalence by the double arrow between the PDF $f_X(x)$ and the Generator $\phi_X(t)$. The sum Z=X+Y in the center is performed on the left PDF side by a *convolutional product* $f_Z(z) = f_X * f_Y[z]$ and on the right Generator side by the *direct product* $\phi_Z(t) = \phi_X(t) \cdot \phi_Y(t)$. The double arrow on the bottom shows $E[e^{Xt}]$ transformation between the two representations for the sum Z. Clearly a direct product is simpler than a convolutional integral provided we can subsequently find the inverse transformation back to an easily interpreted distribution. The following few slides make use of these ideas to obtain the sum PDF by (i) first transforming each RV to the Generator representation $\phi_X(t)$, (ii) taking generator product = $\phi_X(t) \, \phi_Y(t)$, and (iii) transforming back to PDF representation to find $f_Z(z)$. This avoids tedious convolution computations for the sum variable.

2.8.2 Generating Function - Exponential PDF

Generating Function - Exponential PDF

Generating Fcn for Exponential PDF

$$PDF_X = f_X(x) = \begin{cases} \alpha e^{-\alpha x} & x \geq 0 \\ 0 & x < 0 \end{cases} \qquad \Longleftrightarrow \qquad \phi_X(t) = \begin{cases} \dfrac{\alpha}{\alpha - t} & t < \alpha \\ \infty & t \geq \alpha \end{cases}$$

positive PDF requires $\alpha > 0$

$$\phi_X(t) = E[e^{Xt}] = \int_{x=0}^{\infty} e^{xt} \alpha e^{-\alpha x} dx = \int_{x=0}^{\infty} \alpha e^{-(\alpha - t)x} dx = \alpha \left. \frac{e^{-(\alpha - t)x}}{-(\alpha - t)} \right|_{x=0}^{\infty} = \frac{\alpha}{-(\alpha - t)} \cdot \left(\underset{\substack{x \to \infty \\ \alpha > t}}{\lim} e^{-(\alpha - t)x} - \underbrace{e^{-(\alpha - t)0}}_{=1} \right)$$

$$\phi_X(t) = \begin{cases} \dfrac{\alpha}{\alpha - t} & t < \alpha \\ \infty & t \geq \alpha \end{cases} \qquad \text{Generating Fcn is always Eval } \phi(t = 0)$$

so for $(\alpha - t) > 0$ first term $e^{-(\alpha - t)x}$ vanishes!

Use Generating Fcn to Compute Statistics

$$E[X] = \left. \frac{d\phi}{dt} \right|_{t=0} = \left. \frac{\alpha}{(\alpha - t)^2} \right|_{t=0} = \frac{1}{\alpha} \quad ; \quad E[X^2] = \left. \frac{d^2\phi}{dt^2} \right|_{t=0} = \left. \frac{2\alpha}{(\alpha - t)^3} \right|_{t=0} = \frac{2}{\alpha^2} \quad ; \quad Var(X) = \frac{2}{\alpha^2} - \left(\frac{1}{\alpha} \right)^2 = \frac{1}{\alpha^2}$$

Performing the expectation integral for the exponential PDF with parameter $\alpha > 0$ requires evaluation of the expression $e^{-(\alpha - t)x}$ at $x = +\infty$; provided $(\alpha - t) > 0$ or $t < \alpha$ this term vanishes and leads to a well behaved generator function $\phi_X(t) = \alpha / (\alpha - t)$ for $t < \alpha$. Since the generator is only differentiated a number of times and then evaluated at $t=0$ to find the various moments of the distribution, the fact that it becomes infinite when $t \geq \alpha$ is of no consequence and can be ignored.

The generator easily yields the first two moments of the exponential distribution to be $1/\alpha$ and $2/\alpha^2$ by simply differentiating $\phi_X(t)$ and evaluating it at $t=0$ as shown in the slide.

2.8.3 Exponential Sum "Z = X+Y" using Generating Function

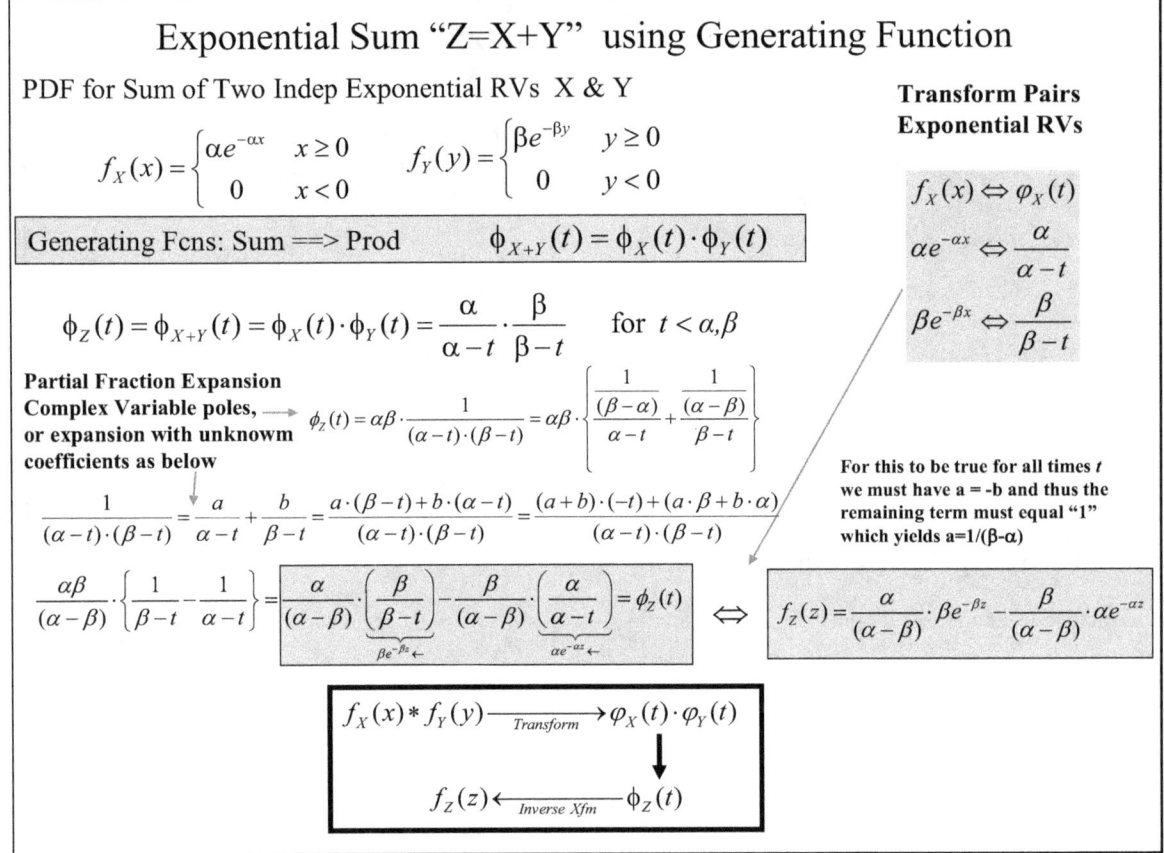

Given two independent exponential distributions with different parameters α and β respectively we may immediately write down their generating functions as $\phi_X(t) = \alpha/(\alpha-t)$ and $\phi_Y(t) = \beta/(\beta-t)$ and obtain the generator for the sum variable Z=X+Y as the direct product $\phi_Z(t) = [\alpha/(\alpha-t)]\cdot[\beta/(\beta-t)]$. Now if we can find a way to transform from Generator representation back to PDF representation $\phi_Z(t) \rightarrow f_Z(z)$ we obtain the PDF for the sum. Note that the "box diagram" shows the path we are taking. The expression for $\phi_Z(t)$ can be expanded using a partial fraction expansion and takes the form

$$\phi_Z(t) = \alpha\beta / [(\alpha-t)(\beta-t)] = [\beta/(\alpha-\beta)]\cdot\{\beta/(\beta-t)\} - [\beta/(\alpha-\beta)]\cdot\{\alpha/(\alpha-t)\}$$

Each of the terms in braces is in the *precise form of a generator for an exponential* and hence the linear combination may be transformed back to PDF to obtain the sum density $f_Z(z)$ given in the shaded boxed equation near the bottom.

It should be noted that the trick is to put the resulting $\phi_Z(t)$ into a form that is *recognizable* for transformation back to PDF domain and this is not always easy to do. Also note that the result of this slide can be written in the form

$$f_Z(z) = \frac{\alpha\beta}{\alpha-\beta}[e^{-\beta z} - e^{-\alpha z}]$$

which agrees with the PDF (Slide#2-21) and CDF (Slide#2-22) methods.

2.8.4 Generating Function - Gaussian

Generating Function - Gaussian

Gaussian PDF
$$f_W(w) = \frac{1}{\sqrt{2\pi}} e^{-w^2/2} \; ; \; -\infty < w < \infty \qquad W \sim N(0,1)$$

Generating Function
$$\phi_W(t) \equiv E[e^{Wt}] = \int_{w=-\infty}^{\infty} dw \frac{1}{\sqrt{2\pi}} e^{-w^2/2} e^{wt} \qquad f_W(w) \xrightarrow{Transform} \phi_W(t)$$

Complete Square
$$-\frac{w^2}{2} + wt \equiv \frac{-(w-t)^2 + t^2}{2}$$

$$\phi_W(t) = e^{t^2/2} \int_{w=-\infty}^{+\infty} dw \frac{e^{-(w-t)^2/2}}{\sqrt{2\pi}} = e^{t^2/2} \int_{w=-\infty}^{+\infty} d(w-t) \frac{e^{-(w-t)^2/2}}{\sqrt{2\pi}} = e^{t^2/2}$$

Normalization = 1

Gaussian

Gaussian Transform Pair $N(0,1)$
$$\frac{1}{\sqrt{2\pi}} e^{-w^2/2} \leftrightarrow e^{t^2/2}$$

General Case: $X \sim N(\mu_X, \sigma_X^2)$ $\qquad W = \dfrac{X - \mu_X}{\sigma_X} \Rightarrow X = \sigma_X W + \mu_X$

$$\therefore \phi_X(t) = E[e^{X \cdot t}] = E[e^{(\sigma_X W + \mu_X) \cdot t}]$$

$$= e^{\mu_X \cdot t} \cdot E[e^{W \cdot (\sigma_X \cdot t)}] = e^{\mu_X \cdot t + (\sigma_X \cdot t)^2/2}$$

Gaussian Transform Pair $N(\mu_X, \sigma_X^2)$
$$\frac{1}{\sqrt{2\pi}} e^{-(x-\mu_X)^2/2\sigma_X^2} \leftrightarrow e^{\mu_X \cdot t + (\sigma_X \cdot t)^2/2}$$

Statistical Moments

$$E[X] = \phi_X'(t)\big|_{t=0} = (\mu_X + \sigma_X^2 \cdot t) e^{\mu_X \cdot t + (\sigma_X \cdot t)^2/2}\big|_{t=0} = \mu_X$$

$$E[X^2] = \phi_X''(t)\big|_{t=0} = [(\mu_X + \sigma_X^2 \cdot t)^2 + \sigma_X^2] e^{\mu_X \cdot t + (\sigma_X \cdot t)^2/2}\big|_{t=0} = \mu_X^2 + \sigma_X^2$$

$$\therefore Var(X) = E[X^2] - E[X]^2$$
$$= (\mu_X^2 + \sigma_X^2) - \mu_X^2 = \sigma_X^2$$

A standardized Gaussian with **zero mean and unity variance** is expressed by the probability density function exp(-w²/2)" /(2π)^{1/2}. The computation of the generating function $\phi_W(t) = E[e^{Wt}]$ results in an integral with $-\frac{1}{2} w^2 + w\,t$ in the exponent and the "square is completed" by adding a subtracting the terms $-t^2/2 + t^2/2$ to yield $-\frac{1}{2}(w-t)^2 + t^2/2$. The term exp(+ t²/2) comes outside of the integral and upon changing the integration variable from w to u=(w-t) the limits for u are (-∞ ,+∞) and the resulting integral is simply the Gaussian normalization integral which leaves $\phi_W(t) = \exp(+ t^2/2)$.

Taking the first and second derivatives of $\phi_W(t)$ with respect to t and evaluating the resulting expression at t=0 yields the 1st and 2nd moments of the Gaussian distribution E[W] =0 and E[W²] =1 so that $\sigma_W^2 = E[W^2] - E[W]^2 = 1 - 0^2 = 1$. The boxed equations show the Transform Pair for the Gaussian probability density; either member of the pair completely characterizes the Gaussian distribution.

The general case for a Gaussian random variable X~N(μX, σX²) with **mean** μX and **variance** σX² is easily obtained from the W~N(0,1) result by the transformation W = (X- μX)/σX ; solving for X and substituting into $\phi_X(t) = E[e^{Xt}]$ we find the generating function for $\phi_X(t) = \exp(\mu_X t + \sigma_X^2 t^2/2)$. Again taking the first and second derivatives of $\phi_X(t)$ with respect to t and evaluating the resulting expression at t=0 yields the 1st and 2nd moments of the Gaussian distribution E[X] = μX and E[W²] = μX² + σX² . The variance is computed to be

$$Var(X) = E[X^2] - E[X]^2 = (\mu_X^2 + \sigma_X^2) - (\mu_X)^2 = \sigma_X^2.$$

2.8.5 Gaussian Sum "U=V+W" using Generating Function

Gaussian Sum "U=V+W" using Generating Function

Sum 2 Indep. Gaussians $N(0,1)$ $U = V + W$; $\phi_U = \phi_V \cdot \phi_W = e^{t^2/2} \cdot e^{t^2/2} = e^{t^2} = e^{\frac{(\sqrt{2} \cdot t)^2}{2}}$

$$E[U] = 2te^{t^2}\Big|_{t=0} = 0 \;\; ; \;\; E[U^2] = 2e^{t^2} + 4t^2 e^{t^2}\Big|_{t=0} = 2 \;\; ; \;\; Var(U) = 2 - 0^2 = 2 \;\; ; \;\; \sigma = \sqrt{2}$$

Generalized Transform Pair :

$$\frac{1}{\sigma \cdot \sqrt{2\pi}} e^{-\frac{1}{2}\left(\frac{x}{\sigma}\right)^2} \leftrightarrow e^{\frac{(\sigma t)^2}{2}} \xrightarrow[\text{Generalizes for } \mu \neq 0]{} \frac{1}{\sigma \cdot \sqrt{2\pi}} e^{-\frac{1}{2}\left(\frac{x-\mu}{\sigma}\right)^2} \leftrightarrow e^{\mu t + \frac{(\sigma t)^2}{2}}$$

Sum n IID Gaussians $N(\mu_X, \sigma_X^2)$: $\mu_Z = n \cdot \mu_X$; $Var(Z) = n \cdot \sigma_X^2$; $\sigma_Z = \sqrt{n} \cdot \sigma_X$

$$Z = \sum_{k=1}^{n} X_k \;\; ; \;\; \phi_Z(t) = E[e^{Zt}] = \underbrace{E[e^{X_1 t}]E[e^{X_2 t}] \cdots E[e^{X_n t}]}_{\text{n factors}} = [\phi_X(t)]^n = e^{\frac{n \cdot t^2}{2}} = e^{\frac{(\sqrt{n} \cdot t)^2}{2}}$$

$$E[Z] = \frac{d}{dt}[\phi_X(t)]^n\Big|_{t=0} = n \cdot [\phi_X(t)]^{n-1} \cdot \phi_X'(t)\Big|_{t=0} = n \cdot \underbrace{[\phi_X(0)]^{n-1}}_{=E[e^{X \cdot 0}]=E[1]=1} \cdot \underbrace{\phi_X'(0)}_{=E[Xe^{X \cdot 0}]=E[X]=\mu_X} = n \cdot E[X] = n \cdot \mu_X$$

$$E[Z^2] = \frac{d^2}{dt^2}[\phi_X(t)]^n\Big|_{t=0} = \frac{d}{dt}\left\{ n[\phi_X(t)]^{n-1} \cdot \phi_X'(t) \right\}_{t=0} = n(n-1)[\phi_X]^{n-2} \cdot (\phi_X')^2 + n[\phi_X]^{n-1} \cdot \phi_X''\Big|_{t=0}$$

$$= n(n-1)\underbrace{[\phi_X(0)]^{n-2}}_{=E[1]^{n-2}=1} \cdot \underbrace{(\phi_X'(0))^2}_{=E[X]^2} + n\underbrace{[\phi_X(0)]^{n-1}}_{=E[1]^{n-1}=1} \cdot \underbrace{\phi_X''(0)}_{=E[X^2]} = n(n-1)E[X]^2 + nE[X^2]$$

$$Var(Z) = n(n-1)E[X]^2 + nE[X^2] - (nE[X])^2 = n\left(E[X^2] - E[X]^2\right) = n \cdot \sigma_X^2$$

Var(Sum "Z") = Sum(Individual Variances)

In the top panel we use the multiplicative property of the generating function for the sum of two independent identically distributed N(0,1) Gaussian RVs to find the generating function for their sum U=V+W as

$$\phi_U(t) = \phi_V(t)\,\phi_W(t) = \exp(+\,t^2/2) \cdot \exp(+\,t^2/2) = \exp(+\,t^2) = \exp(+\,(\sqrt{2} \cdot t)^2/2)$$

Taking the first and second derivatives of $\phi_U(t)$ with respect to t and evaluating the resulting expressions at t=0 yields the 1st and 2nd moments of the Gaussian distribution E[U]=0 and $E[U^2]=2$, so that $\sigma_U^2 = E[U^2] - E[U]^2 = 2 - 0^2 = 2$, or $\sigma_U = \sqrt{2}$. Thus we conclude that the Gaussian means and variances simply add and in this case we have $\mu_U = \mu_V + \mu_W = 0+0=0$ and $\sigma_U^2 = \sigma_V^2 + \sigma_W^2 = 1^2 + 1^2 = 2$. Further note that the equation above for $\phi_U(t)$ can be written in the form $\phi_U(t) = \exp(+\,(\sqrt{2} \cdot t)^2/2) = \exp(+\,(\sigma_U \cdot t)^2/2)$ where σ_U^2 is the variance of the sum variable U, thus forming the *transform pair* shown in the grey box of the top panel.

The bottom panel shows how this result can be extended to sum of n IID Gaussians $N(\mu_X, \sigma_X^2)$ to yield the mean as the sum of n means $\mu_Z = \sum \mu_X = n \cdot \mu_X$, and the variance of the sum is the sum of the individual variances, $\sigma_Z^2 = var(Z) = \sum \sigma_X^2 = n \cdot \sigma_X^2$. These results are derived on the slide in a straight-forward manner and may be further generalized to non-identical Gaussian distributions as indicated by the "word equation" for the variance at the bottom of the slide.

3 Order Statistics

3.1 Discrete Random Variables Min and Max for 4-sided Dice

Discrete Random Variables Min and Max for 4-sided Dice

- Geometry yields the marginal PMFs for L=min(D_1,D_2)   U=max(D_1,D_2)
- However we must go to Original Sample Space to determine Joint PMF

- **L=min(D_1,D_2)**

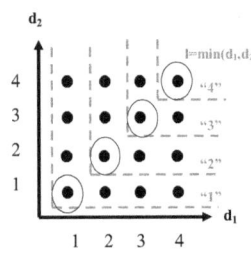

L=min(D_1,D_2)	
L	**Pr(L=l)**
1	7/16
2	5/16
3	316
4	1/16

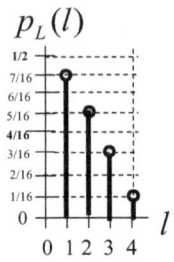

$p_L(l)$

- **U=max(D_1,D_2)**

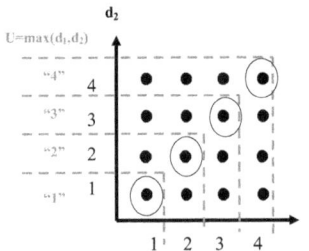

U=max(D_1,D_2)	
U	**Pr(U=u)**
1	1/16
2	3/16
3	5/16
4	7/16

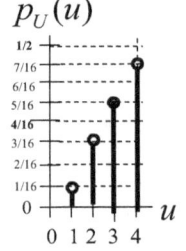

$p_U(u)$

For a pair of 4-sided dice, the coordinate transformation from {D_1, D_2} to lower/upper {L, U} RVs yields a new joint distribution describing the *order statistics* for the minimum (lower) L=min(D_1, D_2) and the maximum (upper) U=max(D_1, D_2) RVs. The marginal PMFs of the underlying joint distribution $p_{LU}(\ell,u)$ are easily found by counting the equally likely points contained in the "corner shapes" defining surfaces of constant "U" or constant "L" as shown in the {D_1, D_2} dice representation of the slide.

In the **upper figure**, we display surfaces of constant L-value as the outward facing corner surfaces shown. The minimum value "L=1" starts on the diagonal point (1,1) contains all points for which one coordinate is fixed at 1, and the other ranges over {1,2,3,4} and hence contains the 7 points {(1,1), (1,2),(1,3),(1,4),(2,1),(3,1),(4,1)}; accordingly the value 7/16 appears in the table and in the PMF plot $p_L(\ell)$ for L=1. Similarly, the surface "L=2" starts at the diagonal point (2,2) and contains all points for which either coordinate is ≥ 1 and hence contains the 5 points {(2,2), (2,3),(2,4),(3,2),(4,2)}; accordingly the value 5/16 appears in the table and in the PMF plot $p_L(\ell)$ for L=2.

In the **lower figure**, we display surfaces of constant U-value as the inward facing corner surfaces shown. The maximum value "U=4" starts on the diagonal point (4,4) contains all points for which either coordinate is ≤ 4 and hence contains the 7 points {(4,4), (4,3),(4,2),(4,1),(3,4),(2,4),(1,4)}; accordingly the value 7/16 appears in the table and in the PMF plot $p_U(u)$ for U=4. Similarly, the surface "U=3" starts at the diagonal point (3,3) and contains all points for which either coordinate is ≤ 3 and hence contains the 5 points {(3,3), (3,2),(3,1),(2,3),(1,3)}; accordingly the value 5/16 appears in the table and in the PMF plot $p_U(u)$ for U=3.

Comparing the PMFs for L and U we see that they contain the same values but have opposite behavior; specifically $p_L(\ell)$ decreases with "ℓ" whereas $p_U(u)$ increases with "u" . Thus by inspection, we immediately have the two marginal probabilities, but recall that we cannot construct the joint PMF $p_{LU}(\ell,u)$ from the two marginals. The next slide constructs the joint PMF directly from the {D_1, D_2} sample space.

3.1.1 Construction of Joint PMF p_UL(u,l) and its Sample Space

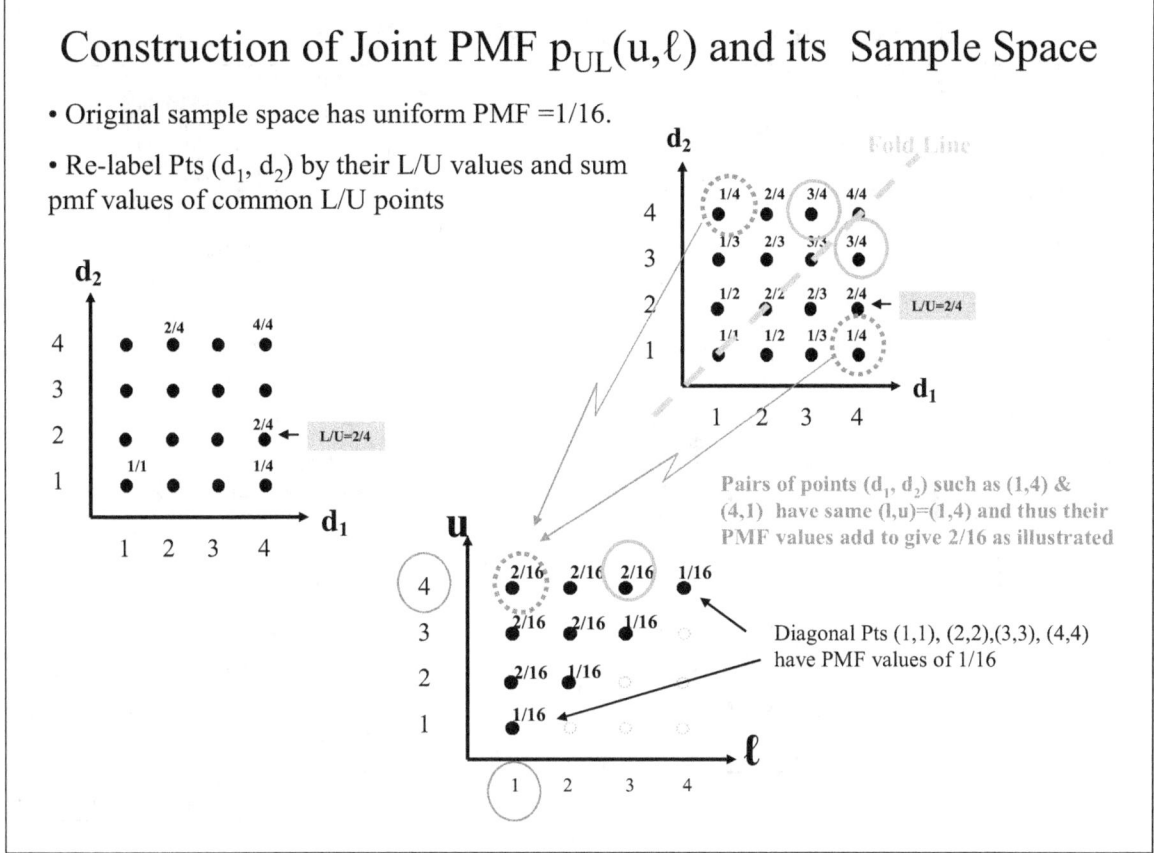

In order to construct the joint PMF $p_{LU}(\ell,u)$, we must start with the original (d_1,d_2) coordinate representation of the sample space and then re-label the points with their min and max values using the new coordinates (ℓ, u). For clarity, the **left top figure** re-labels just a few points in the (d_1, d_2) coordinates using the label notation L/U =2/4 to represent min L=2 and max U=4, *etc.*.

The **right top figure** completes the task by labeling all points in this manner and also shows the fold line about which we have identical values as indicated by the paired values above and below the diagonal. Specifically we note that the paired values $(d_1,d_2)=(1,4)$ and $(d_1,d_2)=(4,1)$ surrounded by dashed (red) circles correspond to the same L/U value with coordinates $(\ell, u)=(1,4)$ and they sum to 2/16 for the single point in the **bottom middle figure**. This transfer of double density is indicated by the broken arrows pointing from the two points in the upper right figure to single point in bottom middle figure. In a similar manner, the pair $(d_1,d_2)=(3,4)$ and $(d_1,d_2)=(4,3)$ surrounded by solid (green) circles correspond to the same L/U value with coordinates $(\ell, u)=(3,4)$ and yield a single point with double density 2/16 in the **bottom middle figure.**

Thus the PMF values in the (ℓ, u) coordinates are doubled to 2/16 for points above the $\ell=u$ diagonal, but remain at 1/16 on the diagonal itself, and are zero below the diagonal as shown in the bottom middle figure for the desired $p_{LU}(\ell, u)$.

On the next slide we show a composite table for joint, marginal, and conditional probabilities, as well as plots of the joint distribution $p_{LU}(\ell, u)$ and its two associated marginals.

3.1.2 Min-Max Joint, Marginal, and Conditional PMFs

Min-Max Joint, Marginal, and Conditional PMFs

l	u	$P_{UL}(u,l)$ Joint	l	$P_L(l)=\sum P_{UL}(u,l)$ Marginal	u	$P_U(u)=\sum P_{UL}(u,l)$ Marginal	given l	u	$P_{UL}(u\|l)$ Conditional
1	1	1/16	1	7/16	1	1/16	1	1	(1/16)/(7/16) = 1/7
	2	2/16						2	(2/16)/(7/16) = 2/7
	3	2/16						3	(2/16)/(7/16) = 2/7
	4	2/16						4	(2/16)/(7/16) = 2/7
2	2	1/16	2	5/16	2	3/16	2	2	(1/16)/(5/16) = 1/5
	3	2/16						3	(2/16)/(5/16) = 2/5
	4	2/16						4	(2/16)/(5/16) = 2/5
3	3	1/16	3	3/16	3	5/16	3	3	(1/16)/(3/16) = 1/3
	4	2/16						4	(2/16)/(3/16) = 2/3
4	4	1/16	4	1/16	4	7/16	4	4	(1/16)/(1/16) = 1

$p_L(l)$ $P_U(u)$ $P_{UL}(u, \ell)$

Conditional Probability $P_{U|L}(u|l) = P_{UL}(u, l) / P_L(l)$

$p_{U|L}(u=\{1,2,3,4\} \mid l=1) = p_{UL}(u=\{1,2,3,4\}, l=1) / p_L(l=1)$
$= \{1/16,2/16,2/16,2/16\} / \{7/16\}=$
$\{1/7,2/7,2/7,2/7\}$

$p_{U|L}(u=\{2,3,4\} \mid l=2) = p_{UL}(u=\{2,3,4\}, l=2) / p_L(l=2)$
$= \{1/16,2/16,2/16\} / \{5/16\}= \{1/5,2/5,2/5\}$

The joint distribution $p_{UL}(u, \ell)$ from the last slide is plotted here again and grouped together with the two marginal distributions $p_L(\ell)$, and $p_U(u)$ in an arrangement that makes their relationships quite intuitive. Starting with the bottom figure for the joint distribution $p_{UL}(u, \ell)$ we see that "projecting and summing" the joint probabilities horizontally left onto the u-axis yields the magnitudes of marginals $p_U(u)$ in the left figure; similarly "projecting and summing" the joint probabilities vertically down onto the ℓ-axis yields the magnitudes of marginals $p_L(\ell)$, in the upper figure. A 3-dimensional stick plot representation of $p_{UL}(u, \ell)$ is also shown on the extreme right.

The large table actually represents a summary of 4 separate probability tables as follows: (i) joint probability density $p_{UL}(u, \ell)$, (ii) marginal density $p_L(\ell)$, (iii) marginal density $p_U(u)$, and (iv) conditional $p_{U|L}(u|\ell)$. The first three tables are taken directly from the three bottom figures. We note that a more natural display of the joint probability would be a 4 x 4 matrix with rows u = 1,2,3,4 and columns ℓ =1,2,3,4, but in order to display the tables together we have only written down the non-zero terms of the joint distribution and have done that in the following linear manner. We first display row "ℓ =1" and columns u={1,2,3,4}, then row "ℓ =2" columns u={2,3,4}, row "ℓ =3" columns u={3,4}, and finally row "ℓ =4" column u={4}. All other values are zero as seen in the plot of joint probability density $p_{UL}(u, \ell)$,.

The last "sub-table" for the conditional probability $p_{U|L}(u|\ell)$ is constructed directly from the joint distribution by using the definition $p_{U|L}(u|\ell) = p_{U,L}(u, \ell)/ p_L(\ell)$. Thus, for ℓ =1, we write in obvious notation

$p_{U|L}(u=\{1,2,3,4\}|\ell =1) = p_{UL}(u=\{1, 2, 3, 4\}, \ell =1) / p_L(\ell =1)$
$= \{1/16, 2/16, 2/16, 2/16\} / \{7/16\}= \{1/7, 2/7, 2/7, 2/7\}$

which are the first four entries of the conditional table; similarly

$p_{U|L}(u=\{2, 3, 4\}|\ell =2) = p_{UL}(u=\{2, 3, 4\}, \ell =2) / p_L(\ell =2) = \{1/16, 2/16, 2/16\} / \{5/16\}= \{1/5, 2/5, 2/5\}$

which are the next three entries.

3.2 DSP Chip with Two Uniform Interrupts - Order Statistics

DSP Chip with Two Uniform Interrupts - Order Statistics

DSP Chip with T= 100 nsec clock
Uniform distribution of interrupts

$$f_T(t) = \begin{cases} 1/100 & 0 \le t \le T = 100\,ns \\ 0 & Otherwise \end{cases}$$

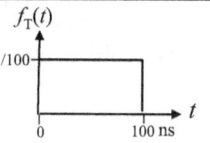

a) Find mean and sigma for the time of interrupts

$$\mu_T = E[t] = \int_{t=0}^{100} \frac{1}{100} \cdot t\, dt = \frac{1}{100} \frac{t^2}{2}\Big|_{t=0}^{100} = 50\,ns \;;\; E[t^2] = \int_{t=0}^{100} \frac{1}{100} \cdot t^2\, dt = \frac{1}{100} \frac{t^3}{3}\Big|_{t=0}^{100} = \frac{100^2}{3}\,ns^2 \;;\; \sigma = \sqrt{Var(t)} = \sqrt{\frac{100^2}{3} - 50^2} = 28.87\,ns$$

b) Consider a "pair" of consecutive interrupts $\{X_1, X_2\}$ Find PDFs for the Min Y & Max W; also E[Y] & E[W]

$$Y = \min(X_1, X_2) \qquad\qquad W = \max(X_1, X_2)$$

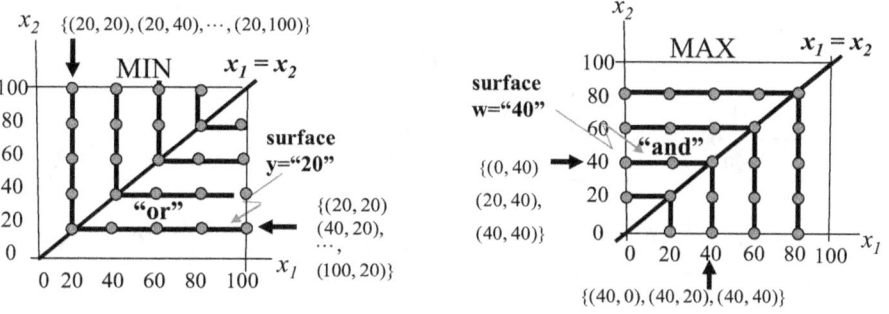

$$y = \begin{cases} x_1 & x_1 \le x_2 \\ x_2 & x_1 > x_2 \end{cases} \qquad\qquad w = \begin{cases} x_2 & x_1 \le x_2 \\ x_1 & x_1 > x_2 \end{cases}$$

Union $\quad \{Y \le y\} = \{X_1 \le y\} \cup \{X_2 \le y\} \qquad \{W \le w\} = \{X_1 \le w\} \cap \{X_2 \le w\}$ Intersection

or **and**

a) In order to have access to a DSP electronic chip it must be pulsed or interrupted during a specified clock cycle. Here we consider that these interrupts are distributed uniformly over a T = 100 ns (nano-second) clock cycle, so that $f_T(t) = 1/100$ as shown in the top figure. Given this PDF we easily compute the mean 50 *ns* and standard deviation 28.9 *ns* statistics characterizing the clock's uniform distribution..

b) Consider two consecutive interrupt "times" $\{X_1, X_2\}$ where both uniform RVs have continuous probability density $f_X(x) = 1/100$ ns and find the PDFs for Y=min$\{X_1, X_2\}$ and W =max$\{X_1, X_2\}$. The surfaces of constant min "Y" and constant max "W" look just like those for the 4-sided dice problem on the previous two slides, except here the corner-shaped "surfaces" are now continuous lines, not discrete points. Note that the red dots have no meaning in the present context; they simply serve to remind us of the discrete analog we have previously solved. They also allow us to "call out" a few of the (infinite in number) continuous pairs such as $\{(20,20), (20,40),..., (20,100)\}$ for the vertical and $\{(20,20), (40,20),..., (100,20)\}$ for the horizontal part of the "Y=20" minimum corner surface shown in the bottom left figure. Now, given any specific pair of values $\{x_1, x_2\}$ drawn from the two independent uniform distributions one of them must be the minimum and the other a maximum; this continuous transformation from coordinates (x_1,x_2) to (y,w) is represented by two functions $y=y(x_1,x_2)$ and $w= w(x_1,x_2)$. More specifically if for a particular sample (x_1, x_2) it happens that $x_1 < x_2$ then we assign the minimum x-value to y, or, $y=x_1$ and assign the maximum x-value to w or $w=x_2$, while for $x_2 < x_1$ the opposite is true, $y=x_2$ and $w=x_1$; when they are equal $(x_1=x_2)$, then $y=w$. These results are summarized mathematically by the functions $y(x_1,x_2)$ and $w(x_1,x_2)$ defined in the equation box in the middle panel.

It will be instructive to compute the joint PDF $f_{YW}(y,w)$ using both the direct PDF Method (Jacobian density transformation) and the CDF Method. Given this joint distribution, we then compute all statistical properties: marginal and conditional densities, means and deviations.

3.2.1 DSP Chip: Marginal Distribution for Minimum Order Statistic

DSP Chip: Marginal Distribution for Minimum Order Statistic

Transformation of Probability Density
Hidden Multiplicity $\alpha=2$

$$f_{YW}(y,w) = \alpha \cdot \frac{f_{X_1X_2}(x_1,x_2)}{|J(y,w)|} = 2 \cdot \frac{f_{X_1}(x_1) \cdot f_{X_2}(x_2)}{|J(y,w)|} = 2 \cdot \frac{(1/100) \cdot (1/100)}{1} = \frac{2}{100^2}$$

$$y = \begin{cases} x_1 & x_1 \le x_2 \\ x_2 & x_1 > x_2 \end{cases} \qquad \frac{\partial y}{\partial x_1} = \begin{cases} 1 \\ 0 \end{cases} ; \quad \frac{\partial y}{\partial x_2} = \begin{cases} 0 \\ 1 \end{cases}$$

$$w = \begin{cases} x_2 & x_1 \le x_2 \\ x_1 & x_1 > x_2 \end{cases} \qquad \frac{\partial w}{\partial x_1} = \begin{cases} 0 \\ 1 \end{cases} ; \quad \frac{\partial w}{\partial x_2} = \begin{cases} 1 \\ 0 \end{cases}$$

$$J(y,w) = \begin{vmatrix} \frac{\partial y}{\partial x_1} & \frac{\partial y}{\partial x_2} \\ \frac{\partial w}{\partial x_1} & \frac{\partial w}{\partial x_2} \end{vmatrix} = \begin{cases} \begin{vmatrix} 1 & 0 \\ 0 & 1 \end{vmatrix} = 1 & x_1 \le x_2 \\ \begin{vmatrix} 0 & 1 \\ 1 & 0 \end{vmatrix} = |-1| & x_1 > x_2 \end{cases}$$

"Since only two RVs X_1 and X_2, if one is a **min**, other is a **max**." Can write this statement as

If $Y = \min(X_1,X_2) = x_1$ then $W = \max(X_1,X_2) = x_2$ & vice-versa $Y = x_2 \Rightarrow W = x_1$

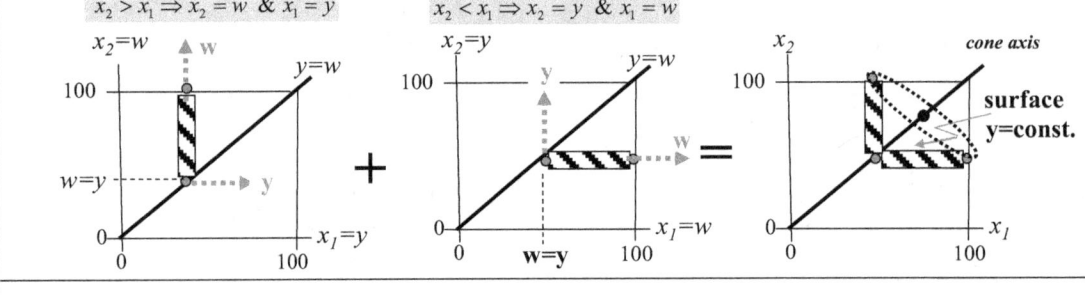

$$f_Y(y) = \underbrace{\int_{w=y}^{w=100} \tilde{f}_{YW}(y,w)\,dw}_{=1/100^2} + \underbrace{\int_{w=y}^{w=100} \tilde{f}_{YW}(y,w)\,dw}_{=1/100^2} = \underbrace{\int_{w=y}^{w=100} f_{YW}(y,w)\,dw}_{=2/100^2} = \int_{w=y}^{w=100} \frac{2}{100^2}\,dw = \boxed{\frac{2}{100^2}(100-y)}$$

We compute the derivatives necessary for the Jacobian and find $J(y,w|x_1,x_2) = 1$ for all (x_1,x_2) and then compute the joint PDF as the product of two independent uniform distributions, divided by J viz., $f_{YW}(y,w) = (1/100)(1/100)/|1| = 1/100^2$. However, this result is incorrect because integrating it over w yields a marginal density $f_Y(y)$ whose normalization integral is off by a factor of two! There is in fact a "hidden multiplicity" $\alpha = 2$ which only becomes obvious if we take a careful look at the "figure equation" in the middle panel. The figure shows that the marginal density calculation must sum the two terms that make up the y=const. "corner surface" one with a vertical strip and the other with a horizontal strip (just as for the discrete 4-sided dice case); these two contributions are shown to be equal and give the needed factor of two for the marginal density $f_Y(y)$. Note that the correct joint density to use in these two contributing integrals is *tilde*-$f_{YW}(y,w) = 1/100^2$ as indicated under the first two integrals. But we must add two *identical* integrals to obtain the correct marginal density $f_Y(y)$; equivalently, we can simply double the joint density using a multiplicity $\alpha = 2$ to write down the corrected density

$$f_{YW}(y,w) = 2/100^2.$$

Note that the transformation from Cartesian (x_1, x_2) to min/max (y, w) coordinates has an unusual type of symmetry because in the 1st term (left figure) for the **vertical** contribution, the **positive y-axis points horizontally to the right** and the **w-axis points up**. However, in 2nd term (center plot) for the **horizontal** contribution, the opposite is true and instead the **y-axis points up** and **w-axis points to right**. The symmetry is about the 45 degree line y=w and we can visualize folding about this line to make the vertical and horizontal strips coincide.

Thus the symmetry doubles the density (effective multiplicity $\alpha = 2$) and upon integrating this true joint density from w="y" to w="100" we obtain the correct marginal density directly from a single integral as

$$f_Y(y) = (2/100^2) \cdot (100-y)$$

3.2.2 DSP Chip: Marginal Distribution for Maximum Order Statistic

DSP Chip: Marginal Distribution for Maximum Order Statistic

Similarly for $f_W(w)$:

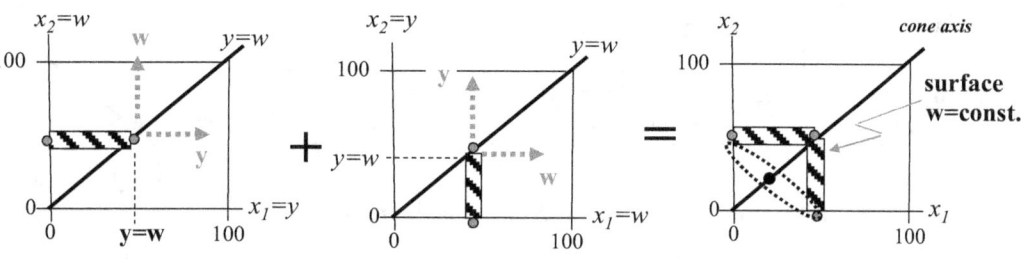

$$f_W(w) = \underbrace{\int_{y=0}^{y=w} \tilde{f}_{YW}(y,w)dy}_{=1/100^2} + \underbrace{\int_{y=0}^{y=w} \tilde{f}_{YW}(y,w)dy}_{=1/100^2} = \underbrace{\int_{y=0}^{y=w} f_{YW}(y,w)dy}_{=2/100^2} = \int_{y=0}^{y=w} \frac{2}{100^2}dy = \boxed{\frac{2}{100^2}w}$$

Check Normalization:

$$\int_{y=0}^{100} f_Y(y)dy = \int_{y=0}^{100} \frac{2}{100^2}(100-y)dy = \frac{2}{100^2}\cdot\left[100y - \frac{y^2}{2}\right]_{y=0}^{100} = \frac{2}{100^2}\cdot\left[100^2 - \frac{100^2}{2}\right] = 1 \quad \checkmark\checkmark$$

$$\int_{w=0}^{100} f_W(w)dw = \int_{y=0}^{100} \frac{2}{100^2}wdw = \frac{2}{100^2}\cdot\left[\frac{w^2}{2}\right]_{w=0}^{100} = 1 \quad \checkmark\checkmark$$

In a similar manner, there are the same two contributions to the marginal density $f_W(w)$ which is obtained by integrating *tilde*-$f_{YW}(y,w)=1/100^2$ from y="0" to y="w" for both horizontal and vertical strips. Again the contributions are identical because the w and y axes are swapped. Both of the marginal densities are summarized here

$$f_W(w) = (2/100^2)\cdot w \quad ; w \; \varepsilon \; [0,100]$$
$$f_Y(y) = (2/100^2)\cdot(100-y) \; ; y \; \varepsilon \; [0,100]$$

The normalization for each of these is verified on the slide by explicit integration over the range [0,100]. We note that the discrete results for U and L on Slide#3-4 follow the trend of the current results: the maximum W increases from zero to 2/100 and the minimum Y decreases from 2/100 to zero; however in the discrete case the probability mass functions are "bunched up" into δ-functions at the discrete points u =1,2,3,4 or l =1,2,3,4 for U and L respectively.

3.2.3 DSP Chip: CDF Method for Maximum Order Statistic

DSP Chip: CDF Method for Maximum Order Statistic

Alternate CDF Method for $f_Y(y)$ & $f_W(w)$ differentiate integrals (Leibnitz):

$$W = \max(X_1, X_2) \quad ; \qquad \text{Event} : W \le w \equiv \{x_1 \le w\} \cap \{x_2 \le w\}$$

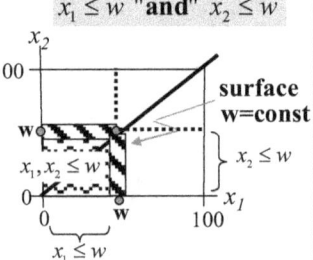

$x_1 \le w$ **"and"** $x_2 \le w$

surface w=const

> **Note:** *Intersection here because if $W = \max(X_1, X_2) \le w = "3.7"$ then the smaller of the two must also be ≤ 3.7 ; thus use $\cap$*

$$F_W(w) = \Pr[W \le w] = \Pr[\{x_1 \le w\} \cap \{x_2 \le w\}] = \Pr[\{x_1 \le w\}] \cdot \Pr[\{x_2 \le w\}]$$

$\{X_1, X_2\}$ Independent RVs

Rewrite in terms of CDFs for $F_{X_1}(x_1 = w) = F_{X_2}(x_2 = w)$

$$\Rightarrow F_W(w) = F_{X_1}(w) \cdot F_{X_2}(w)$$

Independent RVs X_1 & X_2.

Differentiate $F_W(w)$

$$f_W(w) \equiv \frac{d}{dw} F_W(w) = \frac{d}{dw}\left(F_{X_1}(w) \cdot F_{X_2}(w)\right) = \underbrace{\frac{d}{dw}\left(F_{X_1}(w)\right)}_{=f_{X_1}(w)} \cdot F_{X_2}(w) + F_{X_1}(w) \cdot \underbrace{\left(\frac{d}{dw} F_{X2}(w)\right)}_{=f_{X_2}(w)}$$

$$= f_{X_1}(w) F_{X_2}(w) + F_{X_1}(w) f_{X_2}(w) = 2 f_X(w) F_X(w)$$

Order Statistics

$$= 2\frac{1}{100}\frac{w}{100} = \frac{2w}{100^2} \quad \checkmark\checkmark$$

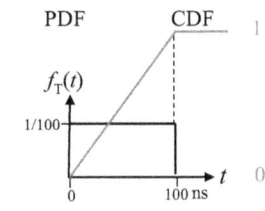

PDF   CDF

$f_T(t)$

$1/100$

On this slide we use the CDF Method to compute the cumulative distribution $F_W(w)$ and differentiate to obtain the probability density $f_W(w)$ and on the next slide we use the same method to compute $f_Y(y)$. In order to do this we again revert to the probability definition of the cumulative distribution function for W, *viz.*, $F_W(w) = \Pr[W \le w]$. The "corner surface" W=w=const. is shown in the x_1-x_2 plane and we need to sum all the probability contributions in the shaded square region determined by the origin (0,0) and the diagonal point $x_2 = x_1 = w$ that is for all (x_1, x_2) points for which $W \le w$. For this to be true x_1 AND x_2 must **both** be $\le w$: as an example, take w=3.7; then if $(x_1, x_2) = (3.5, 3.4)$ the maximum of the two is $x_1 = 3.5$ is ≤ 3.7 then clearly the minimum of the two $x_2 = 3.4$ is also ≤ 3.7. Thus, we conclude that the set of points for which $W \le w$ must be the intersection of the two sets $\{X_1 \le w\}$ AND $\{X_2 \le w\}$, so we have the set equality:

$$W \le w = \{\{X_1 \le w\} \cap \{X_2 \le w\}\}.$$

Since X_1 and X_2 are independent RVs, the probability of their product is the product of their individual probabilities and hence taking the probability of this equality yields

$$\Pr[W \le w] = \Pr[X_1 \le w]\, \Pr[X_2 \le w].$$

Now replacing the probability expressions by their equivalent CDFs yields

$$F_W(w) = F_{X1}(w) \cdot F_{X2}(w) = [F_X(w)]^2$$

where in the last equality uses the fact that the two distributions $F_{X1}(w)$ and $F_{X2}(w)$ are identical. Upon differentiating with respect to w, we find the desired probability density function

$$f_W(w) = d/dw\,[F_W(w)] = d/dw\,[F_X(w)]^2 = 2\,F_X(w)\,d/dw[F_X(w)] = 2\,F_X(w)\,f_X(w)$$

Recall the uniform distribution on X: $f_X(x) = 1/100$ and $F_X(x) = (x/100)$; substituting these two expressions with X=w into the last expression for $f_W(w)$ yields the same results as the Jacobian method, *viz.*,

$$f_W(w) = 2(w/100)\,(1/100) = 2w/(100^2)$$

3.2.4 DSP Chip: CDF Method for Minimum Order Statistic

DSP Chip: CDF Method for Minimum Order Statistic

$$Y = \min(X_1, X_2) \quad ; \quad \text{Event}: Y \le y \equiv \{x_1 \le y\} \cup \{x_2 \le y\}$$

$x_1 \le y \text{ "or" } x_2 \le y$

Note: *Union here because if* $Y = \min(X_1, X_2) \le y = \text{"3.7"}$ *then the larger of the two **may be** ≤ 3.7, **but is not required to be** ≤ 3.7; thus use $\cup$ to take both possibilities into account.*

$$F_Y(y) = \Pr[Y \le y] = \Pr[\{x_1 \le y\} \cup \{x_2 \le y\}]$$

$$= \Pr[\{x_1 \le y\}] + \Pr[\{x_2 \le y\}] - \Pr[\{x_1 \le y\} \cdot \{x_2 \le y\}]$$

$$F_Y(y) = F_{X_1}(y) + F_{X_2}(y) - F_{X_1 X_2}(x_1 \le y, x_2 \le y)$$

$$= 2F_X(y) + [F_X(y)]^2$$

This region is double counted by vertical and horizontal rectangles and thus must subtract their intersection

Differentiate

$$f_Y(y) \equiv \frac{d}{dy} F_Y(y) = \underbrace{\frac{d}{dy} F_{X_1}(y)}_{=f_X} + \underbrace{\frac{d}{dy} F_{X_2}(y)}_{=f_X} - \underbrace{\frac{d}{dy} F_{X_1 X_2}(x_1 \le y, x_2 \le y)}_{=\frac{d}{dx}F_X^2(y) = 2f_X(y)F_X(y)} = 2f_X(y)(1 - F_X(y))$$

Order Statistics

$$\text{Recall}: f_X(y) = 1/100 \quad ; \quad F_X(y) = \int_{x=0}^{x=y} \frac{1}{100} dx = \frac{y}{100}$$

$$f_Y(y) = 2f_X(y)(1 - F_X(y)) = 2\left(\frac{1}{100}\right)\left(1 - \frac{y}{100}\right) = \frac{2}{100^2}(100 - y) \quad \checkmark\checkmark$$

See Next Slide for alternate computation

Again we use the probability definition of the cumulative distribution function for Y, *viz.*, $F_Y(y) = \Pr[Y \le y]$. The "corner surface Y=y=const. is shown in the x_1-x_2 plane and we need to sum all the probability contributions in the shaded square region determined by the diagonal point $x_2 = x_1 = y$ and the point (100,100), that is for all (x_1, x_2) points for which $Y \le y$: for example take y=3.7; then if $(x_1, x_2) = (3.8, 3.4)$ the minimum of the two is $x_2 = 3.4$ is ≤ 3.7 and clearly the maximum of the two $x_1 = 3.8$ is not required to be ≤ 3.7; for another pair such as $(x_1, x_2) = (3.5, 3.4)$ they would both be ≤ 3.7. Thus, *either* x_1 *or* x_2 or both may be $\le y$ and we conclude that the set of points for which $Y \le y$ must be equal to the *union of the two sets* $\{X_1 \le y\}$, $\{X_2 \le y\}$, so we have the set equality:

$$Y \le y = \{\{X_1 \le y\} \cup \{X_2 \le y\}\}.$$

Since X_1 and X_2 are independent RVs, **they must intersect** and hence in taking the probability of their union we must subtract out their intersection in order to avoid "double counting" and we have

$$\Pr[Y \le y] = \Pr[X_1 \le y] + \Pr[X_2 \le y] - \Pr[\{X_1 \le y\}\{X_2 \le y\}]$$

or since X_1 and X_2 are independent RVs

$$\Pr[Y \le y] = \Pr[X_1 \le y] + \Pr[X_2 \le y] - \Pr[X_1 \le y] \cdot \Pr[X_2 \le y]$$

Now replacing the probability expressions by their equivalent CDFs yields

$$F_Y(y) = F_{X1}(y) + F_{X2}(y) - F_{X1}(y) \ F_{X2}(y) = 2 \ F_X(y) - [F_X(y)]^2$$

where in the last equality uses the fact that the distributions for X_1 and X_2 are identical. Upon differentiating with respect to w, we find the desired probability density function

$$f_Y(y) = d/dy \ [F_Y(y)] = d/dy \ \{2 \ F_X(y) - [F_X(y)]^2\} = 2 \ f_X(y) - 2 \ f_X(y) \ F_X(y) = 2 \ f_X(y)[1 - F_X(y)]$$

Once again substituting $f_X(x) = 1/100$ and $F_X(x) = (x/100)$ and setting X=y, yields the correct result for this case

$$f_Y(y) = 2(1/100)[1 - y/100] = (2/100^2) [100 - y]$$

3.2.5 DSP Chip: Differentiation Details

DSP Chip: Differentiation Details

Details for $f_Y(y)$:

Differentiate wrt y:
$$f_Y(y) \equiv \frac{d}{dy} F_Y(y) = \underbrace{\frac{d}{dy} F_{X_1}(y)}_{=f_X} + \underbrace{\frac{d}{dy} F_{X_2}(y)}_{=f_X} - \underbrace{\frac{d}{dy} F_{X_1 X_2}(x_1 \le y, x_2 \le y)}_{=2f_X(y)F_X(y)} = 2f_X(y)(1 - F_X(y))$$

$$\frac{d}{dy} F_{X_1}(y) = \frac{d}{dy} \int_{x_1=0}^{x_1=y} f_{X_1}(x_1)\, dx_1 = f_{X_1}(y) \quad ; \quad \frac{d}{dy} F_{X_2}(y) = \frac{d}{dy} \int_{x_2=0}^{x_2=y} f_{X_2}(x_2)\, dx_2 = f_{X_2}(y)$$

Independent Identical Distributions (IID)

$$\text{IID RVs } X_1, X_2 \Rightarrow \text{ Independent}: f_{X_1 X_2}(x_1, x_2) = f_{X_1}(x_1) \cdot f_{X_2}(x_2)$$

$$\text{Identical}: f_{X_1}(X_1 = y) = f_{X_2}(X_2 = y) = f_X(y)$$

Only y-dep.

$$\frac{d}{dy} F_{X_1 X_2}(x_1 \le y, x_2 \le y) = \frac{d}{dy} \int_{x_1=0}^{\overparen{x_1=y}} dx_1 \int_{x_2=0}^{\overparen{x_2=y}} dx_2 \underbrace{f_{X_1 X_2}(x_1, x_2)}_{=f_{X_1}(x_1) \cdot f_{X_2}(x_2)} = \frac{d}{dy} \underbrace{\int_{x_1=0}^{x_1=y} dx_1 f_{X_1}(x_1) \cdot \int_{x_2=0}^{x_2=y} dx_2 f_{X_2}(x_2)}_{\text{differentiate product of two integrals}}$$

$$= \underbrace{\int_{x_1=0}^{\overparen{x_1=y}} dx_1 f_{X_1}(x_1)}_{=F_{X_1}(y)} \cdot \underbrace{\frac{d}{dy} \int_{x_2=0}^{\overparen{x_2=y}} dx_2 f_{X_2}(x_2)}_{=f_{X_2}(y)} + \underbrace{\frac{d}{dy} \int_{x_1=0}^{x_1=y} dx_1 f_{X_1}(x_1)}_{=f_{X_1}(y)} \cdot \underbrace{\int_{x_2=0}^{x_2=y} dx_2 f_{X_2}(x_2)}_{=F_{X_2}(y)}$$

This slide gives mathematical details for explicit differentiation of the integrals, especially the y-derivative of the joint distribution $d/dy\{F_{X_1 X_2}(x_1 \le y, x_2 \le y)\}$ as developed on the lower half of the slide. The direct method used to perform the derivative on the previous slide is justified by the explicit development of this slide. The definition of the cumulative distribution function $F_X(x)$ as the integral of the density function $f_X(x)$, the converse $f_X(x) = d/dx (F_X(x))$, and Leibnitz's rule for differentiation of an integral are the primary ingredients for the derivations in this slide.

3.2.6 DSP Chip: Interrupt Event Avoidance Regions

DSP Chip: Interrupt Event Avoidance Regions

c) Find Average minimum and maximum E[Y] & E[W]

$$\mu_Y \equiv E[Y] = \int_0^{100} y\, f_Y(y)dy = \int_0^{100} y \cdot \frac{2}{100^2}(100-y)dy = \frac{2}{100^2}\left(100\frac{y^2}{2} - \frac{y^3}{3}\right)_0^{100} = \frac{100}{3} = 33.3$$

$$\mu_W \equiv E[W] = \int_0^{100} w\, f_W(w)dw = \int_0^{100} w \cdot \frac{2w}{100^2}dw = \frac{2}{100^2}\left(\frac{w^3}{3}\right)_0^{100} = \frac{2}{3}100 = 66.7$$

d) Find Prob that the both X_1 and X_2 will be smaller than E[Y] or greater than E[W]

Event A:$\{X_1, X_2: Y \le 33.3\}$ Event B: $\{X_1, X_2: W \ge 66.7\}$

Avoid Pulse Edges Trailing & Leading

$$\Pr[A \cup B] = \Pr[A] + \Pr[B] - \underbrace{\Pr[A \cdot B]}_{=0}$$

$$\Pr[A] = \int_0^{33.3} dx_1 \int_0^{33.3} dx_2 \frac{1}{100^2} = \frac{33.3^2}{100^2} = \frac{1}{9}$$

$$\Pr[B] = \int_{66.7}^{100} dx_1 \int_{66.7}^{100} dx_2 \frac{1}{100^2} = \frac{(100-66.7)^2}{100^2} = \frac{1}{9}$$

$$\Rightarrow \Pr[A \cup B] = \frac{2}{9} = .222$$

Uniform Distribution Allows Geometric Calculation

$\Pr[A \cup B] = [\text{Area}(A) + \text{Area}(B)] / 100^2 = 2(33.3)^2 / 100^2 = .222$

c) Now that we have the marginal distributions in Y and in W we can compute their means to be, 1/3 and 2/3 respectively by performing the integrals..

d) In designing a DSP chip we might want to know how often the pair of interrupts are near the beginning or end of the 100 nsec clock pulse "edges". That is, we need to compute the probability that both interrupts $X_1 X_2$ are smaller than the average minimum $\mu_Y = E[Y] = 33.3$ or greater than the average maximum $\mu_W = E[W] = 66.7$ which corresponds respectively to the two areas A and B in the x_1–x_2 pair interrupt plane. These events are defined by

A: $\{(X_1,X_2) ; Y = \min(X_1,X_2) \le \mu_Y = 33.3\}$ B: $\{(X_1,X_2) ; W = \max(X_1,X_2) \ge \mu_W = 66.7\}$

Taking the union and noting that the two events A and B are mutually exclusive, we need only sum the two probabilities obtained by integrating the uniform joint distribution $f_{X_1X_2}(x_1,x_2) = 1/100^2$ over the two squares to obtain the result 2/9 as shown. Alternately, because the joint density is uniform we only need to compute the areas squares A and B and divide by 100^2. We have by inspection $(33.3^2 + 33.3^2)/100^2 = .222$.

3.3 Order Statistics Formulation

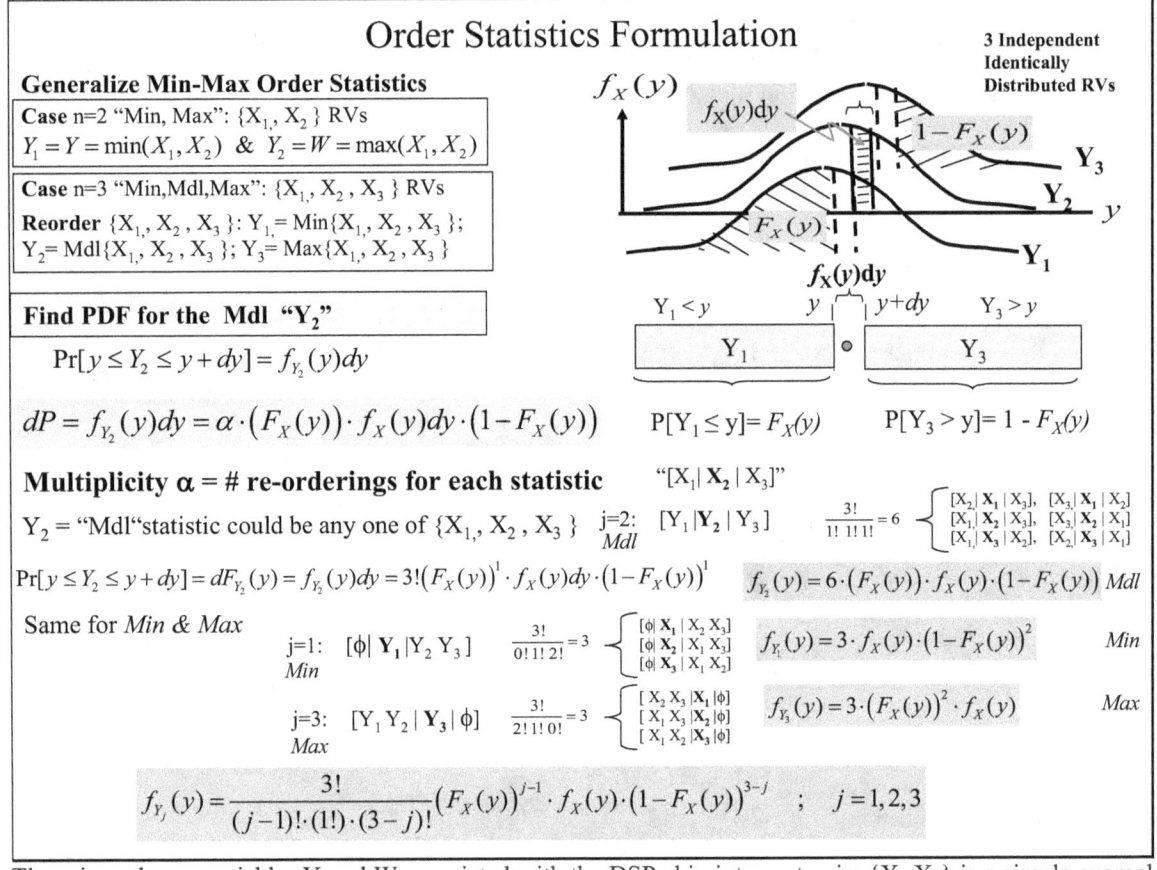

The min and max variables Y and W associated with the DSP chip interrupt pairs $\{X_1, X_2\}$ is a simple example of **order statistics** involving two Independent Identically Distributed (IID) RVs. The concept of order statistics can be extended to triples $\{X_1, X_2, X_3\}$ with associated order statistics {min, middle, max}, and to quadruples $\{X_1, X_2, X_3, X_4\}$ with associated order statistics {min, middle1, middle2, max}, and of course to any number of IID RVs. We first consider triples $\{X_1, X_2, X_3\}$ and denote the order statistics in increasing value within the set of triples $\{Y_1, Y_2, Y_3\}$; each set of triples represents one draw from each of the 3 IIDs and the actual triple $\{x_1, x_2, x_3\} = \{.9, .7, .2\}$ is reordered as $\{y_1, y_2, y_3\} = \{.2, .7, .9\}$ meaning $y_1 = x_3$, $y_2 = x_2$, $y_3 = x_1$ for this particular triple. The order statistic is characterized by the probability density function corresponding to a large number of triples that have been reordered in this manner, so that $f_{Y_2}(y_2)$ is the density of the middle number Y_2 (after re-ordering). The differential probability in the interval between $Y_2 = y$ and $Y_2 = y+dy$ defines the density $f_{Y_2}(y_2)$ via $dP = \Pr[y \le Y_2 \le y+dy] = f_{Y_2}(y_2)\,dy$. If we consider the continuous range of re-ordered Y-values, the small interval from y to y+dy contains the differential probability for the middle value Y_2 given by $f_X(y)dy$; all Y-values less than a particular "y" belong to Y_1 and those greater belong to Y_3 (see inset figure.) The probability to the left of y is $\Pr[Y_1 \le y] = F_X(y)$ and that to the right of y is $\Pr[Y_3 > y] = 1 - \Pr[Y_3 \le y] = 1 - F_X(y)$. The differential probability for Y_2 is just the product of these three terms times a multiplicity factor α,

$$dP = \Pr[y \le Y_2 \le y+dy] = f_{Y_2}(y)\,dy = \alpha\, F_X(y)\, f_X(y)\, [1 - F_X(y)]\,dy = 6\, F_X(y)\, f_X(y)\, [1 - F_X(y)]\,dy$$

The multiplicity factor α results from the number of re-orderings of $\{X_1, X_2, X_3\}$ that yield $Y_2 = y$ and is shown to be $3! = 6$, so we find $f_{Y_2}(y) = 6\, F_X(y)\, f_X(y)\, [1 - F_X(y)]$. The ordering patterns for the triples $\{X_1, X_2, X_3\}$ into left, middle, and right positions describing the reordered triples $\{Y_1, Y_2, Y_3\}$ give the multiplicities as detailed in the slide and the boxed equation summarizes the resulting probability densities for the three statistics Y_1(min), Y_2 (middle), and Y_3 (max).

3.3.1 Order Statistics - General Case n Random Variables

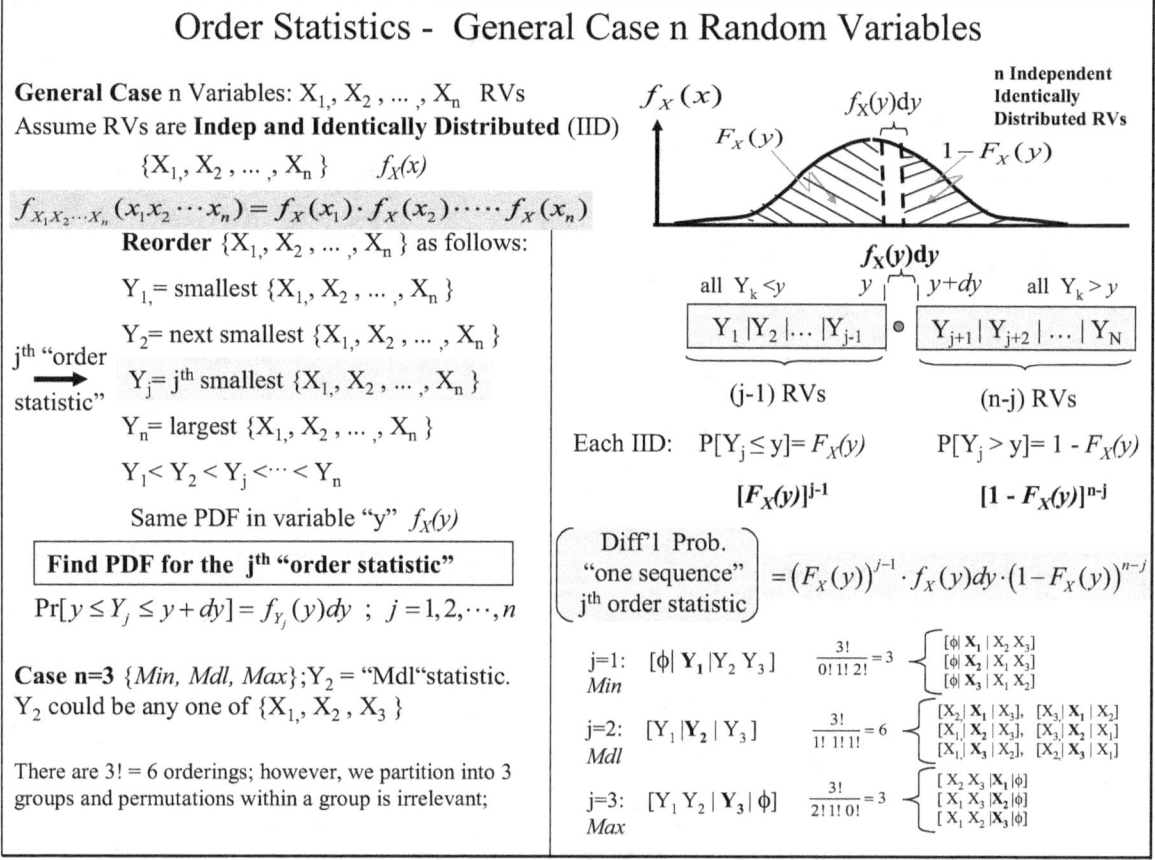

Order Statistics - General Case n Random Variables

General Case n Variables: $X_1, X_2, \ldots, X_n$ RVs
Assume RVs are **Indep and Identically Distributed** (IID)

$$\{X_1, X_2, \ldots, X_n\} \qquad f_X(x)$$

$$f_{X_1 X_2 \cdots X_n}(x_1 x_2 \cdots x_n) = f_X(x_1) \cdot f_X(x_2) \cdot \cdots \cdot f_X(x_n)$$

Reorder $\{X_1, X_2, \ldots, X_n\}$ as follows:

Y_1 = smallest $\{X_1, X_2, \ldots, X_n\}$

Y_2 = next smallest $\{X_1, X_2, \ldots, X_n\}$

j^{th} "order statistic" → Y_j = j^{th} smallest $\{X_1, X_2, \ldots, X_n\}$

Y_n = largest $\{X_1, X_2, \ldots, X_n\}$

$$Y_1 < Y_2 < Y_j < \cdots < Y_n$$

Same PDF in variable "y" $f_X(y)$

Find PDF for the jth "order statistic"

$$\Pr[y \le Y_j \le y+dy] = f_{Y_j}(y)dy \; ; \; j = 1, 2, \cdots, n$$

Case n=3 {*Min, Mdl, Max*}; Y_2 = "Mdl" statistic. Y_2 could be any one of $\{X_1, X_2, X_3\}$

There are 3! = 6 orderings; however, we partition into 3 groups and permutations within a group is irrelevant;

n Independent Identically Distributed RVs

$f_X(x)$ $\qquad$ $f_X(y)dy$

$F_X(y)$ $\qquad$ $1 - F_X(y)$

$f_X(y)dy$

all $Y_k < y$ $\quad$ y $\quad$ $y+dy$ $\quad$ all $Y_k > y$

$\boxed{Y_1 \mid Y_2 \mid \ldots \mid Y_{j-1}} \; \bullet \; \boxed{Y_{j+1} \mid Y_{j+2} \mid \ldots \mid Y_N}$

(j-1) RVs $\qquad\qquad$ (n-j) RVs

Each IID: $\quad P[Y_j \le y] = F_X(y) \qquad P[Y_j > y] = 1 - F_X(y)$

$$[F_X(y)]^{j-1} \qquad\qquad [1 - F_X(y)]^{n-j}$$

$$\begin{pmatrix} \text{Diff'l Prob.} \\ \text{"one sequence"} \\ j^{th} \text{ order statistic} \end{pmatrix} = \left(F_X(y)\right)^{j-1} \cdot f_X(y)dy \cdot \left(1 - F_X(y)\right)^{n-j}$$

j=1: $[\phi \mid \mathbf{Y_1} \mid Y_2 Y_3]$ $\quad \dfrac{3!}{0! \, 1! \, 2!} = 3$ $\begin{cases} [\phi \mid \mathbf{X_1} \mid X_2 X_3] \\ [\phi \mid \mathbf{X_2} \mid X_1 X_3] \\ [\phi \mid \mathbf{X_3} \mid X_1 X_2] \end{cases}$
Min

j=2: $[Y_1 \mid \mathbf{Y_2} \mid Y_3]$ $\quad \dfrac{3!}{1! \, 1! \, 1!} = 6$ $\begin{cases} [X_2 \mid \mathbf{X_1} \mid X_3], \; [X_3 \mid \mathbf{X_1} \mid X_2] \\ [X_1 \mid \mathbf{X_2} \mid X_3], \; [X_3 \mid \mathbf{X_2} \mid X_1] \\ [X_1 \mid \mathbf{X_3} \mid X_2], \; [X_2 \mid \mathbf{X_3} \mid X_1] \end{cases}$
Mdl

j=3: $[Y_1 Y_2 \mid \mathbf{Y_3} \mid \phi]$ $\quad \dfrac{3!}{2! \, 1! \, 0!} = 3$ $\begin{cases} [X_2 X_3 \mid \mathbf{X_1} \mid \phi] \\ [X_1 X_3 \mid \mathbf{X_2} \mid \phi] \\ [X_1 X_2 \mid \mathbf{X_3} \mid \phi] \end{cases}$
Max

Order Statistics for the general case of n IID Random Variables is detailed on this slide. The n IID RVs $\{X_1, X_2, \ldots, X_n\}$ are re-ordered from the smallest Y_1 to the largest Y_n and the jth Y in the sequence Y_j is called the "j^{th} order statistic". Again we fix a value Y=y and consider the continuous range of re-ordered Y-values illustrated in the figure: the small interval from y to y+dy contains the differential probability for the jth order statistic Y_j given by $f_X(y)dy$; all Y-values less than this belong to the Y_1 through Y_{j-1} and those greater belong to Y_{j+1} through Y_n as shown in the inset figure. Now each of the IID Ys on the left have the same probability $\Pr[Y_1 \le y] = F_X(y)$, $\Pr[Y_2 \le y] = F_X(y)$, ... $\Pr[Y_{j-1} \le y] = F_X(y)$, and hence their product yields a total probability $\Pr[Y_{left} \le y] = [F_X(y)]^{j-1}$; similarly on the right we find $\Pr[Y_{right} \le y] = [1-F_X(y)]^{n-j}$. So for the reordered Ys the differential probability is just the product of these three terms multiplied by a multiplicity factor α, *viz.*,

$$dP = \Pr[y \le Y_j \le y+dy] = f_{Yj}(y) \, dy = \alpha \, [F_X(y)]^{j-1} \, f_X(y) \, [1-F_X(y)]^{n-j} \, dy$$

The multiplicity factor α results from the number of re-orderings of $\{X_1, X_2, \ldots, X_n\}$ for each order statistic Y_j; arguments for n=3 and n=4 are illustrated on this slide and the next. These arguments look (in turn) at each order statistic min, middle(s), and max and compute for each statistic the number of distinct arrangements of $\{X_1, X_2, \ldots, X_n\}$ that yield the three groups relative to the "separation point" Y=y and arrive at multinomial forms dependent upon the orderings for each statistic. The specific multiplicity factors for the cases n=3,4 are easily found to be

$$\alpha = {}^3C_{(j-1),1,(3-j)} = 3! \, / \, [(j-1)! \; 1! \; (3-j)!] \;\; ; \; \alpha = {}^4C_{(j-1),1,(4-j)} = 4! \, / \, [(j-1)! \; 1! \; (4-j)!]$$

and the final results for the PDF of the jth order statistic $f_{Yj}(y)$ in these cases are

$$f_{Yj}(y_j) = {}^3C_{(j-1),1,(3-j)} \, [F_X(y_j)]^{j-1} \, f_X(y_j) \, [1-F_X(y_j)]^{3-j} \quad \text{for } j=1,2,3 \quad (n=3)$$

$$f_{Yj}(y_j) = {}^4C_{(j-1),1,(4-j)} \, [F_X(y_j)]^{j-1} \, f_X(y_j) \, [1-F_X(y_j)]^{4-j} \quad \text{for } j=1,2,3,4 \quad (n=4)$$

3.3.2 Order Statistics - Marginal and Joint PDFs

Order Statistics - Marginal and Joint PDFs

Case n=4 {*Min, Mdl₁ , Mdl₂ , Max*}; Y_2 = "*Mdl₁*" statistic
Y_2 could be any one of {X_1, X_2, X_3, X_4}, *etc.*
There are 4! = 24 orderings; however, we partition into 3
groups and permutations within a group is irrelevant;

Tri-nomial
Coeffs

j=1: $[\phi|\ \mathbf{Y_1}|Y_2\ Y_3\ Y_4]$ $\frac{4!}{0!\,1!\,3!}=4$

j=2: $[Y_1|\mathbf{Y_2}|\ Y_3\ Y_4\ ]$ $\frac{4!}{1!\,1!\,2!}=12$

j=3: $[Y_1\ Y_2\ |\ \mathbf{Y_3}|\ Y_4\ ]$ $\frac{4!}{2!\,1!\,1!}=12$

j=4: $[Y_1\ Y_2\ Y_3\ |\ \mathbf{Y_4}\ |\ \phi\ ]$ $\frac{4!}{3!\,1!\,0!}=4$

General Case n: # Sequences $=\dfrac{n!}{(j-1)!\cdot(1!)\cdot(n-j)!}$

Prob all sequences

$$\Pr[y\le Y_j\le y+dy]=dF_{Y_j}(y)=f_{Y_j}(y)dy=\frac{n!}{(j-1)!\cdot(1!)\cdot(n-j)!}\left(F_X(y)\right)^{j-1}\cdot f_X(y)dy\cdot\left(1-F_X(y)\right)^{n-j}$$

The **PDF** for the **jᵗʰ "order statistic"** is therefore

$$f_{Y_j}(y)=\frac{n!}{(j-1)!\cdot(1!)\cdot(n-j)!}\left(F_X(y)\right)^{j-1}\cdot f_X(y)\cdot\left(1-F_X(y)\right)^{n-j}\quad;\quad j=1,2,\cdots,n$$

The **Joint PDF** for all the **"order statistics"**, j=1,2, ⋯,n is defined as follows:

$$\Pr[y_j\le Y_j\le y_j+dy_j,\text{all "j"}]=\Pr[y_1\le Y_1\le y_1+dy_1\cap y_2\le Y_2\le y_2+dy_2\cap\cdots\cap y_n\le Y_n\le y_n+dy_n]$$

$$=f_{Y_1Y_2\cdots Y_n}(y_1y_2\cdots y_n)dy_1dy_2\cdots dy_n\ \ne f_{Y_1}(y_1)dy_1\cdot f_{Y_2}(y_2)dy_2\cdots f_{Y_n}(y_n)dy_n$$

*Y's are not
Independent !*

$$f_{Y_1Y_2\cdots Y_n}(y_1y_2\cdots y_n)=n!\cdot f_X(y_1)\cdot f_X(y_2)\cdots\cdot f_X(y_n)\ \ y_1<y_2\cdots<y_n$$

The PDF requires the ordered RV Vector {$Y_1, Y_2, \ldots, Y_n$} to take on the specific values {$y_1, y_2, \ldots, y_n$}
This must be one of the **n! permutations** of the original sample space {$X_1, X_2, \ldots, X_n$}

$$f_{X_1X_2\cdots X_n}(x_1x_2\cdots x_n)=f_X(x_1)\cdot f_X(x_2)\cdots\cdot f_X(x_n)$$ "X" s are Independent RVs

The specific multiplicity factors for the case for n=4 is detailed in the top portion of this slide using arguments similar to those given for n=3. The n=4 multiplicity coefficients show sufficient structure to write down the general trinomial coefficient to yield the multiplicity factor as $\alpha = {}^nC_{(j-1),1,(n-j)} = n! / [(j-1)!\ 1!\ (n-j)!]$.
The final result for the PDF of the **jᵗʰ order statistic** $f_{Yj}(y)$ is shown on the slide as
$$f_{Yj}(y_j) = {}^nC_{(j-1),1,(n-j)}\ [F_X(y_j)\ ]^{j-1}\ f_X(y_j)\ [1-F_X(y_j)\ ]^{n-j}\quad;\quad j=1,2,\ldots,n$$
Note that the formula for $f_{Yj}(y)$ is left in *pneumonic order* and has four components, namely, (i) the **density term** $f_X(y)$ in the middle, (ii) the **cumulative term** $F_X(y)$ is on the left and raised to a power corresponding to the number of Ys to the left of the point Y=y, *i.e.,* the $(j-1)^{st}$ power, (iii) the **complement of the cumulative term** $[1-F_X(y)\ ]$ is on the right and raised to a power corresponding to the number of Ys to the right of the point Y=y, *i.e.,* the $(n-j)^{th}$ power, and (iv) the **multiplicity factor** $\alpha = {}^nC_{(j-1),1,(n-j)}$ multiplies the whole expression.
Note also for the minimum statistic Y_1, the term $F_X(y)$ disappears j=1: $f_{Y1}(y) = {}^nC_{0,1,(n-1)}\ f_X(y)\ [1-F_X(y)\ ]^{n-1}$, whereas for the maximum statistic Y_n, the term $[1-F_X(y)]$ disappears j=n: $f_{Yn}(y) = {}^nC_{(n-1),1,0}\ [F_X(y)\ ]^{n-1}\ f_X(y)$.
The joint PDF for all the order statistics is obtained using the probability interpretation of joint probability which places all y_j variables in appropriate intervals $[y_j , y_j +dy_j]$ and multiplies by the product of differentials, *viz.,*
$$\Pr[\ y_1\le Y_1\le y_1+dy_1, \ldots , y_n\le Y_n\le y_n+dy_n] = f_{Y1\ldots Yn}(y_1,\ldots, y_n)\ dy_1\ldots dy_n$$
$$f_{Y1\ldots Yn}(y_1,\ldots, y_n) = n!\cdot f_{Y1}(y_1)\cdots\cdot f_{Yn}(y_n)\quad\text{with } y_1\le y_2\le\ldots\le y_n$$
It is important to point out that, even though all the Xs *are independent*, the **Ys are not independent** because of the n! re-orderings of the Xs. This relates back to the n=2 case in which we observed that x_1=min means that x_2 **must be** (dependence) a max which clearly makes X_1 and X_2 dependent. This applies to all order statistics.

3.4 *Order Statistics Applications: DSP Chip Min and Max (n=2)*

Order Statistics Applications: DSP Chip Min and Max (n=2)

Ex.1 DSP Chip - Uniform
Interupts Uniform RVs T_1, T_2

$$f_T(t) = \begin{cases} 1/100 & 0 \le t \le T = 100\,ns \\ 0 & Otherwise \end{cases}$$

$$Y_1 = Y = \min(T_1, T_2)$$
$$Y_2 = W = \max(T_1, T_2)$$

Marginal PDF
(n=2, j=1,2)

$$f_{Y_j}(y_j) = \frac{n!}{(j-1)! \cdot (1!) \cdot (n-j)!} \left(F_T(y_j)\right)^{j-1} \cdot f_T(y_j) \cdot \left(1 - F_T(y_j)\right)^{n-j}$$

Set **n=2, j=1 (Min)** (Remember smallest first)

$$f_{Y_1}(y_1) = 2![F_T(y_1)]^{1-1} f_T(y_1)[1-F_T(y_1)]^{2-1} = 2 \cdot \frac{1}{100} \cdot \left(1 - \frac{y_1}{100}\right) = \frac{2}{100^2} \cdot (100 - y_1) = \frac{2}{100^2} \cdot (100 - y_1)$$

Set **n=2, j=2 (Max)**

$$f_{Y_2}(y_2) = 2![F_T(y_2)]^{2-1} f_T(y_2)[1-F_T(y_2)]^{2-2} = 2 \cdot \frac{y_2}{100} \cdot \frac{1}{100} = \frac{2}{100^2} \cdot y_2 = \frac{2}{100^2} \cdot w$$

Joint PDF
(n=2)

$$f_{Y_1 Y_2}(y_1 y_2) = 2! \cdot f_X(y_1) \cdot f_X(y_2) = 2! \frac{1}{100} \cdot \frac{1}{100} = constant$$

Y_1, Y_2
not indep.
(2!)

Check by integrating:

$$\int_{y_1=0}^{100} dy_1 \int_{y_2=y_1}^{100} dy_2 \cdot f_{Y_1 Y_2}(y_1 y_2) = \int_{y_1=0}^{100} dy_1 \int_{y_2=y_1}^{100} dy_2 \cdot \frac{2!}{100^2} = \frac{2!}{100^2} \int_{y_1=0}^{100} dy_1 \cdot \{y_2\}|_{y_2=y_1}^{100}$$

$$= \int_{y_1=0}^{100} dy_1 \cdot \underbrace{\frac{2}{100^2} \{100 - y_1\}}_{\text{Marginal Density:} f_{Y_1}(y_1)} = \frac{2!}{100^2} \cdot \left\{100 y_1 - \frac{y_1^2}{2}\right\}_{y_1=0}^{100} = 1 \quad ✓✓$$

Note : $f_{Y_1 Y_2}(y_1 y_2) \neq f_{Y_1}(y_1) \cdot f_{Y_2}(y_2) = \frac{2}{100^2} \cdot (100 - y_1) \cdot \frac{2}{100^2} \cdot y_2 = \frac{4}{100^4} \cdot (100 - y) \cdot w$ ✗✗

Marginal PDFs: Let's apply the full order statistics formalism to the n=2 example in which the two DSP chip interrupts are now described by IID uniform RVs X_1 and X_2. We write down the general form of the jth order statistic $f_{Yj}(y)$ from the last slide as

$$f_{Yj}(y_j) = {}^nC_{(j-1),1,(n-j)} [F_X(y_j)]^{j-1} f_X(y_j) [1-F_X(y_j)]^{n-j} \quad ; j=1,2,...,n$$

For the min Y_1 we set n=2, j=1 to obtain directly

$$f_{Y1}(y_1) = {}^2C_{(1-1),1,(2-1)} [F_X(y_1)]^{1-1} f_X(y_1) [1-F_X(y_1)]^{2-1} = {}^2C_{0,1,1} f_X(y_1) [1-F_X(y_1)]$$
$$= 2(1/100) \cdot (1-y_1/100)$$

For the max Y_2 we set n=2, j=2 to obtain

$$f_{Y2}(y_2) = {}^2C_{(2-1),1,(2-2)} [F_X(y_1)]^{2-1} f_X(y_1) [1-F_X(y_1)]^{2-2} = {}^2C_{1,1,0} F_X(y_1) f_X(y_1)$$
$$= 2(y_1/100) \cdot (1/100)$$

Joint PDFs: The general form of the joint distribution is given by

$$f_{Y1...Yn}(y_1,..., y_n) = n! \cdot f_{Y1}(y_1) \cdots f_{Yn}(y_n) \text{ with } y_1 \le y_2 \le ... \le y_n$$

For n=2 we have

$$f_{Y1Y2}(y_1, y_2) = 2! \cdot f_{Y1}(y_1) \cdot f_{Y1}(y_1) = 2 \cdot 1/100 \cdot 1/100 = 2/100^2 \text{ with } y_1 \le y_2$$

and we note that the joint distribution is not the product of the two individual distributions because of the 2! multiplier in front. We check the normalization integral by integrating over all y_1 and y_2 for which $y_1 \le y_2$. This means that the integration over y_2 must start at y_1 so that its value is always greater than or equal to y_1; the integration over y_1 is *unrestricted* over its full range [0,100] nsec. Note that the order of integration affects the limits; integrating over y_1 first requires limits $y_1=0$ to $y_1=y_2$; then integrate over y_2 from $y_2=0$ to $y_2=100$. Both methods integrate to unity as required.

3.4.1 Order Statistics Applications: Middle Statistic Prob[¼ ≤Y$_2$≤ ¾] (n=3)

Order Statistics Applications: Middle Statistic Prob[¼ ≤Y$_2$≤ ¾] (n=3)

Example: $\{X_1, X_2, X_3\}$ are IID RVs Uniform on [0,1]

Find Prob that "middle" value is between ¼ & ¾ , *i.e.*, Prob[¼ ≤Y$_2$≤ ¾]

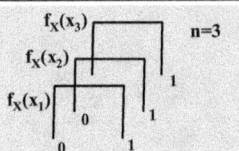

Direct from Order Statistics Formula : n=3, j=2 "middle order statistic"

$$f_{Y_j}(y_j) = \frac{n!}{(j-1)!\cdot(1!)\cdot(n-j)!}\left(F_X(y_j)\right)^{j-1}\cdot f_X(y_j)\cdot\left(1-F_X(y_j)\right)^{n-j}$$

$$f_{Y_2}(y_2) = \frac{3!}{(2-1)!\cdot(1!)\cdot(3-2)!}\left(F_X(y_2)\right)^{2-1}\cdot f_X(y_2)\cdot\left(1-F_X(y_2)\right)^{3-2}$$

$$f_{Y_2}(y_2) = 6\cdot y_2\cdot(1-y_2)$$

$$= 6\cdot \underbrace{F_X(y_2)}_{F_X(y_2)=\int_{x=0}^{x=y_2}\frac{1}{1}dx=y_2}\cdot\left(1-F_X(y_2)\right)\cdot\underbrace{f_X(y_2)}_{=1\,(\text{Uniform on }[0,1])} = 6\cdot y_2\cdot(1-y_2)$$

Normalization:
$$\int_{y_2=0}^{1} f_{Y_2}(y_2)dy_2 = 6\left(\frac{y_2^2}{2}-\frac{y_2^3}{3}\right)\Bigg|_{y_2=0}^{1} = 6\left(\frac{1}{2}-\frac{1}{3}\right)=1$$

$$\Pr[1/4\le y_2 \le 3/4] = F_{Y_2}(3/4) - F_{Y_2}(1/4)$$

$$F_{Y_2}(y_2) = \int_{\zeta=0}^{\zeta=y_2}6(\zeta-\zeta^2)d\zeta = 6\left(\frac{\zeta^2}{2}-\frac{\zeta^3}{3}\right)\Bigg|_{\zeta=0}^{\zeta=y_2} = 6\left(\frac{y_2^2}{2}-\frac{y_2^3}{3}\right)$$

$$\therefore\ \Pr[1/4\le y_2\le 3/4] = \underbrace{6\left(\frac{(3/4)^2}{2}-\frac{(3/4)^3}{3}\right)}_{F_{Y_2}(3/4)} - \underbrace{6\left(\frac{(1/4)^2}{2}-\frac{(1/4)^3}{3}\right)}_{F_{Y_2}(1/4)} = .6875$$

$f_{Y_2}(y_2=1/2)=6(1/2)(1-1/2)$
$=1.5$ Max Value

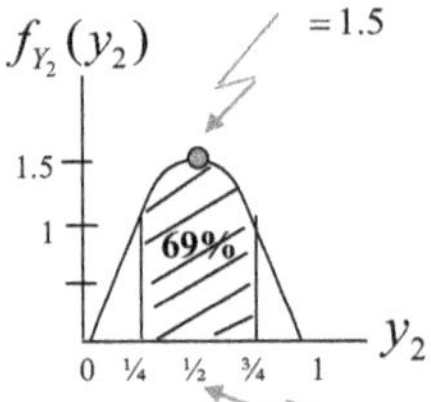

$f_{Y_2}(y_2)$

1.5 ─
1 ─ 69%
   ┼─┼─┼─┼─┼─ y_2
   0 ¼ ½ ¾ 1

$$\frac{df_{Y_2}(y_2)}{dy_2}=6-12y_2=0 \implies y_2=1/2$$

Here is another example with three IID Uniform RVs $\{X_1, X_2, X_3\}$ on [0, 1] and we are asked to find the probability that the "middle" value is between ¼ and ¾; that is, we seek Pr[¼ ≤ Y$_2$≤ ¾]. We first compute the PDF of the middle statistic by setting n=3, j=2 in the general formula

$$f_{Yj}(y_j) = {}^nC_{(j-1),1,(n-j)}\,[F_X(y_j)]^{j-1}\,f_X(y_j)\,[1-F_X(y_j)]^{n-j}\ ;\ j=1,2,...,n$$

to obtain

$$f_{Y2}(y_2) = {}^3C_{(2-1),1,(3-2)}\,[F_X(y_2)]^{2-1}\,f_X(y_2)\,[1-F_X(y_2)]^{3-2}= 6\cdot F_X(y_2)\,f_X(y_2)\,[1-F_X(y_2)] = 6y_2\cdot 1\cdot[1-y_2].$$

The normalization is easily checked by integration over [0, 1] and found to be unity. The cumulative probability function is found by integration to be

$$F_{Y2}(y_2) = 6\,[y_2^2/2 - y_2^3/3]$$

Finally, taking the difference between $F_{Y2}(y_2)$ evaluated at $y_2 = ¾$ and at $y_2 = ¼$ yields

$$\Pr[\,¼ \le Y_2 \le ¾\,].\ = F_{Y2}(y_2=3/4) - F_{Y2}(y_2=1/4) = .6875$$

A sketch of the PDF $f_{Y2}(y_2) = 6y_2(1-y_2)$ shows the probability area between $y_2 = ¼$ and at $y_2 = ¾$ as 69% of the total area; setting the derivative $d/dy_2\,(f_{Y2}(y_2)) = 0$ shows that the PDF peaks at $y_2 = 1/2$; substitution back into the PDF yields the peak magnitude $f_{Y2}(1/2) = 1.5$ and allows us to roughly sketch the distribution as shown.

3.4.2 Order Statistics Applications: Marginal PDFs and Normalization (n=4)

Order Statistics Applications: Marginal PDFs and Normalization (n=4)

Case n=4 {*Min*, *Mdl₁*, *Mdl₂*, *Max*}; all uniformly distributed over [0,1]

The Joint PDF for all the **"order statistics"**, j=1,2, ···,n is defined as follows:

$$f_{Y_1 Y_2 Y_3 Y_4}(y_1, y_2, y_3, y_4) = 4! \cdot \underbrace{f_X(y_1)}_{=1} \cdot \underbrace{f_X(y_2)}_{=1} \cdot \underbrace{f_X(y_3)}_{=1} \cdot \underbrace{f_X(y_4)}_{=1} = 4! \qquad y_1 < y_2 < y_3 < y_4$$

Integrate joint density over all values of y1 , y2 , y3, and y4 to verify normalization

$$\int_{y_4=0}^{1} dy_4 \int_{y_3=0}^{y_3=y_4} dy_3 \int_{y_2=0}^{y_2=y_3} dy_2 \int_{y_1=0}^{y_1=y_2} dy_1 \overbrace{f_{Y_1 Y_2 Y_3 Y_4}(y_1, y_2, y_3, y_4)}^{=4!} = 4! \int_{y_4=0}^{1} dy_4 \int_{y_3=0}^{y_3=y_4} dy_3 \int_{y_2=0}^{y_2=y_3} dy_2 \cdot \underbrace{y_1 \big|_{y_1=0}^{y_1=y_2}}_{=y_2}$$

$\underbrace{\qquad\qquad\qquad\qquad}_{y_1 \,\varepsilon\, [0,y_2]}$ $\underbrace{\qquad}_{\frac{y_2^2}{2}\big|_{y_2=0}^{y_2=y_3} = \frac{y_3^2}{2}}$

$\underbrace{\qquad\qquad\qquad\qquad\qquad}_{y_2 \,\varepsilon\, [0,y_3]}$ $\underbrace{\qquad}_{= \frac{y_4^3}{3\cdot2}}$

$\underbrace{\qquad\qquad\qquad\qquad\qquad\qquad}_{y_3 \,\varepsilon\, [0,y_4]}$

$\underbrace{\qquad\qquad\qquad\qquad\qquad\qquad\qquad}_{y_4 \,\varepsilon\, [0,1]}$ $= 4! \cdot \frac{y_4^4}{4\cdot3\cdot2}\Big|_{y_4=0}^{y_4=1} = 4! \cdot \frac{1^4}{4!} = 1$

alternately,

$$\int_{y_1=0}^{y_1=1} dy_1 \int_{y_2=y_1}^{y_2=1} dy_2 \int_{y_3=y_2}^{y_3=1} dy_3 \int_{y_4=y_3}^{y_4=1} dy_4 \,(4!) = \cdots = 1$$

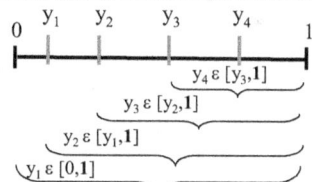

Given 4 IID uniform RVs {X_1, X_2, X_3, X_4} over [0,1] we write down the *joint* PDF as $4! \cdot (1 \cdot 1 \cdot 1 \cdot 1) = 24$. We now integrate this density $f_{Y_1 Y_2 Y_3 Y_4}(y_1, y_2, y_3, y_4)$ over all four dimensions subject to the re-ordering condition on the Ys: $y_1 \leq y_2 \leq y_3 \leq y_4$. Integrating in the order y_1 then y_2 then y_3 then y_4, we assign limits on each variable in accordance with the inequality as follows:

The minimum variable y_1: require $y_1 \leq y_2$; y_1 can start at "0" and go up to "y_2" ; **$y_1 \,\varepsilon\, [0, y_2]$**;
The 1st middle variable y_2: require $y_2 \leq y_3$; y_2 can start at "0" and go up to "y_3" ; **$y_2 \,\varepsilon\, [0, y_3]$**;
Since the 1st step already insures that $y_1 \leq y_2$, this step preserves the order of all three $y_1 \leq y_2 \leq y_3$.
The 2nd middle variable y_3: require $y_3 \leq y_4$; y_3 can start at "0" and go up to "y_4" ; **$y_3 \,\varepsilon\, [0, y_4]$**;
Since already have $y_1 \leq y_2 \leq y_3$ this step preserves the order of all four. $y_1 \leq y_2 \leq y_3 \leq y_4$
The maximum variable y_4 has no restrictions within the full range $y_4 \,\varepsilon\, [0, 1]$.
The multiple integral proceeds in the order y_1, then y_2, then y_3, then y_4, and thus always yields integrands containing the next integration variable as shown; the final result is $24 \cdot y_4^4 / (4\cdot3\cdot2)$ which, upon evaluation between the limits $y_4 = 0$ and $y_4 = 1$ yields the proper normalization of $24/24 = 1$.
Note that if the order of integration is reversed, *i.e.*, y_4, then y_3, then y_2, then y_1, we must assign limits on **y_4** between [y_3, 1] ; on **y_3** between [y_2, 1] ; on **y_2** between [y_1, 1] ; on **y_1** between [0, 1] ; the reader should carefully reason this out and perform the integrations. The final integration over y_1 yields $I = 24 \{(1/6) y_1 - (1/2) (y_1^2/2) + (1/2) (y_1^3/3) - (1/6) (y_1^4/4)\}$ between the limits $y_1=0$ and $y_1=1$ and evaluates to $24\{1/6 -1/4 +1/6 -1/24\} = 1$.

3.4.3 Order Statistics Applications: Two Gaussians (n=2)

Order Statistics Applications: Two Gaussians (n=2)

Ex. 2 - Two IID Gaussian RVs X_1, X_2 with PDF $\qquad f_X(x) = \dfrac{1}{\sigma\sqrt{2\pi}} e^{-\frac{x^2}{2\sigma^2}}$; $\;-\infty < x < \infty$

Find Distrib of larger of two variables $\qquad Y_2 = W = \max\{X_1, X_2\} \qquad$ Set n=2, j=2

$$f_W(w) = 2!\left[F_X(w)\right]^{2-1} f_X(w)\left[1 - F_X(w)\right]^{2-2}$$

$$= 2 \cdot F_X(w) \cdot f_X(w) = 2\left(\int_{x=-\infty}^{x=w} \frac{1}{\sqrt{2\pi}} e^{-\frac{x^2}{2\sigma^2}} dx\right) \cdot \frac{1}{\sqrt{2\pi}} e^{-\frac{w^2}{2\sigma^2}}$$

Normalization

$$\int_{w=-\infty}^{+\infty} f_W(w)\,dw = \int_{w=-\infty}^{+\infty} 2 \cdot F_X(w) \cdot f_X(w)\,dw = \int_{w=-\infty}^{+\infty} 2 \cdot F_X(w) \cdot dF_X(w) = 2\frac{\left[F_X(w)\right]^2}{2}\Bigg|_{w=-\infty}^{w=+\infty} \qquad \checkmark\checkmark$$

$$= \underbrace{\left[F_X(\infty)\right]^2}_{=1} - \underbrace{\left[F_X(-\infty)\right]^2}_{=0} = 1$$

Given to IID Gaussians RVs $\{X_1, X_2\}$ we are asked to find the PDF of the larger of the two variables, i.e., $Y_2 = W = \max\{X_1, X_2\}$; thus set n=2, j=2 in the general formula

$$f_{Yj}(y_j) = {}^nC_{(j-1),1,(n-j)}\ [F_X(y_j)]^{j-1}\ f_X(y_j)\ [1-F_X(y_j)]^{n-j}\ \ ;\ \ j=1,2,\ldots,n$$

to obtain

$$f_{Y2}(y_2) = {}^2C_{(2-1),1,(2-2)}\ [F_X(y_2)]^{2-1}\ f_X(y_2)\ [1-F_X(y_2)]^{2-2} = 2\,F_X(y_2)\,f_X(y_2)$$

Note that the form of the PDF expression automatically insures normalization because of the relation $f_X(y_2) = d/dy_2\{F_X(y_2)\}$; explicitly using this fact, we have the sequence of calculations

$$f_{Y2}(y_2) = 2\,F_X(y_2)\,d/dy_2\{F_X(y_2)\} = 2\,d/dy_2\{[F_X(y_2)]^2/2\} = d/dy_2\{[F_X(y_2)]^2\}$$

The PDF is an exact derivative of the quantity $[F_X(y_2)]^2$, so by the fundamental theorem of calculus its integral is just the integrand evaluated at the limits of the variable y_2 $(-\infty, +\infty)$ and yields

$$\int f_{Y2}(y_2)\,dy_2 = \int d\,[F_X(y_2)]^2 = [F_X(+\infty)]^2 - [F_X(-\infty)]^2 = 1^2 - 0^2 = 1$$

Note that in evaluating F(x) we have used the fact that any cumulative distribution function *starts at 0 and ends at 1*; accordingly, we must always have F($-\infty$) = 0 and F($+\infty$) = 1 *no matter what else it does in between.*

3.4.4 Order Statistics Applications: Gaussian Mean Y_2 Statistic (n=2)

Order Statistics Applications: Gaussian Mean Y_2 Statistic (n=2)

Find mean of W   1) Express as Double Integral   2) Interchange order of Integration

$$E[W] = \int\limits_{w=-\infty}^{+\infty} \{2 \cdot F_X(w) \cdot f_X(w)\} \cdot w\, dw = 2 \int\limits_{w=-\infty}^{+\infty} dw\, f_X(w) \cdot w \underbrace{\left\{ \int\limits_{\xi=-\infty}^{\xi=w} d\xi\, f_X(\xi) \right\}}_{\substack{=F_X(w) \\ \text{strip "A"}}} = 2 \int\limits_{\xi=-\infty}^{+\infty} d\xi\, f_X(\xi) \underbrace{\left\{ \int\limits_{w=\xi}^{w=+\infty} dw\, f_X(w) \cdot w \right\}}_{\text{strip "B"}}$$

$$= 2 \int\limits_{\xi=-\infty}^{+\infty} d\xi\, \frac{1}{\sigma\sqrt{2\pi}} e^{-\frac{\xi^2}{2\sigma^2}} \int\limits_{w=\xi}^{w=+\infty} dw\, \frac{1}{\sigma\sqrt{2\pi}} e^{-\frac{w^2}{2\sigma^2}} \cdot w$$

$$= 2 \int\limits_{\xi=-\infty}^{+\infty} d\xi\, \frac{1}{\sigma\sqrt{2\pi}} e^{-\frac{\xi^2}{2\sigma^2}} \int\limits_{w=\xi}^{w=+\infty} \sigma^2 d(\underbrace{\frac{w^2}{2\sigma^2}}_{=u})\frac{1}{\sigma\sqrt{2\pi}} \underbrace{e^{-\frac{w^2}{2\sigma^2}}}_{=e^{-u}} = \frac{1}{\pi} \int\limits_{\xi=-\infty}^{+\infty} d\xi\, e^{-\frac{\xi^2}{2\sigma^2}} \left\{ -e^{-\frac{w^2}{2\sigma^2}} \right\}\Big|_{w=\xi}^{+\infty}$$

$$= \frac{4}{2\pi \cdot \sigma^2} \int\limits_{\xi=0}^{\infty} d\xi \cdot e^{-\frac{\xi}{2\sigma^2}} \int\limits_{w=\xi}^{w=\infty} \sigma^2 \cdot d\left(\frac{w^2}{2\sigma^2}\right) \cdot e^{-\left(\frac{w^2}{2\sigma^2}\right)} = \frac{2}{\pi} \int\limits_{\xi=0}^{\infty} d\xi \cdot e^{-\frac{\xi^2}{2\sigma^2}} \cdot \left[-e^{-\left(\frac{w^2}{2\sigma^2}\right)} \right]_{w=\xi}^{w=\infty} = \frac{2}{\pi} \int\limits_{\xi=0}^{\infty} d\xi \cdot e^{-2\cdot\frac{\xi^2}{2\sigma^2}}$$

Define $u = \xi^2/\sigma^2$; $du = 2\xi d\xi/\sigma^2 \Rightarrow$

$$d\xi = \sigma^2 du/(2\xi) = \sigma^2 du/(2\sigma \cdot u^{1/2})$$

$$\therefore\; E[W] = \frac{2}{\pi} \cdot \frac{\sigma}{2} \int\limits_{u=0}^{\infty} du \cdot u^{-1/2} e^{-u}$$

$$= \frac{\sigma}{\pi} \cdot \Gamma(1/2) = \frac{\sigma}{\pi} \cdot \sqrt{\pi} = \boxed{0.564 \cdot \sigma}$$

The mean value of the maximum statistic $Y_2 = W$ is defined by the integral $E[W] = \int f_W(w) \cdot w\, dw$; substitution of $f_W(w)$ in terms of the underlying Gaussian PDF $f_X(x)$ and CDF $F_X(x)$ then requires over dw first and then over $d\xi$ making sure to cover the same doubly infinite shaded region above the line $\xi = w$. evaluation of the integral $\int 2 \cdot w \cdot F_X(w) \cdot f_X(w) dw$. Since $F_X(w)$ is itself an integral of $f_X(\xi)$ over the "dummy variable" ξ from $\xi = -\infty$ to $\xi = w$, we can rewrite $E[W]$ as the **double integral** on the top line of the slide.
Note that the region of integration for the double integral variables dw $d\xi$ is the **doubly infinite region above the line $\xi = w$** as indicated by the grey region in the figure. We have identified the integral over the dummy variable ξ by the horizontal strip labeled A in the ξ-w plane as shown in the figure; the second integration over dw sums all such strips from $w = -\infty$ to $+\infty$ thereby covering the entire shaded region. Now a standard "trick" is to change the *order of integration* by integrating over dω first and then over dξ and the integral can be done. Thus, changing the order, first integrate dw over the vertical **strip B** from the line $w="\xi"$ to $+\infty$ and then integrate dξ over all such strips emanating from the line at points ranging from $\xi = -\infty$ to $\xi = +\infty$. This trick allows the first integral to be evaluated exactly as $-\exp(-w^2/2\sigma^2)$ between the limits $w="\xi"$ and $w = +\infty$ which yields $\exp(-\xi^2/2\sigma^2)$; subsequent integration over dξ from $\xi = -\infty$ to $\xi = +\infty$ is just a Gaussian normalization integral yielding unity, and hence the desired mean of the larger random variable $Y_2 = W$ is $\mu_W = E[W] = \sigma/(\pi)^{1/2} \cdot 1 = .564\, \sigma$. Thus the mean of the *larger* random variable W increases in proportion to the spread σ of the underlying zero-mean Gaussian distribution for the two IID RVs X_1 and X_2.

4 Random Processes

Random Processes

4.1 *Random Processes Overview*

Random Processes Overview

- Time Series Data = Physical Measurements in *time*
- Random Process = Sequence of random variable realizations
    - Geiger Counter Sequence of "detections" - Poisson Process
    - Communication Binary Bit Stream - Bernoulli Process " 01001···"
    - E&M Propagation Phase (I-Q components) - Gaussian Process
- Arrival Event: Success ="arrival" (of an event in time)
- Interarrival Times for Random Processes
    - Not only interested in how many successes K (" arrivals") there are
    - But also interested in *"specific time of arrivals,"* e.g., T_K = time of kth arrival
    - **DSP Chip Interrupts:**
        - Time between interrupts
        - used for data processing
    - **Waiting on Telephone:**
        - "you are 10th customer in line and …
        - wait will be approximately "7 min"
    - **Experimental Bernoulli Trials:**
        - Large but finite number of trials
        - Statistics of Regrouped Data
        - Number and timing of arrivals

Physical Random Process	Number of Arrivals	Interarrival Times
Geiger Counter $p \ll 1$, $n \to \infty$	Poisson	Exponential
Binary Bit Stream $p\varepsilon[0,1]$, $n \to \infty$	Bernoulli	Geometric
Bernoulli Trial Grouping $p\varepsilon[0,1]$, n fixed	Binomial	Negative Binomial

Observations of physical processes produce measurements over time that almost always have components described by a random process. Some examples are Geiger counter detections (Poisson Process), Binary bit streams (Bernoulli Process) and Electromagnetic wave I, Q Phase components (Gaussian Process).

Because, these processes take place over time, the notion of a "success" is translated to an "arrival" at a specific time. Moreover, we are not only interested in how many successes K there are, but also their specific arrival times, *e.g.*, the time of the kth arrival T_k. This has application to many physical processes such as the timing of DSP chip interrupts relative to their "clock cycles" and the queuing of customers in a telephone answering system. In both cases the system must handle the "load" in an appropriate manner; for the DSP chip it is important to minimize the number of times the interrupt is near the *leading or trailing edges* of the timing pulse where pulse shape distortion causes errors, while for the telephone answering service, the 10th customer, would like to know how long he or she must wait in the queue before being served. For the general analysis of a fixed number n of Bernoulli trials, the re-grouping of such experimental data results in a Binomial process shown in the third row of the table. Slides# 4-10 to 4-12 describe the analysis and trade-offs involved.

The table summarizes some physical random processes in terms of the distributions characterizing their two complementary aspects, *i.e.*, number and timing of arrivals. The 1st column names the processes and gives restrictions (if any) on the single trial probability p and on the number of trials n for each. The 2nd and 3rd columns labeled "Number of Arrivals" and "Interarrival Times" display the complementary process pairs as {Poisson, Exponential}, {Bernoulli, Geometric}, and {Binomial, Negative Binomial} representing the distributions for {number arrivals, times of arrival}. It is interesting to note that the Poisson distribution is discrete while its complementary aspect, the Exponential distribution, is continuous. On the next few slides, the relationships and limiting forms of these processes will be described in more detail.

4.1.1 Relations between Random Processes

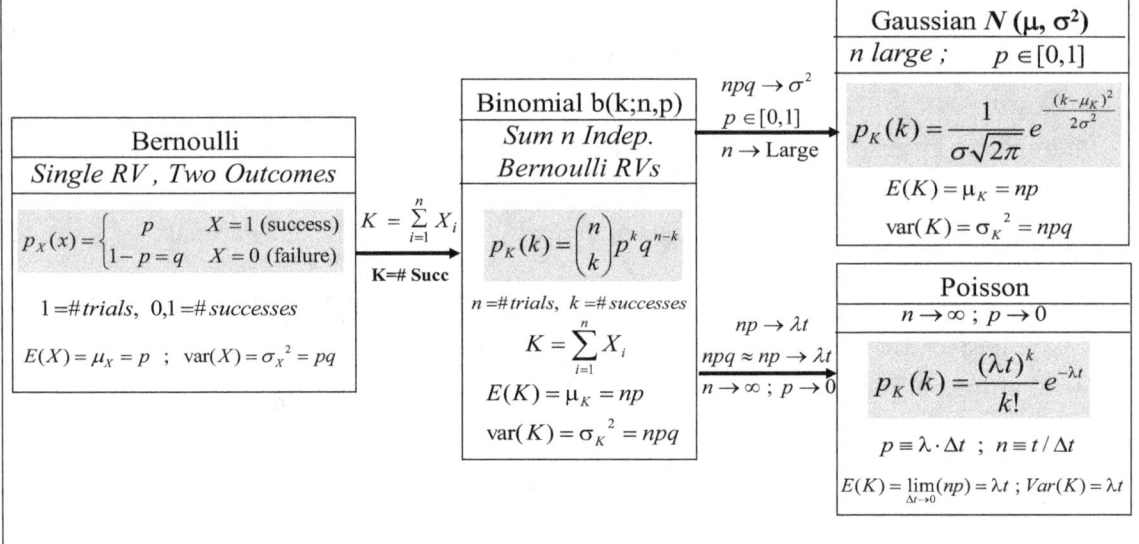

Relations Between Random Processes

- Progression: Bernoulli to Binomial to Gaussian & Poisson
- Framework for Introduction of "time"& approximations
- Problem Types A and B

The Bernoulli process is the fundamental random process underlying a number of other important random processes; the figure illustrates the flow from Bernoulli to Binomial to the two limiting forms: Gaussian and Poisson. The **sum of n independent Bernoulli processes** results in the Binomial distribution b(k;n,p) representing k-successes in n-trials each having a *single trial probability of success* "p". This is true because the Bernoulli process takes on a value of "1" with probability p for success and "0" with probability q=1-p for failure; thus the sum of n Bernoulli trials acts as a "success indicator" since it adds "1" to the sum for each success and "0" to the sum for each failure. This is precisely the Binomial distribution b(k;n,p).

As the number of **Bernoulli variables n becomes large**, the computation of the discrete binomial distribution terms $^{n}C_k \, p^k \, q^{n-k}$ becomes problematic because of the "factorial explosion" (*e.g.*, n!=10,000!) and more convenient approximate representations are obtained by the two limiting forms Gaussian (continuous) and Poisson (discrete) as shown at the extreme right.

Gaussian Limit: Single trial probability "p" can be *small or large within limits* [0, 1], and as the number of trials "n" becomes large the product $np \to \mu_K$, the mean, and $npq = np(1-p) \to \sigma_K^2$, the variance. The Gaussian form is used to approximate the discrete number of Binomial successes K=k and is left written in terms of the variable "k" instead of the more usual notation with the variable "x". It is important to note that the variable "k" in the Gaussian is not only continuous now, but can also take on any positive or negative value in $[-\infty, +\infty]$. Even though such "fractional successes and negative successes" have no meaning, the Gaussian approximation to the Binomial does yield valid results as shown in the Central Limit Theorem slides and examples of Section 8 (see Slides#8-8 to 8-11, 8-14, 8-15.)

Poisson Limit: Single trial probability "p" *must be small*, and as the number of trials "n" becomes large, the product $np \to \lambda t = \mu_K$ = "a", the mean, and $npq = np(1-p) \approx np \to \lambda t = \sigma_K^2$, the variance of the Poisson distribution. Note that the Poisson PMF is discrete (not continuous) and the Poisson parameter a = $\lambda \cdot t$ (rate * time) is equal to both the mean μ_K and the variance σ_K^2. An explicit derivation is given in Slide#5-3 and application examples are given in Slide#5-6 to 5-9.

4.1.2 Bernoulli, Geometric, Binomial and Negative Binomial PMFs

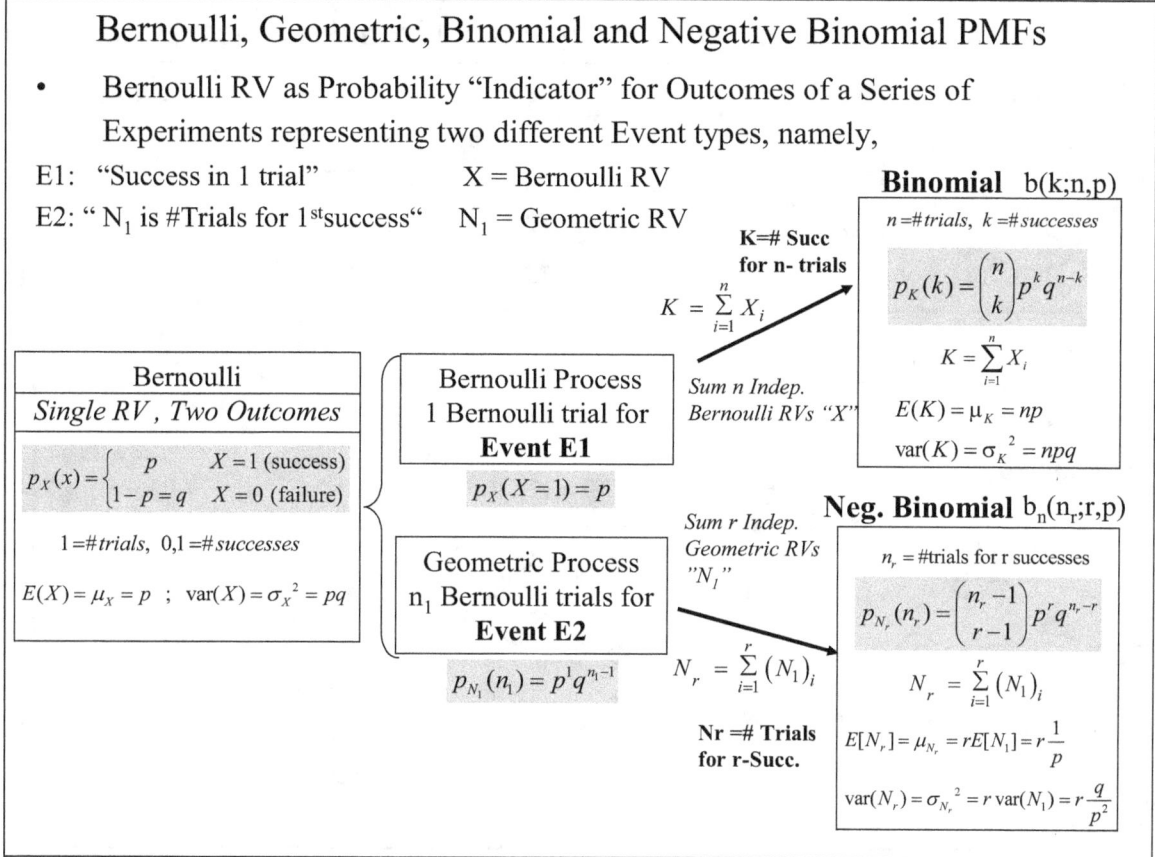

Bernoulli, Geometric, Binomial and Negative Binomial PMFs

- Bernoulli RV as Probability "Indicator" for Outcomes of a Series of Experiments representing two different Event types, namely,

E1: "Success in 1 trial" $\quad$ X = Bernoulli RV

E2: " N_1 is #Trials for 1st success" $\quad N_1$ = Geometric RV

Binomial b(k;n,p)

n =#trials, k =#successes

$$p_K(k) = \binom{n}{k} p^k q^{n-k}$$

$$K = \sum_{i=1}^{n} X_i$$

$E(K) = \mu_K = np$

$\text{var}(K) = \sigma_K^2 = npq$

K=# Succ for n- trials

$$K = \sum_{i=1}^{n} X_i$$

Bernoulli

Single RV , Two Outcomes

$$p_X(x) = \begin{cases} p & X=1 \text{ (success)} \\ 1-p=q & X=0 \text{ (failure)} \end{cases}$$

1 =#trials, $0,1$ =#successes

$E(X) = \mu_X = p$; $\text{var}(X) = \sigma_X^2 = pq$

Bernoulli Process 1 Bernoulli trial for **Event E1**

$$p_X(X=1) = p$$

Sum n Indep. Bernoulli RVs "X"

Geometric Process n_1 Bernoulli trials for **Event E2**

$$p_{N_1}(n_1) = p^1 q^{n_1-1}$$

Sum r Indep. Geometric RVs "N_1"

$$N_r = \sum_{i=1}^{r} (N_1)_i$$

Nr =# Trials for r-Succ.

Neg. Binomial $b_n(n_r;r,p)$

n_r = #trials for r successes

$$p_{N_r}(n_r) = \binom{n_r-1}{r-1} p^r q^{n_r-r}$$

$$N_r = \sum_{i=1}^{r} (N_1)_i$$

$E[N_r] = \mu_{N_r} = rE[N_1] = r\dfrac{1}{p}$

$\text{var}(N_r) = \sigma_{N_r}^2 = r\,\text{var}(N_1) = r\dfrac{q}{p^2}$

The event E_1: success in a single Bernoulli trial and the event E_2: the number of trials required for one success represent the complementary aspects of the Bernoulli process (see Slide#4-2). The Bernoulli RV "X" is the basic building block for other RVs ("atomic" RV) and has a PMF with only two outcomes: X=1 with probability p and X=0 with probability q=1- p. We have seen that n such Bernoulli variables when added yield a Binomial PMF {b(x;n,p), x=0,1,2,...,n} which gives the "#successes "x" for "n" trials. We have also seen that this **Binomial PMF** can be understood by repeatedly appending the Bernoulli tree graph to each of its nodes (repeated independent trials) thereby constructing a tree with 2^n outcomes corresponding to the n Bernoulli trials, each with two possible outcomes.

Alternately, the **Geometric PMF** can be constructed by repeatedly appending a Bernoulli tree graph to itself an infinite number of times, but this time only to the *failure node*, thereby constructing a tree with an infinite number of outcomes which correspond to *exactly "1" success* and *"x-1" failures* where x=1,2, ...,∞. Just as the Bernoulli tree graph is a building block for the Binomial tree graph, the infinite Geometric tree graph is a building block for the Negative Binomial. The Negative Binomial tree graph for r=2 successes is constructed by appending a Geometric tree graph to itself an infinite number of times, but this time only to the *success nodes*, resulting in a doubly infinite tree graph corresponding to *exactly "2" successes* and *"x-2" failures* where x= 2,3, ..., ∞. Repeating this process r-times yields the r-fold infinite tree graph corresponding to *exactly "r" successes* and *"x-r" failures*, where x= r, r+1, ..., ∞. The mathematical transformations relating Bernoulli, Binomial, Geometric, and Negative Binomial are shown in this slide and the corresponding tree progressions are shown on the next slide.

4.1.3 Tree Progression from Bernoulli to: Binomial, Geometric, Neg. Binomial

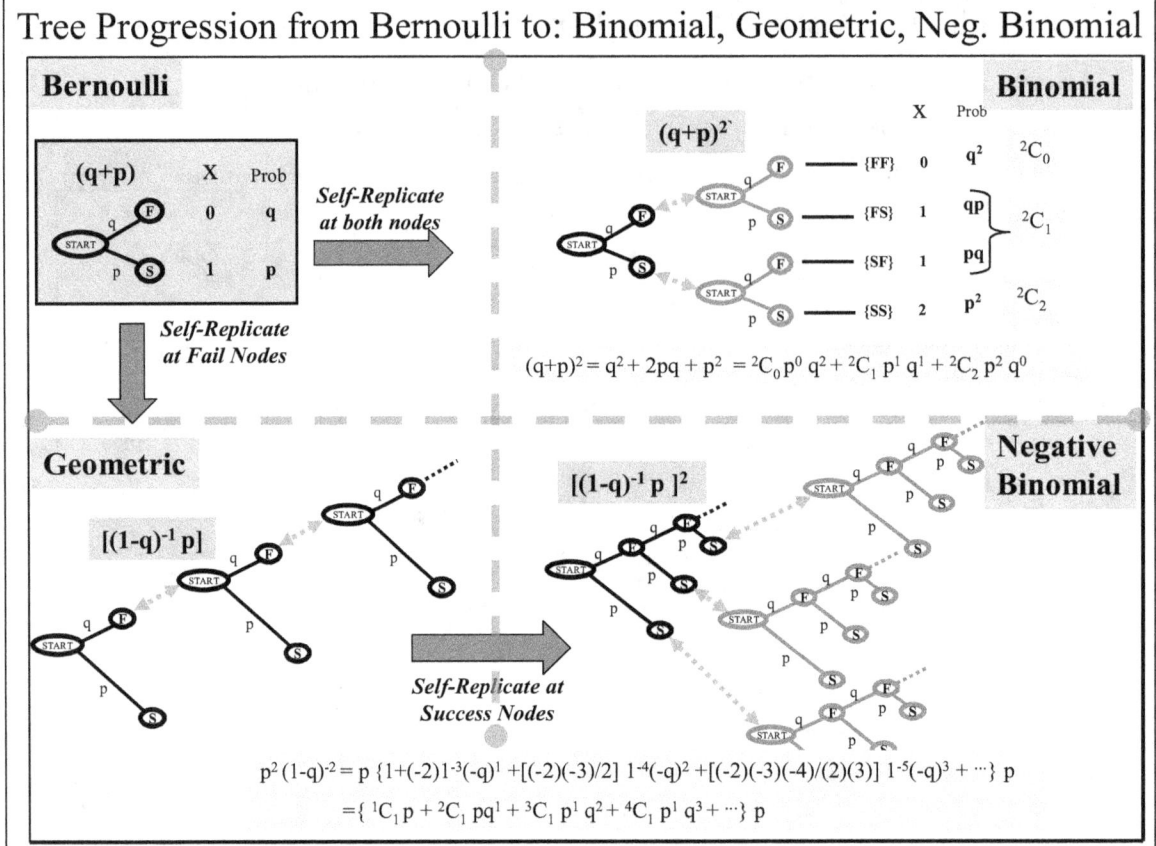

This slide illustrates a progression from the fundamental Bernoulli process tree to Binomial and Geometric and then from Geometric to Negative Binomial with transitions indicated by the *dark arrows*.

Bernoulli to Binomial: This transition is shown by the horizontal arrow on the top where the n=2 Binomial is obtained by appending a second Bernoulli tree to each output node of the first, thus yielding the 4 output states {{FF}, {FS}, {SF}, {SS}}. Furthermore, we see that this structure for n=2 is generated algebraically by the expansion $1 = (q+p)^2 = q^2 + 2q^1 p^1 + p^2$. This can be continued for n=3, 4, ... by repeatedly appending a binary Bernoulli tree to each new node; expanding the identity $1 = (q+p)^n$ obviously yields the appropriate Binomial expansion terms for n Bernoulli trials.

Bernoulli to Geometric: This tree is constructed from an infinite number of Bernoulli trials by successively appending an additional Bernoulli tree to *each failure node* as shown by the down transition arrow and the geometric tree on the bottom left. The 1st Bernoulli trial yields either success S or failure F; this failure node spawns a 2nd Bernoulli trial, and the process continues indefinitely. It accurately describes all possible probability paths for a single success in 1, 2, 3,...,∞ #trials and is algebraically generated by the expression $1 = [(1-q)^{-1} p]$ which expands to give an infinite sequence $[1 + q^1 + q^2 + q^3 + ...]\cdot p$, corresponding to exactly 0, 1, 2, 3,..."failures before a single success."

Geometric to Negative Binomial: The Negative Binomial tree structure is obtained by appending the basic Geometric tree to itself at each success S-node (infinite number) as shown by the horizontal transition arrow and tree on the bottom right. This yields a doubly infinite tree structure for the r=2 Negative Binomial representing the number of trials X =x required for r=2 successes. We can verify the first few terms in the Negative binomial by tracing directly through the tree. Alternately, direct expansion of the algebraic identity yields

$$1^2 = [(1-q)^{-1} p]^2 = \{ {}^1C_1 p + {}^2C_1 pq^1 + {}^3C_1 p^1 q^2 + {}^4C_1 p^1 q^3 + \cdots \} p,$$

in agreement with terms of the r=2 Negative Binomial PMF. This process may be extended to r=3, 4, ... successes by repeatedly appending the Geometric tree to each success *node*; the algebraic generator $1^r = [(1-q)^{-1} p]^r$ for the exponent r yields the correct result for the general Negative Binomial. Note that the "Negative" modifier to Binomial is a natural designation in view of the $(1-q)^{-1}$ term in its algebraic structure.

4.1.4 Type A and B Problems: Number and Timing of Successes

Type A and B Problems: Number and Timing of Successes

- Two Types of Problems

— **Type A) Bernoulli => Binomial ~ Poisson, Gauss**

  - **k** successes (**varies**); n trials (**fixed**) $n=1$ Bernouli
    $n>1$ Binomial

$$p_K(k;n,p) = b(k:n,p) = \binom{n}{k}p^k q^{n-k}$$

$$\Pr[k-\text{succ},n-\text{trials}] \Rightarrow \Pr[k_{min} \le K \le k_{max}] = \sum_{k=k_{min}}^{k_{max}} \binom{n}{k}p^k q^{n-k} = \alpha$$

$$\underset{\#succ}{}$$

$p_K(k)$

*Bernoulli **Inter-arrival** times leads to:*

| 5 Params: Given 4, find 5th |
Find	Given
kmin	kmax,p,n,alpha
kmax	kmin,p, n, alpha
n	kmin,kmax,p, alpha
p	kmin,kmax, n,alpha
alpha	**kmin,kmax,p, n**

A) $p_K(k)$
B) $p_{Nr}(n_r)$ α

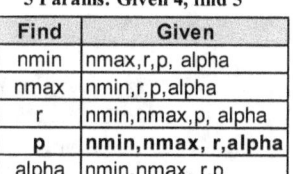

— **Type B) Geometric => Neg Binomial**

  - **n** trials (**varies**); r successes (**fixed**) $r=1$ Geometric
    $r>1$ Neg. Binomial
  - exchange roles *fixed ↔ vary*

$$p_{N_r}(n;r,p) = \begin{cases} \binom{n-1}{r-1}p^r q^{n-r} & ; \ n=r,r+1,\cdots \quad \binom{n}{r}=0 \ \ for \ \ n<r \\ 0 & ; \ \text{otherwise} \end{cases}$$

$$\Pr[r^{th}\text{succ. occurs on} n^{th} \text{trial}] \Rightarrow$$

$$\Pr[n_{min} \le N_r \le n_{max}] = \sum_{n=n_{min}}^{n_{max}} \binom{n-1}{r-1}p^r q^{n-r} = \alpha$$

$$\underset{\substack{\#trials \\ n \text{ "varies"}}}{}$$

| 5 Params: Given 4, find 5th |
Find	Given
nmin	nmax,r,p, alpha
nmax	nmin,r,p,alpha
r	nmin,nmax,p, alpha
p	**nmin,nmax, r,alpha**
alpha	nmin,nmax, r,p

We have seen that the Bernoulli RV "X" is the basic building block for the Binomial and Geometric discrete processes. The sum of n Bernoulli RVs corresponds to appending the Bernoulli tree to itself n times and leads a *finite tree* with 2^n outputs; the tree paths generate the Binomial PMF probability structure corresponding to "k successes in n trials". On the other hand, appending Bernoulli trials *only to the failure nodes* until a single success occurs yields an *infinite tree* whose outputs generate the Geometric PMF; its terms give the probability that "a single success occurs after n trials". Thus, the **Binomial** considers a **fixed number of trials n** and finds the probability of **k successes**, while the **Geometric PMF** considers a **single success** and finds the probability that it **occurs after n trials**. Since trials take place within a fixed interval of time Δt, the number of trials in fact represents an **"arrival time for the first success"**, $t_n = n \Delta t$.

Thus the Geometric process can be thought of as an arrival process answering a question about when in "time" a success occurs as opposed to the Bernoulli process which simply asks whether or not the trial is successful. Sums of Bernoulli RVs and of Geometric RVs yield Binomial and Negative Binomial processes respectively:

(i) The sum of **"n" Bernoulli RVs yields the Binomial** distribution $b(k;n,p)$ for the number of successes k in n-trials (with single trial probability of success p)

(ii) The sum of **"r" Geometric RVs yields the Negative Binomial** distribution $p_{Nr}(n;r,p)$ for the number of trials N_r required for r-successes. See Slide#4-20 for summing r= 2 or 3 geometric RV's.

The two processes are complementary in the sense that the Binomial answers a question about "the number of successes" while the Negative Binomial answers a question about the "timing of the successes." This distinction yields two general types of problems with parameters displayed in the slide equations and tables

A) *Parameters*: $\{k_{min}, k_{max}, n, p, \alpha\}$;   *Equation* stated as $\Pr[k_{min} \le K \le k_{max}] = \alpha$, and

B) *Parameters*: $\{n_{min}, n_{max}, r, p, \alpha\}$;   *Equation* stated as $\Pr[n_{min} \le N \le n_{max}] = \alpha$.

The general set up is given by the type A or B equations and all problems reduce to a specification of 4 of these 5 parameters and a determination of the 5th.

4.1.5 Negative Binomial Sums over Number of Trials and of Successes

Negative Binomial Sums over Number of Trials and of Successes

Note 1: Sum over # *trials 'n'* ==> (Normalization = 1)

$$\sum_{n=r}^{\infty} p_{N_r}(n;r,p) = \sum_{n=r}^{\infty} \binom{n-1}{r-1} p^r q^{n-r} = \underbrace{\binom{r-1}{r-1} p^r q^{(r)-r}}_{n=r} \overset{q^0}{} + \underbrace{\binom{(r+1)-1}{r-1} p^r q^{(r+1)-r}}_{n=r+1} \overset{q^1}{} + \cdots + \underbrace{\binom{(r+i)-1}{r-1} p^r q^{(r+i)-r}}_{n=r+i} \overset{q^i}{} \cdots$$

$$= p^r \cdot \left\{ 1 + rq + \frac{r(r+1)}{2!} q^2 + \frac{r(r+1)(r+2)}{3!} q^3 + \cdots \right\} = p^r \cdot \frac{1}{(1-q)^r} \left(=1\right) \quad \checkmark\checkmark$$

Note 2: Sum over # *successes 'r'* ==> (Principle of Total Probability for Single Trial = p)

$$\sum_{r=1}^{n} p_{N_r}(n;r,p) = \sum_{r=1}^{n} \binom{n-1}{r-1} (p^{r-1} \cdot p^1) q^{n-r} = p^1 \cdot \underbrace{\sum_{r=1}^{n} \binom{n-1}{r-1} p^{r-1} q^{(n-1)-(r-1)}}_{=(p+q)^{(n-1)}} = p \cdot \underbrace{(p+q)^{(n-1)}}_{=1} = p$$

Interpretation for Note 2:

•**Prob of success on trial #n** = sum of Prob that either the 1st, the 2nd, ..., or the nth success (arrival) occurs on trial #n

•This exhausts all possibilities ; *i.e.*, you cannot have any more than **n successes** in **n trials**.

•By independence, probability of success p is same for *all trials*, so it is p for n^{th} trial as well

$$\begin{pmatrix} \text{Prob of} \\ \text{succ. on} \\ \text{trial \#n} \end{pmatrix} = p = \underbrace{p_{N_1}(n;1,p)}_{1^{st} \text{ Arrival on } \textbf{Trial \#n}} + \underbrace{p_{N_2}(n;2,p)}_{2^{nd} \text{ Arrival on } \textbf{Trial \#n}} + \underbrace{p_{N_3}(n;3,p)}_{3^{rd} \text{ Arrival on } \textbf{Trial \#n}} + \cdots + \underbrace{p_{N_n}(n;n,p)}_{n^{th} \text{ Arrival on } \textbf{Trial \#n}}$$

Sum of Prob over a ME & CE set of outcomes for the nth trial

We are already familiar with the fact that the **finite sum** of Binomial terms b(k;n,p) from k=1,2,...,n must be unity; similarly, the **infinite sum** of the Negative Binomial terms $p_{Nr}(n;r,p)$ over n= r,r+1,...,∞ must also be unity if it is to be a valid distribution. Moreover, for a fixed number of trials n, the sum of a Negative Binomial, over the number of successes, r = 1,2, ..., n, turns out to be "p", the single trial probability of success. The two notes below discuss the proofs given in this slide.

Note 1 verifies the normalization requirement by explicitly writing out several terms in the sum, factoring out the common factor p^r, and recognizing that the binomial coefficients $^rC_{r-1} = {^rC_1} = r$, *etc.*, yield the terms on the 2nd line. The terms within the braces {} are precisely the expansion of $1/(1-q)^r$; this leaves the product $p^r \cdot \{1/(1-q)^r\} = p^r/p^r = 1$ (Q.E.D.)

Note 2 shows that if we instead fix the number of trials "n" and take the finite sum of the terms $p_{Nr}(n;r,p)$ over the number of successes r =1,2,...,n the result is the single trial probability of success. The interpretation of this result is that in n=5 trials (say) the ME and CE ways to have a success on the 5th trial are: for it to be the $r=1^{st}$ success, or $r=2^{nd}$ success, ..., or $r=5^{th}$ success; this exhausts all possibilities since we cannot have more than 5 successes in 5 trials. Now since these are all independent trials, the probability of success in any given trial is always the same, namely "p". In particular, for the $n=5^{th}$ trial, the probability of success is also "p" and so given that the 5th trial is uniquely composed of the five outcomes listed above, it is clear that the sum of $p_{Nr}(5;r,p)$ over r=1,2,3,4,5 must yield "p." The next slide gives an analytic proof of this result.

4.1.6 Negative Binomial $p_{Nr}(n)$ Sum over Number of Arrivals "r"

Negative Binomial $p_{N_r}(n)$ Sum over Number of Arrivals "r"	
Identity : $\displaystyle\sum_{r=1}^{n} p_{N_r}(n) = \underbrace{p_{N_1}(n)}_{\substack{\text{n trials}\\\text{1 succ.}}} + \underbrace{p_{N_2}(n)}_{\substack{\text{n trials}\\\text{2 succ.}}} + \cdots + \underbrace{p_{N_n}(n)}_{\substack{\text{n trials}\\\text{n succ.}}} = p$ **Neg Binom Term :** $\quad p_{N_r}(n) = \binom{n-1}{r-1} p^r q^{n-r}$	**Interpretation** **1) Prob of success on trial #n** is the same as for any other trial namely "p" **2)** This success event is broken up into a **ME and CE set of outcomes** for the nth trial, *i.e.,* that the 1st, the 2nd, ..., or the nth success (arrival) occurs on trial #n

Case n=1: $\quad \displaystyle\sum_{r=1}^{1} p_{N_r}(n) = p_{N_1}(1) = p =$ "single trial prob succ."

Case n=2:

$$\sum_{r=1}^{2} p_{N_r}(2) = p_{N_1}(2) + p_{N_2}(2) = \sum_{r=1}^{2}\binom{2-1}{r-1} p^r q^{n-r} = \binom{2-1}{1-1} p^1 q^{2-1} + \binom{2-1}{2-1} p^2 q^{2-2} = pq + p^2 = p\underbrace{(q+p)}_{=1} = p$$

Case n=3: $\quad \displaystyle\sum_{r=1}^{2} p_{N_r}(3) = p_{N_1}(3) + p_{N_2}(3) + p_{N_3}(3) =$

$$= \sum_{r=1}^{3}\binom{3-1}{r-1} p^r q^{3-r} = \binom{3-1}{1-1} p^1 q^{3-1} + \binom{3-1}{2-1} p^2 q^{3-2} + \binom{3-1}{3-1} p^3 q^{3-3}$$

$$= pq^2 + 2p^2 q + p^3 = p\underbrace{(q^2 + 2pq + p)}_{=(p+q)^2=1} = p$$

Gen. Case n:

$$\sum_{r=1}^{n} p_{N_r}(n) = \sum_{r=1}^{n}\binom{n-1}{r-1} p^r q^{n-r} \underset{r\to r+1}{=} \sum_{r=0}^{n-1}\binom{n-1}{r} p^{r+1} q^{n-r-1} = p\cdot\sum_{r=0}^{n-1}\binom{n-1}{r,\,(n-1)-r} p^r q^{(n-1)-r} = p\underbrace{(p+q)}_{=1}^{n-1} = p$$

We have seen that for the binomial PMF the sum over all possible values of n_r = r, r+1, r+2, ...,∞ must naturally yield unity, while the sum over all possible successes r = 1,2, ..., n leads to the single trial probability of success p. The combinatorial derivation for the latter was demonstrated on previous slide by considering the ME and CE possible results for n trials: r=1 success in n trials, r=2 successes in n trials, ..., r = n successes in n trials and we found that $p_{N1}(n) + p_{N2}(n) +... p_{Nn}(n) = p$, where we have dropped the parameters in favor of the more compact notation $p_{N1}(n) = p_{N1}(n;1,p)$.

Here we give an alternate and more explicit proof that, for a fixed number of trials n, the sum of the Negative Binomial PMF $p_{Nr}(n) = {}^{n-1}C_{r-1} p^r q^{n-r}$ over the number of arrivals r = 1,2, ..., n, yields the single trial probability of success *p*.

The proof is by demonstration; the trivial case n=1 is obvious; we then explicitly write out the sums for n =2, 3 on the slide to verify that they indeed sum to p. Finally, taking the case for arbitrary n, we find that re-indexing r→ r+1 converts the sum to a range r=0 to r=n-1 and that it takes on a form equivalent to the expansion of $(p+q)^{n-1}$ times p; noting that q=1-p, this reduces to $(p+1-p)^{n-1}\cdot p = p$, thus completing the proof.

4.1.7 Type A and Type B Problem Examples

Type A and Type B Problem Examples

Type A: Trials "fixed" Successes "vary"   Time $t = n\Delta t$ is fixed   ***Binomial***

Find the total probability 'α' for 15 to 20 successes (arrivals) out of 5000 events given the single trial probability $p = 0.3$.

> **Given**: $k_{min}=15$, $k_{max}=20$, n =5000, p =0.3 ; **Find**: 'α'
>
> $$\alpha = \sum_{k=15}^{20} \binom{5000}{k} p^k (1-p)^{5000-k} \qquad \text{Easy Sum of 6 terms}$$

Type B: Trials "vary" Successes "fixed"   Time $t = n\Delta t$ varies   ***Negative Binomial***

Find the single trial probability 'p' so that with probability $\alpha \leq 0.5$, the r=5th error (arrival) for a 10,000 bit data packet does not occur until after 5000 bits have already been sent.

> **Given**: $n_{min}=5000+1=5001$, $n_{max}=10,000$, r=5, $\alpha =0.5$; **Find**: 'p'
>
> $$\sum_{n_5=5001}^{10,000} \binom{n_5 -1}{5-1} p^5 (1-p)^{n_s -5} \leq 0.5 \qquad \begin{array}{l}\text{This is a polynomial of degree 10,000 in 'p'}\\ \text{We need an approximation technique!!}\end{array}$$

Trial and Error or Approximations

Here are two examples for Type A (#succ. k in n trials) and Type B (#trials n needed for r succ.) problems; in each case 4 of the 5 parameters are specified and we are asked to find the 5th .

Type A: *We are asked a question involving the number of successes in a fixed number of trials, that is, for a given "run of data".* Specifically, find the probability α that out of a run of n=5000 trials we will have between k_{min} =15 and k_{max} =20 successes, given a single trial probability $p = 0.3$. The set-up is straight forward and simply requires the sum of 6 Binomial terms k=15,16, ..., 20. This problem can also be framed in terms of the cumulative distribution for the binomial $F_X(x) \rightarrow F_K(k;n,p)$ as $\alpha = F_K(20;5000,0.3) - F_K(14;5000,0.3)$.

An alternate problem would be to specify the probability $\alpha = 0.02$ (2%) and find the single trial probability of success p; this is more difficult because the set-up would be essentially the same except the 6 terms we sum $p^{15}(1- p)^{5000-15} +p^{16}(1- p)^{5000-16} +...+p^{20}(1- p)^{5000-20} = .02$ now requires the solution of the polynomial of degree 5000 in "p" which requires numerical root-finding techniques or a technique involving the Gaussian approximation to the Binomial.

Type B: *We are asked a different type of question concerning the timing of an event, i.e., its "arrival" within the a fixed length of data: how many trials "n" ($t_n =n\Delta t$) are needed to yield a given number of successes "r" (errors)?* Again the specific question can be framed in a number of ways; the example asks us to find the single trial probability "p" such that there is a 50% chance that the 5th error ("arrival") in a 10,000 bit packet of digital data occurs after the n=5000th bit in the packet. Thus, we frame the problem in terms of Negative Binomial RV "N_5", the number of trials for 5 successes ("errors"), and write down the sum of Negative Binomial terms between the $N_5 = n_{min}$ =5001 and $N_5 = n_{max}$ = 10,000 "time bins" as shown. We then set this equal to probability $\alpha = 0.5$ (50%) and need to solve a polynomial of degree 10,000 in "p".

We could take a brute force trial and error approach and make an initial guess of p =.001 (say) and then compute the sum of 5000 terms repeatedly adjusting the value of p until we obtain 0.5 as the output. This is not an unreasonable approach given the power of modern computers; however, in Section 8, we develop some special approximation techniques that allow a more efficient solution to this type of problem.

4.2 Bernoulli Random Process

Bernoulli Random Process

Single Bernoulli Trial
1-*Trial*
X=x succ.
"0" or "1" successes

$$p_X(x) = \begin{cases} p & X = 1 \text{ (success)} \\ 1-p = q & X = 0 \text{ (failure)} \end{cases}$$

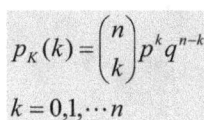

$$E[X] = 0 \cdot (1-p) + 1 \cdot p = p$$

$$E[X^2] = 0^2 \cdot (1-p) + 1^2 \cdot p = p$$

$$\phi_K(s) = E[e^{Xs}] = \sum_{x=0}^{1} e^{xs} p_X(x) = e^{0 \cdot s} \underbrace{p_X(0)}_{=q} + e^{1 \cdot s} \underbrace{p_X(1)}_{=p} = \boxed{q + pe^s}$$

$$\text{var}(X) = p - p^2 = p(1-p) = pq$$

n-Independent Bernoulli Trials = "Bernoulli Process"
PMF: k - succ. in n-trials

$$K = \sum_{i=1}^{n} x_i \quad \text{"Binomial RV"}$$

$$p_K(k) = \binom{n}{k} p^k q^{n-k}$$

$$k = 0, 1, \cdots n$$

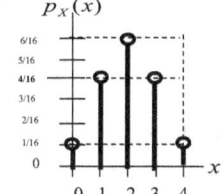
$p_X(x)$

$$E[X] = \sum_{x=0}^{n} x \binom{n}{x} p^x q^{n-x} = n \cdot p$$

$$\text{var}(X) = n \cdot pq$$

$$\phi_K(s) = E[e^{Ks}] = \sum_{k=0}^{n} e^{ks} \binom{n}{k} p^k q^{n-k} = \sum_{k=0}^{n} \binom{n}{k} \cdot \left(pe^s\right)^k q^{n-k} = \boxed{\left(pe^s + q\right)^n}$$

The **Bernoulli random variable** X is completely characterized by either its PMF or its moment generating function $\phi_X(s) = p \cdot e^s + q$. The mean $\mu_X = p$ and a variance $\sigma_X^2 = pq$ (and all other moments) can be obtained directly from the PMF by computing the expectation of its 1st and 2nd moments as shown on the slide; alternately, they can be calculated by differentiating the generator with respect to s several times and evaluating each derivative at s=0. The Bernoulli PMF has a *symmetric* shape when p=q=.5 as illustrated in the upper figure; it becomes *asymmetric* for all other values of p.

The **Binomial random variable** in the 2nd panel is the sum of n Independent Identically Distributed (IID) Bernoulli RVs $K = X_1 + X_2 + ... + X_n$ and has the PMF resulting from the product of the binomial coefficient ^nC_k with the k-Bernoulli successes p^k and n-k failures q^{n-k} as shown. The sum variable has a new mean and variance that is simply n times that of a single Bernoulli RV, *viz.*, $\mu_K = n \cdot p$ and $\sigma_K^2 = n \cdot (pq)$; the generating function for the sum random variable K is just the product of n Bernoulli generators, *viz.*,

$$\phi_K(s) = \phi_X(s) \cdot \phi_X(s) \cdot ... \cdot \phi_X(s) = (p \cdot e^s + q)^n.$$

The Binomial PMF for n=5 has a symmetric shape when p=.5 as illustrated in the lower figure; it becomes asymmetric for all other values of p

4.2.1 Statistical Analysis of Experimental Data

Statistical Analysis of Experimental Data

Bernoulli: Single Trial
Two outcomes

$$p_X(x) = \begin{cases} p & X = 1 \text{ (success)} \\ 1 - p = q & X = 0 \text{ (failure)} \end{cases}$$

Binomial: n indep Bernoulli Trials
Each with two outcomes
p= single trial prob of success

$$p_X(x) \equiv b(x; n, p) = \binom{n}{x} p^x q^{n-x}$$

$$x = 0, 1, \cdots n$$

Bernoulli = Binomial for n=1: $\quad b(x; "1", p) = \binom{1}{x} p^x q^{1-x}$

Experiment: N_B -independent *Bernoulli trials* (say N_B =10,000)

X: 0 1 1 0 0 1 0 0 0 ••• 1 1 1

Bern. Trial #: 1 2 3 4 5 6 7 8 9

1 2 3

3 "Bern. Trials/ Sample

$N_S = N_B / n = 10^4 / 3 = 3333$

n=3
ξ =#succ

2 1 0 ••• 3

Compare Expt'l frequency of occurrence $f(\xi)$ in 3333 samples to theoretical PMF

Consider an "experiment" consisting of N_B=10,000 independent Bernoulli trials yielding data that is either a "0" or "1" in the 10,000 data bins as illustrated. The subsequent analysis of such data depends on precisely how the experiment was structured. For example, if all the data was taken in a single experiment "flipping a coin" repeated 10,000 times, then the correct analysis is to compare the results to a Bernoulli distribution with success probability p illustrated in the top figure. Thus, we would compute an experimental value of the Bernoulli success probability p by dividing the number of "1"s by 10,000; because we have so many samples we would feel pretty confident about this experimental value of p_{exptl} = 2544/10000 =.2544 (say) and conclude that the underlying distribution is Bernoulli and that the *coin is unfair* (p≠.5).

Alternately, suppose that the data was taken in 3333 different sub-experiments in which three people simultaneously "flip a coin" and the data was recorded as a single concatenated sequence (with 1 bad 10,000th datum which we discard). In this case we would more naturally analyze the data in terms of 3333 samples (bottom figure) representing the sum of "group successes" which take on values {0,1,2,3} corresponding to 0,1,2,3 heads. Thus, we would compare the experimental results with a Binomial distribution b(k; 3, p) by computing the frequency-of-occurrence for k=0,1,2,3 successes, this time dividing the number count for each by 3333. The experimental value of p_{exptl} is not directly computable in this case since we now have 4 possible outcomes not just the 2 as in a single Bernoulli trial. We could estimate p_{exptl} by adjusting p in the binomial until the theoretical probabilities most "closely match" the experimental results (frequencies-of-occurrence); but this requires a careful definition of "closely match" which is the subject of statistical testing. The main point here is that the same data can be analyzed differently depending on how it was actually collected.

4.2.2 Trade-Off: Sample Averaging *vs.* Number of Samples

Trade-Off: Sample Averaging vs. Number of Samples

Break up into arbitrary sample sizes n and the compare with appropriate binomial distribution with p=q=1/2

---

n=1: $N_S = 10^4$ samples

Compare sample frequency to Binomial b(x; **n=1**, p=1/2)=**Bernoulli** for x=0,1

Compare sample mean to theoretical mean (**Bernoulli**) E[X]= 1·p=1(1/2)=.5

Compare sample Var to theoretical (**Bernoulli**) Var(X) =1· pq=1(1/2)(1/2)=.25

Too Granular
No Averaging
of "pixels"

---

n=3: $N_S =10^4/3 = 3333$ samples

Compare sample frequency to Binomial b(x; **n=3**, p=1/2) for x=0,1,2,3

Compare sample mean to theoretical (**sum 3 Bernoullis**) E[X]=np=3(1/2)=1.5

Compare sample Var to theoretical (**sum 3 Bernoullis**) Var(X)=npq=3(1/2)(1/2)=.75

Possible
Trade-off
strategy"

---

n=5000: $N_S = 10^4/5000 = 2$ samples

Compare sample frequency to Binomial b(x; **n=5000**, p=1/2) for x=0,...,5000

Compare sample mean to theoretical (**sum 5000 Bernoullis**) E[X]=np=5000(1/2)=2500

Compare sample Var to theoretical (**sum 5000 Bernoullis**) Var(X)=npq=5000(1/2)(1/2)=1250

Poor Statistics
only 2 samples!

Must trade-off sample averaging *versus* number of samples

Another important aspect of data analysis is to consider the trade-off between "sample averaging" which removes unwanted "experimental noise" and the *reduction in the total number of samples leading to less statistical significance of the result*. We shall consider grouping the data into sums of n samples (equivalent to averaging without dividing by n) and compare the sample mean and variance with the theoretical values computed from an assumed Binomial distribution b(x; n, p). (Below, we also assume p=q=1/2.)

For n=1 (no averaging), there are N_s = 10,000 samples and the Binomial is b(x; 1, p=1/2)with x=0,1 (Bernoulli); we compare against the theoretical values E[X]= n·p = 1·.5=.5 and var(X)=npq=1·.5·(1-.5)=.25 . The 10,000 samples yield good "statistical significance" but are subject to random noise variations that can distort the underlying data; thus the sample variance should be good but the sample mean (consisting of a single point) may be corrupted by the noise and yield a poor estimate.

For n=3 (3-sample averaging), there are N_s = 3333 samples (1 is dropped) and the Binomial is b(x; 3, p=1/2) with x=0,1,2,3; we compare against the theoretical values E[X]= n·p = 3·.5=1.5 and var(X) = npq = 3·.5·(1-.5) =.75. The 3333 samples still yield good "statistical significance" and now the random noise variations have been reduced by averaging 3 samples at a time; thus both the sample mean and sample variance should be "good" estimates.

For n=5000 (5000-sample averaging), there are now just Ns = 2 samples, and Binomial b(x; 5000,p=1/2) with x=0,1,2,3, ...,5000; we compare against the theoretical values E[X]= n·p = 5000·.5=2500 and var(X)=npq=5000·.5·(1-.5)= 1250. The 2 samples have virtually no "statistical significance" and although averaging almost completely eliminates the random noise variations, we only have two samples and we do not expect the resulting estimates to be meaningful.

4.2.3 Statistical Analysis of N=16 Bernoulli Trials

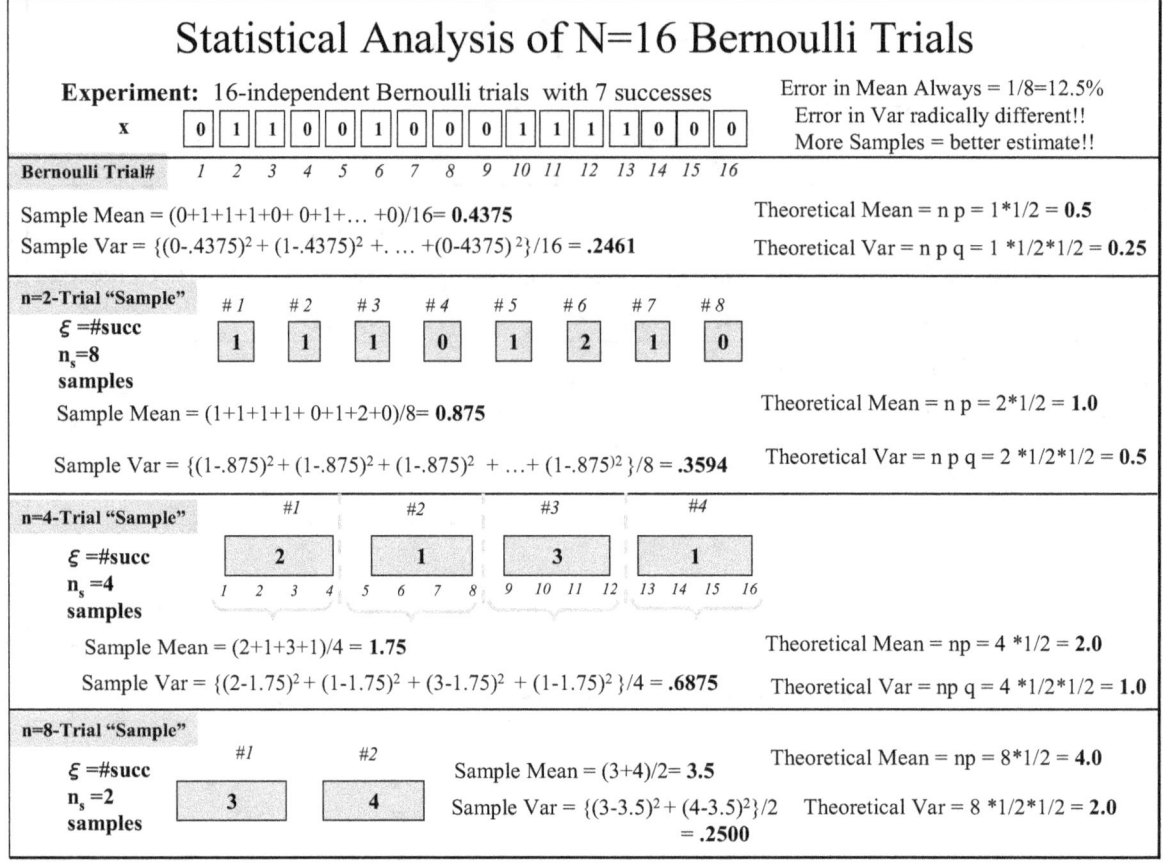

Statistical Analysis of N=16 Bernoulli Trials

Experiment: 16-independent Bernoulli trials with 7 successes

x

| 0 | 1 | 1 | 0 | 0 | 1 | 0 | 0 | 0 | 1 | 1 | 1 | 1 | 0 | 0 | 0 |

Error in Mean Always = 1/8=12.5%
Error in Var radically different!!
More Samples = better estimate!!

Bernoulli Trial#   1 2 3 4 5 6 7 8 9 10 11 12 13 14 15 16

Sample Mean = (0+1+1+1+0+ 0+1+... +0)/16= **0.4375**

Sample Var = {(0-.4375)^2 + (1-.4375)^2 +. ... +(0-4375)^2}/16 = **.2461**

Theoretical Mean = n p = 1*1/2 = **0.5**

Theoretical Var = n p q = 1 *1/2*1/2 = **0.25**

n=2-Trial "Sample"

ξ =#succ

n_s=8

samples

| #1 | #2 | #3 | #4 | #5 | #6 | #7 | #8 |
| 1 | 1 | 1 | 0 | 1 | 2 | 1 | 0 |

Sample Mean = (1+1+1+1+ 0+1+2+0)/8= **0.875**

Sample Var = {(1-.875)^2 + (1-.875)^2 + (1-.875)^2 +...+ (1-.875)^2}/8 = **.3594**

Theoretical Mean = n p = 2*1/2 = **1.0**

Theoretical Var = n p q = 2 *1/2*1/2 = **0.5**

n=4-Trial "Sample"

ξ =#succ

n_s =4

samples

| #1 | #2 | #3 | #4 |
| 2 | 1 | 3 | 1 |

1 2 3 4   5 6 7 8   9 10 11 12   13 14 15 16

Sample Mean = (2+1+3+1)/4 = **1.75**

Sample Var = {(2-1.75)^2 + (1-1.75)^2 + (3-1.75)^2 + (1-1.75)^2 }/4 = **.6875**

Theoretical Mean = np = 4 *1/2 = **2.0**

Theoretical Var = np q = 4 *1/2*1/2 = **1.0**

n=8-Trial "Sample"

ξ =#succ

n_s =2

samples

| #1 | #2 |
| 3 | 4 |

Sample Mean = (3+4)/2= **3.5**

Sample Var = {(3-3.5)^2 + (4-3.5)^2}/2 = **.2500**

Theoretical Mean = np = 8*1/2 = **4.0**

Theoretical Var = 8 *1/2*1/2 = **2.0**

Here is an explicit trade-off analysis for averaging n= 1, 2, 4, and 8 data samples given the data from 16 Bernoulli trials shown. For each case, we compute the experimental sample mean and variance and compare them with those for the assumed underlying Binomial b(x; n, 0.5) distribution. Since no noise was actually added to this data, we can only describe the benefits of averaging as if noise were actually present.

For n=1 we have 16 Bernoulli trials and compare *sample with theoretical statistics* to find: means .4375 *vs.* 0.5, and variances .2461 *vs.* 0.25. This clearly shows reasonably good results for both. (However, these single trial results provide no noise reduction and will not be so good in the presence of noise.)

For n=2 we have 8 samples after averaging and compare *sample with theoretical statistics* to find: means .875 *vs.* 1.0, and variances .3594 *vs.* 0.5. This clearly shows some degradation of results for both. (However, there will be an improvement in noise reduction because 2 trials are averaged.)

For n=4 we have 4 samples after averaging and compare *sample with theoretical statistics* to find: means .1.75 *vs.* 2.0, and variances .6875 *vs.* 1.0. This clearly shows increased degradation of results for both. (However, the improvement in noise reduction will be correspondingly greater because 4 trials are averaged.)

For n=8 we have 2 samples after averaging and compare *sample with theoretical statistics* to find: means 3.5 *vs.* 4.0, and variances .2500 *vs.* 2.0. With only 2 samples not much is expected in either comparison. (However, this would have the best improvement in noise reduction because 8 trials are averaged.)

Finally, we note that the error in mean (1/8=12.5%) is the same in all cases, but this will be different if we actually add noise to the data. On the other hand, the error in the computed variance increases as the number of samples decreases because more samples are needed to obtain a good estimate of the statistical variability of the actual data.

4.3 *Useful Binomial Distribution Identities*

Useful Binomial Distribution Identities

1) $b(k;n,p) \equiv \begin{pmatrix} n \\ k,(n-k) \end{pmatrix} p^k q^{n-k} = \begin{pmatrix} n \\ (n-k),k \end{pmatrix} q^{n-k} p^k = b(n-k;n,q)$ **Symmetry of Binomial "Taken" & "Not Taken"**

2) $\underbrace{b(k;n+1,p)}_{\text{k succ., n+1 trials}} = \underbrace{b(k-1;n,p)}_{\text{k-1 succ., n trials}} \cdot \underbrace{p}_{\text{1 succ.}} + \underbrace{b(k;n,p)}_{\text{k succ., n trials}} \cdot \underbrace{q}_{\text{1 failure}}$ **Decomposition of Event "k-successes in (n+1)-trials"**

3) $\displaystyle\sum_{k=r}^{n} b(k;n,p) = 1 - \sum_{k=n-r+1}^{n} b(k;n,q)$ **Decomposition of Normalization Identity**

Proof of 3):   Break up identity sum: $1 = \displaystyle\sum_{k=0}^{r-1} b(k;n,p) + \sum_{k=r}^{n} b(k;n,p)$

$\displaystyle\sum_{k=r}^{n} b(k;n,p) = 1 - \sum_{k=0}^{r-1} \begin{pmatrix} n \\ k,(n-k) \end{pmatrix} p^k q^{n-k} \xrightarrow{k=n-k'} = 1 - \sum_{k'=n}^{k'=n-(r-1)} \begin{pmatrix} n \\ n-k',k' \end{pmatrix} q^{k'} p^{n-k'} = 1 - \sum_{k'=n}^{k'=n-r+1} b(k';n,q)$

Q.E.D.

Alternately, expand sum explicitly

$\displaystyle\sum_{k=r}^{n} b(k;n,p) = 1 - \left[\begin{pmatrix} n \\ 0,\boxed{(n-0)} \end{pmatrix} p^0 q^{n-0} + \begin{pmatrix} n \\ 1,(n-1) \end{pmatrix} p^1 q^{n-1} + \cdots + \begin{pmatrix} n \\ (r-1),\boxed{(n-r+1)} \end{pmatrix} p^{(r-1)} q^{(n-r+1)} \right]$

$\xleftarrow{\qquad\qquad}$ **Sum terms backwards**

$\displaystyle\sum_{k=r}^{n} b(k;n,p) = 1 - \sum_{\boxed{k=n-r+1}}^{n} \begin{pmatrix} n \\ k,(n-k) \end{pmatrix} q^k p^{n-k} = 1 - \sum_{k=n-r+1}^{n} b(k;n,q)$

Q.E.D.

Identity 1 states that b(k;n,p) = b(n-k;n,q), *i.e.*, the binomial term is unchanged under the simultaneous swap k↔(n-k) and p↔q . This is obvious when the binomial term is written in symmetric form as shown.

Identity 3 may be verified by breaking up the Binomial normalization sum into the first "r" terms and the remaining "n-r" terms and setting their sum to unity. Moving the term $\sum_{k=r}^{k=n} b(k;n,p)$ to the LHS leaves $1 - \sum_{k=0}^{k=r-1} b(k;n,p)$ as shown on the 2nd line of the proof. Changing the summation index to k′= k - n yields the second equality which is immediately identified as $1 - \sum_{k'=n}^{k'=n-r+1} b(k';n,q)$ where the binomial terms are actually being summed backwards. Clearly, reversing the summation order leaves the sum unchanged, so the theorem is proved. Alternately, the terms of the sum on the RHS can be written out explicitly and summed backwards which changes the roles of p and q (as indicated on the slide) and we arrive at the desired result.

An alternate proof sums **Identity 1** from k=r to k=n to obtain

$$\sum_{k=r}^{k=n} b(k;n,p) = \sum_{k=r}^{k=n} b(n-k;n,q)$$

Now letting m=n-k in the RHS sum above shifts its range to be "m=n-r" to "m=n-n =0" and yields

$$\sum_{k=r}^{k=n} b(k;n,p) = \sum_{m=n-r}^{m=0} b(m;n,q) = \sum_{m=0}^{m=n-r} b(m;n,q)$$

where the order of summation is reversed in the last equality. Finally using the fact that normalization sum must be unity we can write the RHS as one minus the rest of the sum, or

$$\sum_{k=r}^{k=n} b(k;n,p) = \sum_{m=0}^{m=n-r} b(m;n,q) = 1 - \sum_{m=n-r+1}^{m=n} b(m;n,q),$$

thereby proving Identity 3 (with "dummy" sum index m rather than k)

4.3.1 Binomial Identities -Derivations

Binomial Identities -Derivations

Proof of 2):
$$\Pr\begin{bmatrix} \text{k succ.} \\ \text{n trials} \end{bmatrix} = \Pr\begin{bmatrix} \text{k-1 succ.} \\ \text{n-1 trials} \end{bmatrix} \cdot \Pr\begin{bmatrix} \text{succ.} \\ \text{next time} \end{bmatrix} + \Pr\begin{bmatrix} \text{k succ.} \\ \text{n-1 trials} \end{bmatrix} \cdot \Pr\begin{bmatrix} \text{fail} \\ \text{next time} \end{bmatrix}$$

$$b(k;n,p) = b(k-1;n-1,p) \cdot p + b(k;n-1,p) \cdot q$$

Write out Explicitly
$$\binom{n}{k,(n-k)} p^k q^{n-k} = \binom{n-1}{k-1,(n-k)} p^{k-1} q^{n-k} \cdot p^1 + \binom{n-1}{k,(n-k-1)} p^k q^{n-k-1} \cdot q^1$$

Common factor
$$p^k q^{n-k} \left\{ \binom{n}{k,(n-k)} = \binom{n-1}{k-1,(n-k)} + \binom{n-1}{k,(n-k-1)} \right\} \implies$$
Yields binomial identity
$$\boxed{\binom{n}{k} = \binom{n-1}{k-1} + \binom{n-1}{k}}$$

Recall Interpretation: Consider a unique single object & (n-1) more:

Note: Whether or not you include the unique object you still exclude it from further consideration *i.e.,* (n-1) remain

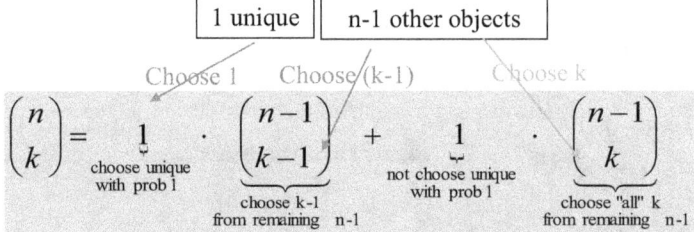

Identity #2 is easily proved by decomposing the event into the union of two ME and CE events as $E = E_1 \cup E_2$, where the events are described as follows: $E = \{k\text{-successes in n trials}\}$,
$E_1 = \{(k\text{-}1)\text{-successes in }(n\text{-}1)\text{-trials \textbf{and} success in the last trial}\}$ a joint event
$E_2 = \{k\text{-successes in }(n\text{-}1)\text{-trials \textbf{and} failure in the last trial}\}$ a joint event
Taking the probabilities over this union of disjoint events yields just two terms $P(E) = P(E_1)+P(E_2)$. Now success or failure of each trial is independent, so the probability of the joint event E_1 is simply the probability product of its constituent events, and we have $P(E_1)= P((k\text{-}1)\text{-succ in }(n\text{-}1)\text{-trials}) \times P(\text{succ in next-trial})$; similarly $P(E_2)= P(k\text{-succ in }(n\text{-}1)\text{-trials}) \times P(\text{fail in next-trial})$. The top panel contains a word equation describing the above events which is immediately transcribed into the desired Identity#2 written in terms of the Binomial *functional forms* b(k;n,p) and the success and failure probabilities p and q. We prove its validity by substituting explicit forms for the binomial terms, thereby reducing the identity to the binomial coefficient identity in the shaded box.

A **combinatorial argument** for this coefficient identity is illustrated in the lower half of the slide and hinges on the fact that any event space S is simply the union of a set A which contains a *unique object* and its complement A^c which *does not contain that object*. Because A contains the object by definition, there remain n-1 other objects to choose from; and since we have already made one selection (the "object") we only need choose k-1 more, *i.e.*, we have $^{n-1}C_{k-1}$ choices for the set A. By definition, the complement A^c does not contain the unique object; hence it must be removed from the original set of n leaving n-1 elements to choose from and we need to make all k choices from this set, *i.e.*, $^{n-1}C_k$. Since these two sets are ME and CE, $A \cup A^c = S$, there is no overlap and the sum of the number of outcomes for these two sets must equal the number of outcomes for "n take k" ^nC_k in the original sample space S; thus, we have shown $^nC_k = {}^{n-1}C_{k-1} + {}^{n-1}C_k$.

4.4 Bernoulli Process Nuances - Interarrival Times and Geometric PMF

Bernoulli Process Nuances - Interarrival Times and Geometric PMF

1st order Interarrival Time: #Trials $N_1 = k$ for 1st Success

If first arrival occurs after $N_1 = k$ trials, there must have been "k-1" failures before 1st success

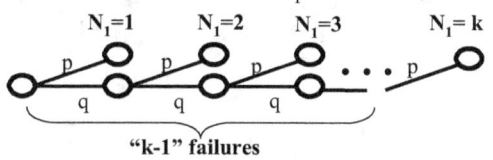

$$= q^{k-1} \cdot p^1$$

First Success at trial "k"

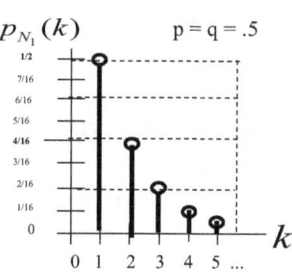

$p_{N_1}(k)$ $p = q = .5$

$$\Pr[N_1 > k - 1] = q^{k-1}$$

Prob #trials for 1st Succ. N_1 exceeds k-1 equals Prob [k-1 failures]

$$CDF(k) = \Pr[N_1 \leq k]$$

$$= 1 - \Pr[N_1 > k] = 1 - q^k$$

Cumulative Distribution

Geometric PMF $N_1 = \#$ trials needed to yields 1st success

$$\Pr[N_1 = k] = p_{N_1}(k) = q^{k-1} \cdot p \qquad \text{for } k = 1,2,3, \ldots \quad \substack{\text{k-1 failures} \\ \text{next trial =success}}$$

Properties $E[N_1] = \dfrac{1}{p}$ $\text{var}(N_1) = \dfrac{q}{p^2}$ $\phi_{N_1}(s) = E[e^{sN_1}] = \dfrac{pe^s}{1 - qe^s}$

Proof:

$$\phi_{N_1}(s) = E[e^{sN_1}] = \sum_{N_1=1}^{\infty}(p \cdot q^{N_1-1}) \cdot e^{s \cdot N_1} = (p/q)\sum_{N_1=1}^{\infty}(qe^s)^{N_1} = (p/q)\frac{qe^s}{1 - qe^s} = \frac{pe^s}{1 - qe^s}$$

The geometric PMF results from concatenating the Bernoulli process an infinite number of times as shown in the figure. Starting at the left-most failure node (F) the Bernoulli trial fails with probability q and succeeds with probability p; the next Bernoulli tree is attached to the failure or F-node only and again results in F with probability q and in S with probability p; this process of adding Bernoulli trees continues indefinitely. Obviously, there must be at least one trial in order to have a success, so the 1st success can occur after k=1 trial and this event occurs with probability p; it can also occur after 2 trials with probability qp, or after 3 trials with probability q^2p, ...   Thus the individual terms of the geometric distribution give the probability for "the number of trials $N_1 = k$ required for a single success" and clearly this requires (k-1)-failures before the 1st success, so we write the kth term of the Geometric PMF as

$$\Pr[N_1 = k] = p_{N1}[k] = q^{k-1} p \quad k=1,2, 3, \ldots$$

The PMF for the case p = q = 0.5 is illustrated in the plot; note that there are an infinite number of terms in this PMF which decay very rapidly in magnitude. The mean or expected value of this discrete distribution is $\mu_{N1}=E[N_1] = 1/p$ which says that the average number of trials required for a single success is just the inverse of the single trial probability of success; so for instance if p=1/3 the average number of trials needed for one success is 3. The variance is $\text{var}(N_1) = q/p^2$ and the generator is $\phi_{N1}(s) = pe^s/(1-qe^s)$.

4.4.1 Memoryless Property of Geometric PMF

Memoryless Property of Geometric PMF

Conditional PMF: remaining # trials "k" for 1st success given "m" prior failures

$$\Pr[\underbrace{N_1 > m+k}_{\substack{\text{more than "k"}\\ \text{additional trials}}} | \underbrace{N_1 > m}_{\substack{\text{given "m"}\\ \text{failures means}\\ \text{1st succ} > m}}] = \frac{\Pr[N_1 > m+k, N_1 > m]}{\Pr[N_1 > m]} = \frac{\Pr[N_1 > m+k]}{\Pr[N_1 > m]} = \frac{q^{m+k}}{q^m} = q^k$$

Memoryless Property: "k" remaining trials before 1st success is indep. of prior "m" failures

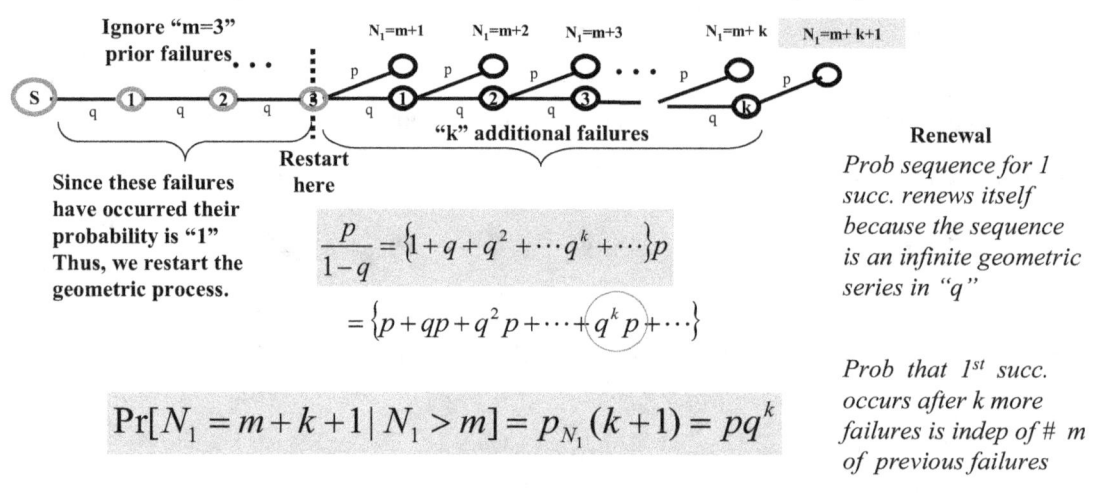

Since these failures have occurred their probability is "1" Thus, we restart the geometric process.

Renewal
Prob sequence for 1 succ. renews itself because the sequence is an infinite geometric series in "q"

$$\frac{p}{1-q} = \left\{1 + q + q^2 + \cdots q^k + \cdots\right\}p$$

$$= \left\{p + qp + q^2 p + \cdots + q^k p + \cdots\right\}$$

$$\Pr[N_1 = m+k+1 \mid N_1 > m] = p_{N_1}(k+1) = pq^k$$

Prob that 1st succ. occurs after k more failures is indep of # m of previous failures

Since every trial is independent and has the same probability of success *p* (or failure q=1-p) it should not be surprising that the process has *no memory of its prior failures.* (If it did have memory of its prior failures, then the probability of failure could not have the same value q for each trial and the trials would not be independent!) This means that the probability of k failures given m previous failures is still q^k (not q^{m+k}) as shown in the first equation of the slide. This result is formally obtained by using the definition of the conditional probability of m+k failures given m prior failures, *viz.*, $\Pr[N_1 > m+k \mid N_1 > m]$. This probability is defined to be the joint probability $\Pr[N_1 > m+k, N_1 > m]$ = q^{m+k} divided by the conditioning event of m prior failures $\Pr[N_1 > m] = q^m$ to find $q^{m+k} / q^m = q^k$. (Note that $\Pr[N_1 > m]$ means that a success can occurs for $N_1 = m+1$ or m+2 or m+3 which is another way of stating that there have been "m" failures.)

This becomes clearer if we look at the renewal properties of geometric (infinite) tree in which we assume m=3 prior failures (red circles labeled 1,2,3) have occurred since the original start (red circle on left). The dashed vertical line in the tree indicates the "new start" from which the tree continues indefinitely with the same fail-succeed pattern and nothing is changed by "cutting off its tail." We are given that m= 3 prior failures have occurred so the restarted tree structure has the same generator, namely, p/(1-q) and the probability of a single success after k more failures is given by the term $q^k p$ (see bottom equation). Note that the m= 3 prior failures have occurred with certainty, *i.e.*, with probability "1"; what happens after m=3 is completely independent of these *known* prior failures.

4.4.2 Geometric RV Example: Urn with Replacement

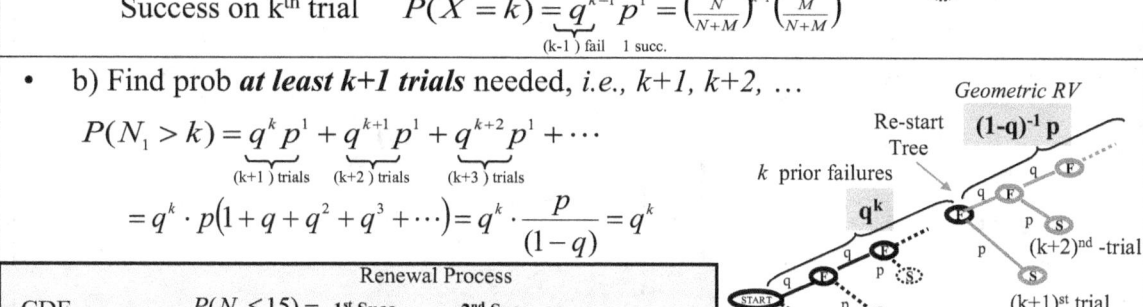

Geometric RV Example: Urn with Replacement   ***#Trials "x" for "1" Succ.***

- Urn with "N" White (W) and "M" Black (B) balls
  - Many independent trials (***w/replacement***) until B is drawn
  - ***Single trial prob succ.*** $p = M/(M+N)$; $q = (1-p) = N/(M+N)$
  - "2" outcomes: W = fail, B = success
- a) Find prob exactly "***k***" trials needed "means"

  Failure on 1st (k-1)-trials

  Success on kth trial $P(X=k) = \underbrace{q^{k-1}}_{(k-1)\text{ fail}} \underbrace{p^1}_{1 \text{ succ.}} = \left(\frac{N}{N+M}\right)^{k-1}\left(\frac{M}{N+M}\right)$

- b) Find prob ***at least k+1 trials*** needed, *i.e.*, *k+1, k+2, …*

$$P(N_1 > k) = \underbrace{q^k p^1}_{(k+1)\text{ trials}} + \underbrace{q^{k+1} p^1}_{(k+2)\text{ trials}} + \underbrace{q^{k+2} p^1}_{(k+3)\text{ trials}} + \cdots$$

$$= q^k \cdot p\left(1 + q + q^2 + q^3 + \cdots\right) = q^k \cdot \frac{p}{(1-q)} = q^k$$

Geometric RV

Re-start Tree $(1-q)^{-1} p$

k prior failures

q^k

$q^k \{(1-q)^{-1} p\}$

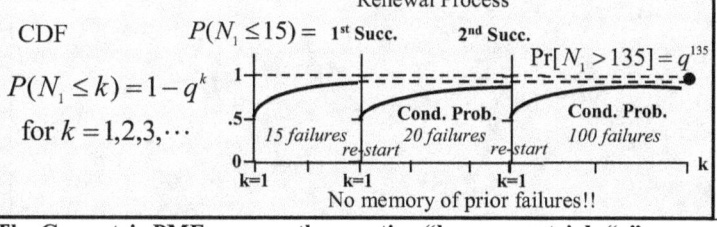

Renewal Process

CDF $P(N_1 \le 15) =$ 1st Succ.   2nd Succ.

$\Pr[N_1 > 135] = q^{135}$

$P(N_1 \le k) = 1 - q^k$
for $k = 1,2,3,\cdots$

Cond. Prob.   Cond. Prob.
15 failures   20 failures   100 failures
re-start   re-start

No memory of prior failures!!

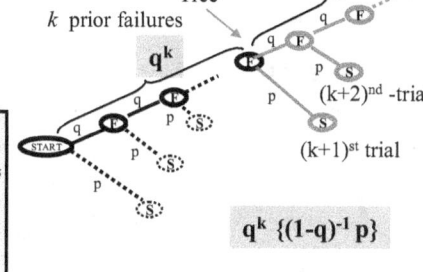

The Geometric PMF answers the question "how many trials "x" are needed for "1" success?"
Given an urn containing N white (W) and M black (B) balls, one ball is drawn and then placed back into the urn until a black ball is drawn. Because of replacement, the number of W and B balls in the urn is always the same; thus each draw is *independent* and has the same "single trial probability of success (selecting B)," namely, $p=M/(M+N)$ [and $q=1-p= N/(M+N)$.] The answer to the question *"how many trials "x" are needed for exactly "1" success"* is characterized by a Geometric RV "X" with PMF, $p_X(x) = q^{x-1} p^1$, with x=1,2,3,...denoting the number of trials needed for a single success.
a) **Find probability "exactly k trials"** are needed: $P(X=k) = p_X(x=k) = (N/(M+N))^{k-1} (M/(M+N))^1$
b) **Find probability "at least k+1 trials"** are needed: $P(X>k) = q^k p^1 + q^{k+1} p^1 + q^{k+2} p^1 + q^{k+3} p^1 + ...$, which factors to
$$P(X>k) = q^k p^1 (1 + q^1 + q^2 + q^3 + ...) = q^k p / (1-q) = q^k p / p = q^k$$
Note that "greater than k trials" requires a sum over an infinite number of possible paths to success, starting with k-failures prior to success on trial k+1, or (k+1)-failures prior to success on trial k+2, *ad infinitum*. The calculation above shows that we have the same infinite series, even after k prior failures, so the result is *independent of prior failures*. The infinite sum always yields the same value $1/[1-q]$ which cancels the "p" term in the numerator leaving q^k as the resulting probability. This can be visualized as appending an entire geometric tree $\{(1-q)^{-1} p\}$ to the tree representing the first k-failures as illustrated in the slide. This means that the conditional probability of *k more failures* given *k failures have already occurred* is still q^k and not $q^{2(k)}$; this is called a "renewal process" since it has no memory of the previous k-failures.
Alternately, the nature of this renewal process may be somewhat more intuitive by considering the cumulative distribution function (CDF) for success in k trials as follows: Setting p=q=0.5, the CDF for success in k trials can be written as
$$F(k) \equiv P(X \le k) = 1 - P(X>k) = 1 - (q)^k = 1 - (.5)^k$$
The CDF curve $F(k) = 1 - q^k = 1 - (.5)^k$ starts at 0 for k=0 rises to .5 for k=1 and increases to .75 for k=2, .875 for k=3, … and asymptotically approaches 1 as $k \to \infty$ as shown in the *first segment* of the plot. However, if a success occurs at k=16, the whole process re-starts by resetting the failures to zero and starting again with F(1)=.5 at k=1 as shown in the *second plot segment*; now the 2nd success may occur after, say 20 failures, at k= 21 and the process restarts once more at k=1 and F(1)=.5, *ad infinitum*.

4.4.3 Sum of r-independent Geometric RVs – Neg. Binomial (Pascal) PMF

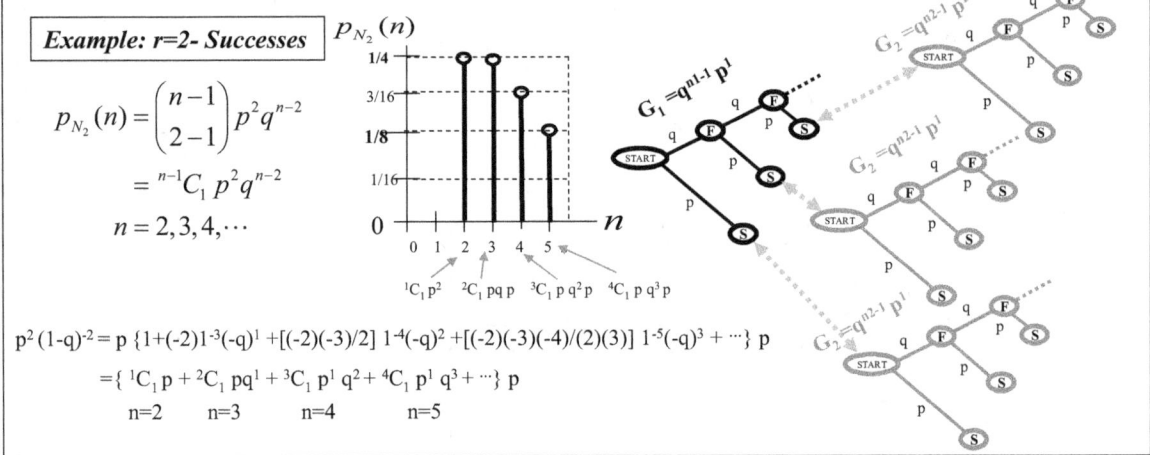

In the top panel of the slide we give the general expression for the Negative Binomial PMF

$$p_{Nr}(n) = {}^{n-1}C_{r-1}p^r q^{n-r} \text{ with } n_r = r, r+1, r+2, ...$$

which is the probability that N_r trials are required for the rth success. Since N_r is the sum of r IID Geometric RVs, it is no surprise that the mean and variance of N_r are r times the same quantities for a single geometric RV

$$\mu_{Nr} = E[N_r] = r \cdot (1/p) \quad ; \quad var(N_1) = r \cdot (q/p^2)$$

The geometric tree structure gives the "timing structure" for all paths that end with only one success; however, we are often interested in paths that lead to two or more successes. Paths which generate two successes are obtained by re-applying the geometric tree structure to each 1st success node as shown in the bottom panel of the slide. This is equivalent to summing two geometric RVs $N_2 = (N_1)_1 + (N_1)_2$ to obtain the Negative Binomial distribution for the sum variable N_2. The resulting PMF $p_{N2}(n)$ gives the probability that $n = 2,3,4,5, ...$ trials are required for $r=2$ successes. As an example, the case $p=q=.5$ is illustrated by the PMF plot in the bottom panel; the magnitudes of the probabilities $p_{N2}(n)$ are determined by the binomial coefficients ${}^{n-1}C_{r-1}p^2 q^{n-2} = {}^{n-1}C_1 p^2 q^{n-2}$ for $n=2,3,4,...$. The equation at the bottom of the slide is an expansion of the algebraic structure $p^2(1-q)^{-2}$ which generates all the terms of this discrete PMF; the first four $n=2,3,4,5$ are shown explicitly. The first term gives $N_2=n=2$ as the number of trials for exactly 2 successes; the tree has a unique path SS that goes diagonally down to the 1st success S-node with probability p and continues directly to the 2nd S-node again with probability p giving the term $({}^1C_1 \, p)p$. The second term $({}^2C_1 \, pq)p$ has two node paths, namely FSS and SFS; the 3rd and 4th path sequences are left as an exercise for the reader.

4.4.4 Sum of Geometric Random Variables $N_2=G_1+G_2$ & $N_3=G_1+G_2+G_3$

Sum of Geometric Random Variables $N_2=G_1+G_2$ & $N_3=G_1+G_2+G_3$

Sum Z=G_1 + G_2
2nd arrival r=2

$$p_{G_1}(n_1)=p^r q^{n_1-r} \quad ; \quad p_{G_2}(n_2)=p^r q^{n_2-r} \quad ; \quad \boxed{Z=G_1+G_2}$$

$$p_Z(z)=\sum_{n_1+n_2=z} p_{G_1}(n_1)\cdot p_{G_2}(n_2)=\sum_{n_1+n_2=z} p^r q^{n_1-r}\cdot p^r q^{n_2-r}$$

$$\underbrace{1+1+\ldots+1}_{(z-1)\text{ terms}}=z-1$$

Let n_1= k
n_2 = z-k

$$p_Z(z)=\sum_{k=1}^{z-1} p^1 q^{k-1}\cdot p^1 q^{z-k-1}=p^2 q^{z-2}\cdot\left(\underbrace{\sum_{k=1}^{z-1} 1}\right)=p^2 q^{z-2}\cdot(z-1)$$

Setting Z=N_2
$$p_{N_2}(n_2)=p^2 q^{n_2-2}\cdot(n_2-1)=\binom{n_2-1}{2-1}\cdot p^2 q^{n_2-2}$$

Sum Z=G_1+ N_2
3rd arrival r=3

$$p_{G_1}(n_1)=p^r q^{n_1-r} \quad ; \quad p_{N_2}(n_2)=\binom{n_2-1}{2-1}\cdot p^2 q^{n_2-2} \quad ; \quad \boxed{Z=G_3+N_2}$$

Let n_1= k
n_2 = z-k

$$p_Z(z)=\sum_{n_1+n_2=z} p_{G_1}(n_1)\cdot p_{G_2}(n_2)=\sum_{n_1+n_2=z} p^1 q^{n_1-1}\cdot\binom{n_2-1}{2-1}\cdot p^2 q^{n_2-2}=\sum_{k=1}^{z-2} p^1 q^{k-1}\cdot\binom{z-k-1}{2-1}p^2 q^{z-k-2}$$

$$=p^3 q^{z-3}\cdot\sum_{k=1}^{z-2}\binom{z-k-1}{2-1}=p^3 q^{z-3}\cdot\sum_{k=1}^{z-2}(z-k-1)=p^3 q^{z-3}\cdot\{(z-1)\sum_{k=1}^{z-2} 1-\sum_{k=1}^{z-2} k\}$$

$$=p^3 q^{z-3}\cdot\{(z-1)(z-2)-(z-2+1)(z-2)/2\}=\frac{(z-1)(z-2)}{2}p^3 q^{z-3}$$

Setting Z=N_3
$$p_{N_3}(n_3)=p^3 q^{n_3-3}\cdot\frac{(n_3-1)\cdot(n_3-2)}{2}=\binom{n_3-1}{3-1}\cdot p^3 q^{n_3-3}$$

In the upper panel we add two identical independent (IID) geometric RVs $Z=G_1+G_2$ using the convolution sum technique (see Discrete Probability Slide#7-17 or Slide#2-25 for Continuous Probability). We first write down the probability densities $p_{G_1}(n_1)$ and $p_{G_2}(n_2)$, take their product to form the joint distribution $p_{G_1 G_2}(n_1,n_2)=p_{G_1}(n_1)\cdot p_{G_2}(n_2)$, and finally take their sum for constant $z=n_1+n_2$. Setting n_1=k and n_2=z-k and summing from k=1 to k=z-1, which yields $p_Z(z)=p^2 q^{z-2}(z-1)$; since Z represents the sum of the original two geometric variables, we can identify Z=N_2 and thus

$$p_Z(z)\Rightarrow p_{N_2}(n_2)=p^2 q^{n_2-2}\cdot(n_2-1)=\binom{n_2-1}{2-1}\cdot p^2 q^{n_2-2}$$

In the lower panel we add three identical independent (IID) geometric RVs $Z=G_1+(G_2+G_3)=G_1+N_2$ following the same steps. First we form the joint distribution $p_{G_1 N_2}(n_1,n_2)=p_{G_1}(n_1)\cdot p_{N_2}(n_2)$, and sum for constant $z=n_1+n_2$; again setting set n_1=k and n_2=z-k in order to sum from k=1 to k=z-2 (leaving two "slots" for two arrival times of n_2) we find

$$p_{N_3}(n_3)=\binom{n_3-1}{3-1}\cdot p^3 q^{n_3-3}.$$

Having established the cases r=1 and r=2, the general case is proved by *assuming* it is true for r and *showing* it is true for the next value r+1 (induction).

4.4.5   Negative Binomial Example: Multi-User Digital Communication "CDMA"

Neg. Binomial Example: Multi-User Digital Communication "CDMA"

- Two signals s_1 , s_2 . Decode s_1 or s_2 in ***given time slot***
- *a priori Prob:* $P[s_1]=3/4$; $P[s_2]=1/4$
- Decoding Statistics:

decoded "1" : $P[1|s_1]=2/3$; $P[1|s_2]=2/3$

not decoded "0" : $P[0|s_1]=1/3$; $P[0|s_2]=1/3$

N_r time slots ("trials")	$p_{N_r}(n) = \binom{n-1}{r-1} p^r q^{n-r}$
r-Decodes of s_1 $p_1=q_1=1/2$	

1) Pr[1st decode in 4th slot] $\Pr[N_1 = k] = p_{N_1}(k) = q^{k-1}p^1 \Rightarrow \Pr[N_1 = 4] = p_{N_1}(4) = \left(\frac{1}{2}\right)^{4-1}\left(\frac{1}{2}\right)^1 = \frac{1}{16}$

2) Pr[4th decode in 10th slot | 3 decodes

   in 1st 6 time slots]

No memory - slots 7 to 10

$\Pr[N_1 = 4] = p_{N_1}(4) = q^3 p = \left(\frac{1}{2}\right)^3\left(\frac{1}{2}\right)^1 = \frac{1}{16}$

3) Pr[2nd decode in 4th slot] $\Pr[N_r = n] = p_{N_r}(n) = \binom{n-1}{r-1} p^r q^{n-r} \Rightarrow \Pr[N_2 = 4] = p_{N_2}(4) = \binom{4-1}{2-1}p^2 q^{4-2} = 3\left(\frac{1}{2}\right)^4 = \frac{3}{16}$

4) Pr[2nd decode in 4th slot | no decodes   No memory of failures in slots 1 & 2

   in 1st 2 time slots ("means" $N_2>2$)] $\Pr[N_2 = 2] = p_{N_2}(2) = p^2 = \left(\frac{1}{2}\right)^2 = \frac{1}{4}$

$$\Pr[N_2 = 4] = \binom{4-1}{2-1} \cdot p^1 q^2 \cdot p = 3 \cdot \left(\frac{1}{2}\right)^4 = 3/16$$

$$\Pr[N_2 = 4 | N_2 > 2] = \frac{\Pr[N_2 = 4, N_2 > 2]}{\Pr[N_2 > 2]} = \frac{p_{N_2}(4)}{1 - p_{N_2}(2)} = \frac{(3/16)}{1-(1/4)} = \frac{1}{4}$$

In a multiuser environment the digital signals from multiple transmitters can occupy the same signal processing time slot so long as they can be distinguished by their modulation characteristics. Code Division Multiple Access (CDMA) uses a pseudorandom code unique to each user in order to "decode" the proper signal source. Two signals s_1 and s_2 are being processed in the same time slot with *a priori* "system usage" $P[s_1] = ¾$ and $P[s_2] = ¼$; further let "1" denote successful and "0" denote unsuccessful decodes respectively. Given that each signal has the same 2/3 probability of a successful decode $P[1|s_1]=P[1|s_2]=2/3$, we can use the tree to find the single trial probability of success for decoding each signal. The end state $\{s_1, 1\}$ represents a successful decode for signal s_1 and has a probability $p_1=1/2$; all the other states $\{s_1, 0\}$, $\{s_2, 1\}$, $\{s_2, 0\}$ combined represent a failure to decode signal s_1 and has probability $q_1 = 1/4+1/6 +1/12 = 1/2$. Similarly, for signal s_2, we see that the end state $\{s_2, 1\}$ represents a successful decode of s_2 and has $p_2 =1/6$; all the other states $\{s_2, 0\}$, $\{s_1, 1\}$, $\{s_1, 0\}$ combined represent a failure to decode signal s_2 and has probability $q_2 =1/12+1/2+1/4=10/12=5/6$. The successive decodes of s_1 are independent trials with probability of success $p_1=1/2$; thus, the probability of having r-successful decodings of signal s_1 in N_r signal processing slots ("trials") is given by the Negative Binomial PMF $p_{Nr}(n) = {}^{n-1}C_{r-1}p_1^r q_1^{n-r}$ with $n_r = r, r+1, r+2, ...$, with $p_1=q_1=1/2$. The following four examples decode signal s_1 with $p_1= q_1=1/2$ for a number of different conditions [Note: For signal s_2 we would use $p_2= 1/6$, $q_2=5/6$ instead.]

1) Pr of 1st decode (r=1) in 4th slot (N_1 =4) is $p_{N1}(4) = {}^{4-1}C_{1-1}p_1^1 q_1^{4-1} = 1(1/2)^4 = 1/16$

2) Pr of 4th decode (r=4) in 10th slot (N_4 =10) *given 3 previous decodes in 1st 6 slots* is found by "restarting the process at slot #7." Thus using slots #7, 8, 9, 10, we need one decode (r =1) in 4 slots, *i.e.*, N_1 =4, which is identical to part 1) and yields $\Pr[N_4 = 10 | N_3=6] = p_{N1}(4) = {}^{4-1}C_{1-1}p_1^1 q_1^{4-1} = 1(1/2)^4 = 1/16$

3) Pr of 2nd decode (r=2) in 4th slot (N_2 =4) is $p_{N2}(4) = {}^{4-1}C_{2-1}p_1^2 q_1^{4-2} = 3(1/2)^4 = 3/16$

4) Pr of 2nd decode (r=2) in 4th slot *given 1st two slots were not decoded* is found by "restarting the process at #3, and using slots #3,4" so we need r=2 decodes in the two remaining N_2 =2 slots (two decodes in two trials), and we compute $p_{N2}(2)= {}^{2-1}C_{2-1}p_1^2 q_1^{2-2} = 1(1/2)^2= 1/4$.

5 The Poisson Process

5.1 Role of Time in the Poisson Process

Role of Time in the Poisson Process

- **Continuous time variable**

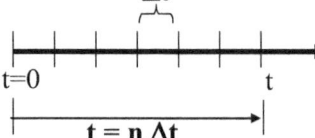

- Physical Examples - rare occurrences over long periods of time
  - *Radioactive Decay:* # particles emitted within a *fixed observation interval*
  - *Incoming Tel. Calls:*# calls made within a *fixed observation interval*
  - *Self Noise Generation*: # photons emitted within a *fixed observation interval*
  - *Earth Quake Prediction*: # earthquakes that occur in a *fixed observation intrvl*
- Limit of Bernoulli Process (Binomial PMF) as $\Delta t \to 0$, $n \to \infty$
- Assumptions:
  - Δt is small; no more than 1 arrival in Δt
  - $t = n\,\Delta t$ is the fixed observation interval
  - One Bernoulli Trial takes place in subinterval Δt:
  - $p = \lambda\,\Delta t$ is the Single Trial Prob of success:
  - $n = t/\Delta t$ is the number of independent trials in interval $[0, t]$
  - Number of successes in interval $[0, t]$ is

$$K(t) = \sum_{j=1}^{n} X_j \qquad \begin{aligned} K &= K(t) \\ &= \#\,succ.\ in\ interval\ t \end{aligned}$$

Sum of n Bernoulli RVs

The Poisson Random Process describes a series of random events that occur infrequently over *fixed interval of time* as indicated in the timeline at the top. Thus, many trials yield only a few successes and the average probability of success is small. The Poisson process is usually described by two parameters, namely (i) an average arrival (success) rate λ and (ii) a fixed interval of time t; their product $a = \lambda \cdot t$ is the dimensionless Poisson parameter characterizing the process. Common examples include the number of alpha particles emitted in a year for radioactive material, the number of incoming phone calls in an hour, the number of "noise" photons emitted in a one-second interval, and the number of Earth quake tremors in a week. The specific call-out of time intervals {a year, an hour, a second, a week} is important in these Poisson process examples because the same physical rate λ yields different PMFs depending on the value of the time interval.

The Poisson process is the limiting form of the Bernoulli process consisting of n Bernoulli trials over a fixed interval of time t as indicated in the lower timeline sketch. The Bernoulli "trial interval" Δt is chosen so that the physical process (*e.g.*, alpha decay, telephone calls, Earthquakes) can have at most one arrival within Δt. The total duration for these n Bernoulli trials is accordingly $t = n\,\Delta t$; taking the limit $\Delta t \to 0$ insures that Δt is small enough to contain *at most* a single arrival and taking the limit as $n \to \infty$ insures that there are a large number of observations (Bernoulli trials) in the fixed Poisson time interval *i.e.*, $\lim_{n \to \infty} \lim_{\Delta t \to 0} (n \cdot \Delta t) = $ finite duration "t".

If λ is the parameter describing the average arrival rate, then, since a single trial takes place in a small interval Δt, the single trial probability of success is the product $p = \lambda \cdot \Delta t$. The total number of successes is simply the sum of the n Bernoulli RVs, *viz.*, $K(t) = X_1 + X_2 + ... + X_n$, where we write "$K(t)$" to emphasize that the sum variable K is a function of chosen observation interval "t. Clearly, if Δt is held fixed, increasing the interval to $t_{new} = 10 \cdot t$ will increase the number of Bernoulli trials (RVs) to $n_{new} = 10 \cdot (t/\Delta t) = 10 \cdot n$, and hence allow for a larger number of arrivals $K(t_{new})$.

5.1.1 Derivation of Poisson PMF from Binomial

Derivation of Poisson PMF from Binomial

Binomial with Single Trial Prob Succ: $\mathbf{p = \lambda\, \Delta t}$

$$\lim_{n\to\infty} b(k;n,p) = \lim_{\substack{n\to\infty \\ \Delta t\to 0 \\ \lambda\cdot\Delta t\to p}} \binom{n}{k}\left(\lambda\cdot\underbrace{\Delta t}_{=(t/n)}\right)^{k}\left(1-\lambda\cdot\Delta t\right)^{n-k}$$

$$= \lim_{\substack{n\to\infty \\ \Delta t\to 0 \\ \lambda\cdot\Delta t\to p}} \frac{\overbrace{n(n-1)\cdots(n-k+1)}^{\cong n^{k}}}{k!}\cdot\frac{(\lambda\cdot t)^{k}}{n^{k}}\cdot\underbrace{\left(1-\frac{\lambda\cdot t}{n}\right)^{n}}_{\to e^{-\lambda t}}\cdot\underbrace{\left(1-\frac{\lambda\cdot t}{n}\right)^{-k}}_{\to 1} \quad\to\quad \frac{(\lambda\cdot t)^{k}}{k!}e^{-\lambda t}$$

$$\lim_{n\to\infty}\left(1-\frac{\lambda\cdot t}{n}\right)^{n} \underset{\substack{\text{Binomial}\\\text{Expansion}}}{=} \lim_{n\to\infty}\left\{1^{n}+n\cdot 1^{n-1}\left(-\frac{\lambda\cdot t}{n}\right)^{1}+\frac{n(n-1)}{2!}\cdot 1^{n-2}\left(-\frac{\lambda\cdot t}{n}\right)^{2}+\cdots+\binom{n}{n-k}\cdot 1^{n-k}\left(-\frac{\lambda\cdot t}{n}\right)^{k}+\cdots+\binom{n}{0}\cdot 1^{0}\left(-\frac{\lambda\cdot t}{n}\right)^{n}\right\}$$

$$= \lim_{n\to\infty}\left\{1+(-\lambda t)+\frac{(-\lambda t)^{2}}{2!}+\cdots+\frac{(-\lambda t)^{k}}{k!}+\cdots\right\}\cong e^{-\lambda t} \qquad \lim_{n\to\infty} 1\cdot 1^{0}\cdot\left(\frac{\lambda\cdot t}{n}\right)=0$$

Poisson PMF - Discrete PMF related to a time interval t and a rate λ

$$p_{K}(k;\lambda,t)=\frac{(\lambda\cdot t)^{k}}{k!}e^{-\lambda\cdot t} \quad t=n\Delta t \;;\; p=\lambda\Delta t \;;\; k=0,1,2,\cdots$$

$$E[X]=\lambda t \qquad \operatorname{var}(X)=\lambda t \qquad \phi_{K}(s)=e^{\lambda\cdot t(e^{s}-1)}$$

$$\lambda\cdot t=\lim_{\substack{n\to\infty\\p\to 0}}(n\cdot p)=(\text{aver. arrival rate})*\text{time}$$

We have previously stated that the Binomial becomes a Poisson PMF if we identify the single trial probability of success "p" for the Binomial with $\lambda\cdot\Delta t$, the product of the Poisson rate and the small interval during which the "trial" takes place. Making the substitution $p = \lambda\,\Delta t$ and $n = t/\Delta t$ in the binomial expression b(k;n,p) and taking the limit as $n \to \infty$ yields the Poisson PMF: $p_K(k;\lambda,t) = \{(\lambda\, t)^k / k!\}e^{\lambda t}$ for k=0,1,2,3,.... The Poisson parameter $a = \lambda\cdot t$ is the product of the arrival rate λ and a time t representing the duration of the data run.

The **Poisson parameter "a"** is dimensionless so the units of λ must be the inverse of those of t. More specifically, the rate λ can represent decays per sec per gram, customers per minute, bit errors per 10,000 bits, defects per square inch, Earth quakes per week, *etc.*; the corresponding "time" t must have *appropriate units* to make the product $a = \lambda\cdot t$ dimensionless. If the result is to be a pure number "a" cannot contain any dimensions! The **mean and variance** are both equal to the Poisson parameter a $(= \lambda\cdot t)$; this equality means that the larger the mean, the more spread out the distribution is.

Any random process which has very few successes over a given "run of data" and whose "average rate of success" λ, is known can be described as a Poisson process with arrival rate λ and Poisson parameter $a = \lambda\cdot t$ where t is the duration of the data run. For a new data run taken over a *different interval* t_{new}, we can immediately write down a new Poisson parameter as $a_{new} = \lambda\cdot t_{new}$. Thus, for example, if we know from past measurements that the average rate of Earthquakes (of magnitude > 2) is 4 per week, the arrival rate is λ=4/week; for a 3 week period we have a=(4/wk)·(3 wks) = 12; for a new 5 week period we have the new value a=(4/wk)·(5 wks) =20, *etc.*. Although the Poisson process is described by the same arrival rate λ, the resulting Poisson PMFs are different because of the different parameters "a".

The Poisson Process

5.1.2 Poisson PMF Properties

Poisson PMF Properties

Normalization

$$\sum_{k=0}^{\infty} p_K(k;\lambda,t) = e^{-\lambda \cdot t} \cdot \left\{ \frac{(\lambda \cdot t)^0}{0!} + \frac{(\lambda \cdot t)^1}{1!} + \frac{(\lambda \cdot t)^2}{2!} + \cdots \right\} = e^{-\lambda \cdot t} \cdot \underbrace{\sum_{k=0}^{\infty} \frac{(\lambda \cdot t)^k}{k!}}_{\text{Taylor Series for } e^{+\lambda \cdot t}} = 1$$

Mean

$$E[K] = e^{-\lambda \cdot t} \sum_{k=0}^{\infty} k \cdot \frac{(\lambda \cdot t)^k}{k!} = (\lambda \cdot t) \cdot e^{-\lambda \cdot t} \underbrace{\sum_{k=1}^{\infty} \frac{(\lambda \cdot t)^{k-1}}{(k-1)!}}_{=1} = \lambda \cdot t \qquad \boxed{E[X] = \lambda t}$$

Variance

$$E[K(K-1)] = e^{-\lambda \cdot t} \sum_{k=0}^{\infty} k(k-1) \cdot \frac{(\lambda \cdot t)^k}{k!} = (\lambda \cdot t)^2 \cdot e^{-\lambda \cdot t} \underbrace{\sum_{k=2}^{\infty} \frac{(\lambda \cdot t)^{k-2}}{(k-2)!}}_{=1} = (\lambda \cdot t)^2$$

"Trick"

$$\underbrace{E[K(K-1)]}_{=(\lambda t)^2} = \underbrace{E[K^2] - E[K]}_{=\lambda t} \Rightarrow E[K^2] = (\lambda t)^2 + (\lambda t)$$

$$\therefore Var(K) = E[K^2] - E[K]^2 = \left\{ (\lambda t)^2 + (\lambda t) \right\} - (\lambda t)^2 = \lambda t \qquad \boxed{var(X) = \lambda t}$$

Moment Generating Fcn

$$\phi_K(s) = E[e^{Ks}] = e^{-\lambda \cdot t} \sum_{k=0}^{\infty} (e^s)^k \cdot \frac{(\lambda \cdot t)^k}{k!} = e^{-\lambda \cdot t} \underbrace{\sum_{k=0}^{\infty} \frac{(\lambda \cdot t \cdot e^s)^k}{k!}}_{= e^{\lambda \cdot t \cdot e^s}} = e^{\lambda \cdot t(e^s - 1)} \qquad \boxed{\phi_K(s) = e^{\lambda \cdot t(e^s - 1)}}$$

Some important properties of the Poisson PMF are listed and proved in this slide.

Normalization: Since each Poisson contribution corresponds to a discrete value of k and a single term in the Taylor expansion of the exponential $e^{+\lambda t}$, the normalization sum over the infinite number of terms clearly leads to $e^{-\lambda t} \cdot e^{+\lambda t} = 1$.

Mean: The expectation value involves an infinite sum of terms of the form $k \cdot [e^{-\lambda t} \cdot (\lambda \cdot t)^k / k!]$; since the k=0 term yields zero contribution, we may omit it and sum from k=1 to ∞ instead; canceling k into k! leaves (k-1)! in the denominator of the summand and upon taking out a factor of $\lambda \cdot t$ the summand becomes $(\lambda \cdot t) \cdot [e^{-\lambda t} \cdot (\lambda \cdot t)^{k-1} /(k-1)!]$. By changing the summation index k-1 → k, we recognize the normalization sum leaving $E[K] = (\lambda \cdot t)$ as the mean.

Variance: Instead of computing the 2nd moment $E[K^2]$, it is more convenient to compute the expression $E[K(K-1)]$ which zeros out two terms in the sum as the product k(k-1) vanishes for both k=0 and k=1. Thus omitting these terms we sum from k=2 to ∞ instead; the summand now starts at the k=2 term and has the form $(\lambda t)^2 \cdot [e^{-\lambda t} \cdot (\lambda \cdot t)^{k-2} /(k-2)!]$. By changing the summation index k-2 → k, we recognize the normalization sum, leaving the result $E[K(K-1)] = E[K^2] - E[K] = (\lambda \cdot t)^2$. Using $E[K] = (\lambda \cdot t)$ and solving for the 2nd moment we find $E[K^2] = (\lambda \cdot t)^2 + (\lambda \cdot t)$; the variance is the easily found by subtracting $E[K]^2 = (\lambda \cdot t)^2$ to give var(K) = $(\lambda \cdot t)$ (variance equals mean!).

Moment Generating Function: We use the notation $\phi_K(s)$ with parameter "s" now for the generating function since "t" is taken to be the data time span. The term $e^{-\lambda t}$ factors out of the sum leaving $e^{-\lambda t}$ $\sum \{(\lambda \cdot t \cdot e^s)^k / k!\}$ which yields $e^{-\lambda t} \cdot \exp(\lambda \cdot t \cdot e^s) = \exp(-\lambda \cdot t) \cdot \exp(\lambda \cdot t \cdot e^s) = \exp(\lambda \cdot t \cdot [e^s - 1])$.

5.1.3 Useful Properties for Sums of Poisson RVs

Useful Properties for Sums of Poisson RVs

Recursive Computation of Poisson

$$p_K(k;\lambda,t) = \frac{(\lambda \cdot t)^k}{k!} e^{-\lambda \cdot t} = \frac{\lambda \cdot t}{k} \left\{ \frac{(\lambda \cdot t)^{k-1}}{(k-1)!} e^{-\lambda \cdot t} \right\}$$

$$p_K(k;\lambda,t) = \frac{\lambda \cdot t}{k} \cdot p_K(k-1;\lambda,t)$$

$$p_K(k=0;\lambda,t) = \frac{(\lambda \cdot t)^0}{0!} e^{-\lambda \cdot t} = e^{-\lambda \cdot t}$$

$$p_K(k=1;\lambda,t) = \frac{\lambda \cdot t}{1} \cdot p_K(k=0;\lambda,t)$$

$$p_K(k=2;\lambda,t) = \frac{\lambda \cdot t}{2} \cdot p_K(k=1;\lambda,t), \cdots$$

Sum of Two Independent Poissons:

$$Z = X + Y \quad ; \quad X = p_X(x;\lambda_X,t) \quad ; \quad Y = p_Y(y;\lambda_Y,t)$$

$$\varphi_Z(s) = \varphi_X(s) \cdot \varphi_Y(s) = e^{\lambda_X \cdot t(e^s - 1)} \cdot e^{\lambda_Y \cdot t(e^s - 1)} = e^{(\lambda_X + \lambda_Y) \cdot t(e^s - 1)} \rightarrow p_Z(z; \underbrace{\lambda_X + \lambda_Y}_{\text{"sum rate params"}}, t)$$

Mean and Variance

$$\mu_Z = \mu_X + \mu_Y$$

$$\text{var}(Z) = \sigma_Z^2 = \sigma_X^2 + \sigma_Y^2$$

The Poisson distribution can be written down in a recursive form $p_K(k;\lambda,t) = (\lambda t/k) \cdot p_K(k-1;\lambda,t)$ for k=1,2, ... with the start value $p_K(k=0;\lambda,t) = e^{-\lambda t}$. This nested form minimizes numerical round-off errors.

In the bottom panel, the sum of two independent Poisson RVs X and Y with rates λ_X and λ_Y respectively, is easily obtained using the product of their generating functions to find that the rates just add $\lambda_Z = \lambda_X + \lambda_Y$; transforming back yields a PMF $p_Z(z;(\lambda_X + \lambda_Y),t)$ for the sum random variable Z=X+Y. The mean and variance of the sum Poisson RV Z are also given and note the variance is equal to the mean. This is easily generalized to the sum of any number of independent Poisson RVs and will be useful for approximations to binomial sums for the case in which p<<1. Examples are given in Section 8 on approximations (see Slide#8-12 to 8-15)

5.2 *Poisson Examples*

Poisson Examples

- Customer Arrival: $\lambda =50/\text{hr}$;
  time interval $t=[0, 0.5]$ hr $\lambda\, t = 25$

- At Least 17 Customers $\Pr[K \geq 17] = 1 - \Pr[K \leq 16] = 1 - e^{-25}\underbrace{\sum_{k=0}^{16}\frac{(25)^k}{k!}}_{=3.775e-2} = 96.23\%$

- More than 8 Customers
  $$\Pr[K > 8] = 1 - \Pr[K \leq 8] = 1 - e^{-25}\underbrace{\sum_{k=0}^{8}\frac{(25)^k}{k!}}_{=7.548e-5} = 99.99\%$$

- More than 3 Customers
  $$\Pr[K > 3] = 1 - \Pr[K \leq 3] = 1 - e^{-25}\underbrace{\sum_{k=0}^{3}\frac{(25)^k}{k!}}_{=4.087e-8} = 99.999996\%$$

- Nesting

$$e^{-25}\sum_{k=0}^{3}\frac{(25)^k}{k!} = e^{-25}\left(1+\frac{25}{1}\left(1+\frac{25}{2}\left(1+\frac{25}{3}\right)\right)\right) = 4.087\cdot 10^{-8}$$

$$\therefore\ \ 1 - 4.087\cdot 10^{-8} = 0.99999996$$

The next two slides give some typical applications of the Poisson distribution. If information is gathered over a 3 hour period and it is found that 150 customers walk through the door, then their average rate of arrival is 150/3hrs and the customer arrival rate is $\lambda = 50/\text{hr}$. If we now wish to consider a duration of $t= \frac{1}{2}$ hr, the Poisson parameter $a = 50/\text{hr}\cdot \frac{1}{2}$ hr $=25$ and the probability of any given number of customers being in the store during the $\frac{1}{2}$ hour period is answered by evaluating specific terms of the Poisson PMF

$$p_K(k;\lambda=50/\text{hr}, t=1/2\text{ hr}) = p_K(k; a=25) = e^{-25}(25)^k/k!\ \ \text{for } k=0,1,2,3, \ldots$$

The probability of at least 17 customers would require an infinite sum from 17 to ∞ so instead we compute $\Pr[K\geq 17] =1- \Pr[K\leq 16]$ to obtain 96.23%. The probability of more than 8 customers or more than 3 customers is computed the same way. Note that in the latter case we are taking the difference of two numbers that are nearly equal and finite digit arithmetic will be subject to subtractive cancelation and loss of significant digits. The nesting procedure shown for calculating the sum reduces the numerical round-off error.

If we wish to find the probability that there are between 3 and 8 customers during this period we could just sum the terms k=3,4,5,6,7,8, or if we have a tabulation of cumulative probabilities we could simply take the difference $\Pr[K\leq 8] - \Pr[K\leq 2]$; the 1st probability contains terms k=0,1,2,...,8 and we subtract the 2nd probability containing terms k=0,1,2, leaving terms k=3 through 8 as required. (Note the result is not $\Pr[K\leq 8] - \Pr[K\leq 3]$ as you might at first think since this difference excludes k=3.)

5.2.1 More Poisson Examples

More Poisson Examples

- **Typographical Errors:** 320 errors/500pages ;

Error Rate: $\lambda = .64$ err/page

Data Run: 4 pages $\lambda t = .64*4 = 2.56$

Prob 4 pages are error-free:

$$\Pr[K = 0] = p_K(k = 0; t = 4, \lambda = .64)$$
$$= \frac{(2.56)^0}{0!} e^{-2.56} = .077$$

- **Moon Craters**

Error Rate: $\lambda = 900$ craters/km^2 = .0009 craters/m^2

Data Run: diam=50 m; Area= $\pi\, 50^2/4$ m^2

a)Prob no craters in circle diam d= 50 m

$$\Pr[K = 0] = p_K(k = 0; t = \pi(50)^2/4, \lambda = .0009)$$
$$= \frac{(1.767)^0}{0!} e^{-1.767} = .1708$$

---

b) Find diam d for which

Prob[no craters]=.90

$$\Pr[K = 0] = p_K(k = 0; t = \pi d^2/4, \lambda = .0009)$$
$$= \frac{(.0009 \cdot \pi d^2/4)^0}{0!} e^{-.0009 \cdot \pi d^2/4} = e^{-(7.069e-4)d^2} = .90$$

$$p_K(k; a) = \frac{(a)^k}{k!} e^{-a} \quad ; \quad a = \underset{\text{rate}}{\lambda} \cdot \underset{\text{"run"}}{t}$$

$$\Rightarrow d = \sqrt{\frac{-\ln(.90)}{7.069e-4}} = 12.21\ m$$

Typographical errors in a book are found to be 320 errors in a 500 pages or at a rate $\lambda = .64$ err/pg. If we now take a look at just 4 pages of the book what is the probability that there are no errors? The data run is 4 pages, so the Poisson parameter is a = .64 err/pg $\cdot$ 4 pg = 2.56 and the k= 0 Poisson term is computed to be .077.

Craters on the Moon are found to occur with "rate" $\lambda = 900$ craters/km^2 = .0009 craters/m^2 and so if we consider a "data run" described as a circular region of diameter d = 50 m , what is the probability that no craters are found? The Poisson parameter is computed to be 1.767 and the k=0 term becomes p_K(k=0; a=1.767) = .1708 which is approximately 17%.

Another type of question is "for what diameter d is there a 90% probability of finding no craters?" In this case the Poisson parameter a is unknown and we write

$\qquad$ p_K(k=0; a=.0009 $\pi d^2/4$) = exp{-[.0009$\pi d^2/4$]} [.0009$\pi d^2/4]^0/0!$ = .90

and solve to find d = 12.21 m.

Clearly, there are many variations of these "Type A" problems for the Poisson PMF, but they are all handled in the same manner.

5.2.2 Poisson with "Erasures"- Some Arrivals NOT Detected

Poisson with "Erasures"- Some Arrivals NOT Detected

- Erasures in the Poisson Process (Bernoulli Trial Conditioned on Poisson Arrival)
- Given an arrival, make Bernoulli Trial: "1"=detect ; "0" =no detect
- Ex. Optical detection
  - **Photons Arrive** according to **Poisson** process rate λ
  - **Successful Detection** of arrival is a **Bernoulli** process p, q=1-p
  - RV K = # arrivals   Not all arrivals are detected and counted!
  - RV D = # Detects   They must arrive K and must also be detected D
  - K and D independent RVs ➜ Joint distribution is just their product

$$\Pr[D=d] = \sum_{k=d}^{\infty} \Pr[D=d, K=k] = \sum_{k=d}^{\infty} \Pr[D=d \mid K=k] \cdot \Pr[K=k]$$

  Num Detected   *Sum Joint Probability over all 'k'*   *Detected Cond. on Arrivals*   *Num of Arrivals*

  - PMF for counts

$$p_D(d) = \sum_{\substack{k=d \\ arrivals}}^{\infty} p_{DK}(d,k) = \sum_{k=d}^{\infty} \underbrace{p_{D|K}(d \mid k)}_{Bernoulli} \; p_K(k)$$

  Poisson

  Condition "detects" D on "arrivals" K

$$= \sum_{k=d}^{\infty} \underbrace{\binom{k}{d} p^d q^{k-d}}_{\substack{K \; Bernoulli \\ trials}} \cdot \underbrace{e^{-\lambda t} \frac{(\lambda t)^k}{k!}}_{Poisson}$$

  Arr. ≥ #Detects

  D=#succ in K trials (arrivals) yields Binomial

The arrivals of many physical processes are predicted by the Poisson process with an arrival rate λ over a data run t in terms of the PMF: $p_K(k;\lambda, t) = e^{-\lambda t} (\lambda t)^k/k!$ for k=0,1,2,3, ...; however, these arrivals must be "detected" or they will not be recorded as an arrival. If we assume that each arrival is subjected to a Bernoulli trial with probability p of detection, then the number of **detected arrivals D** (given that there are k arrivals!) is just the sum of K independent Bernoulli trials and is therefore given by the Binomial distribution

$\quad b_D(d;p,k) = {}^kC_d \, p^d \, q^{k-d}$ for d = 0,1,2, ..., k   (d $\leq$ k means #detections cannot exceed arrivals)

Thus, in order to describe the "detected arrivals D" we must consider the joint PMF of the two independent RVs D and K which is simply their product

$\quad p_{DK}(d,k) = b_D(d;p,k) \cdot p_K(k;\lambda, t) = {}^kC_d \, p^d \, q^{k-d} \cdot e^{-\lambda t} (\lambda t)^k/k!$ k=0,1,2... and d = 0,1,2, ..., k

(Note the joint distribution above actually represents the *conditional* (Binomial) times the *a priori* (Poisson) *probability densities*, viz., $p_{DK}(d,k) = p_{D|K}(d|k) \cdot p_K(k)$.)

Summing this joint distribution over all k≥d yields the desired marginal distribution $p_D(d)$ which turns out to be the Poisson distribution again, but with a rate reduced by the Bernoulli trial probability p, viz., $\lambda_{new} = p\,\lambda < \lambda$. The resulting probability density for the detection random variable D gives the number of arrivals that are actually detected

$\quad p_D(d) = p_D(d; (p\lambda), t) = e^{-p\lambda t} (p\lambda t)^d/d!$ for d=0,1,2,3, ...

and is called a "Poisson PMF with erasures." The result is derived on the next slide.

5.2.3 Poisson with "Erasures"- Details

Poisson with "Erasures"-Details

Photon Detection Probability

$$p_D(d) = \sum_{k=d}^{\infty} \binom{k}{d} p^d q^{k-d} \cdot e^{-\lambda t} \frac{(\lambda t)^k}{k!}$$

Changing the sum index to $\ell = k - d$ leads to

$$p_D(d) = \sum_{l=0}^{\infty} \binom{l+d}{d} p^d q^l \cdot e^{-\lambda t} \frac{(\lambda t)^{l+d}}{(l+d)!}$$

$$= e^{-\lambda t} p^d (\lambda t)^d \sum_{l=0}^{\infty} \frac{(l+d)!}{d!\,l!} (\lambda t)^l q^l \cdot \frac{1}{(l+d)!}$$

$$= \frac{(p\lambda t)^d}{d!} e^{-\lambda t} \underbrace{\sum_{l=0}^{\infty} \frac{(q\lambda t)^l}{l!}}_{=e^{q\lambda t}} = \frac{(p\lambda t)^d}{d!} \underbrace{e^{-\lambda t} e^{q\lambda t}}_{=e^{-\lambda t(1-q)}}$$

Erasures yield Poisson with "Reduced Rate" $\qquad \lambda \rightarrow p\lambda \qquad \boxed{p_D(d) = \frac{(p\lambda t)^d}{d!} e^{-(p\lambda t)}}$

The proof of the reduced Poisson rate for "detected arrivals D" follows by the substitution l = k-d which transforms the sum from k = d to ∞ to a sum over l from l =0 to ∞. Writing out the binomial coefficient explicitly provides some cancelation as indicated by the red slashes; the terms $e^{-\lambda t} p^d (\lambda t)^d$ can be taken outside the sum to leave a sum of terms that is precisely $e^{q\lambda t}/d!$ Thus collecting terms in the product $e^{-\lambda t} p^d (\lambda t)^d \cdot e^{q\lambda t} /d! = e^{-\lambda t(1-q)} (p\lambda t)^d /d!$ and using the identity p=1-q in the exponential yields the final result for the reduced rate Poisson process $e^{-p\lambda t}(p\lambda t)^d /d!$ given in the boxed equation. The PMF for the detection random variable D is still a Poisson process, but has its arrival rate reduced by the probability of detection p, *viz.*, λ_{new} = p·λ .

5.3 PDF for Time of the 3rd Poisson Arrival T_3

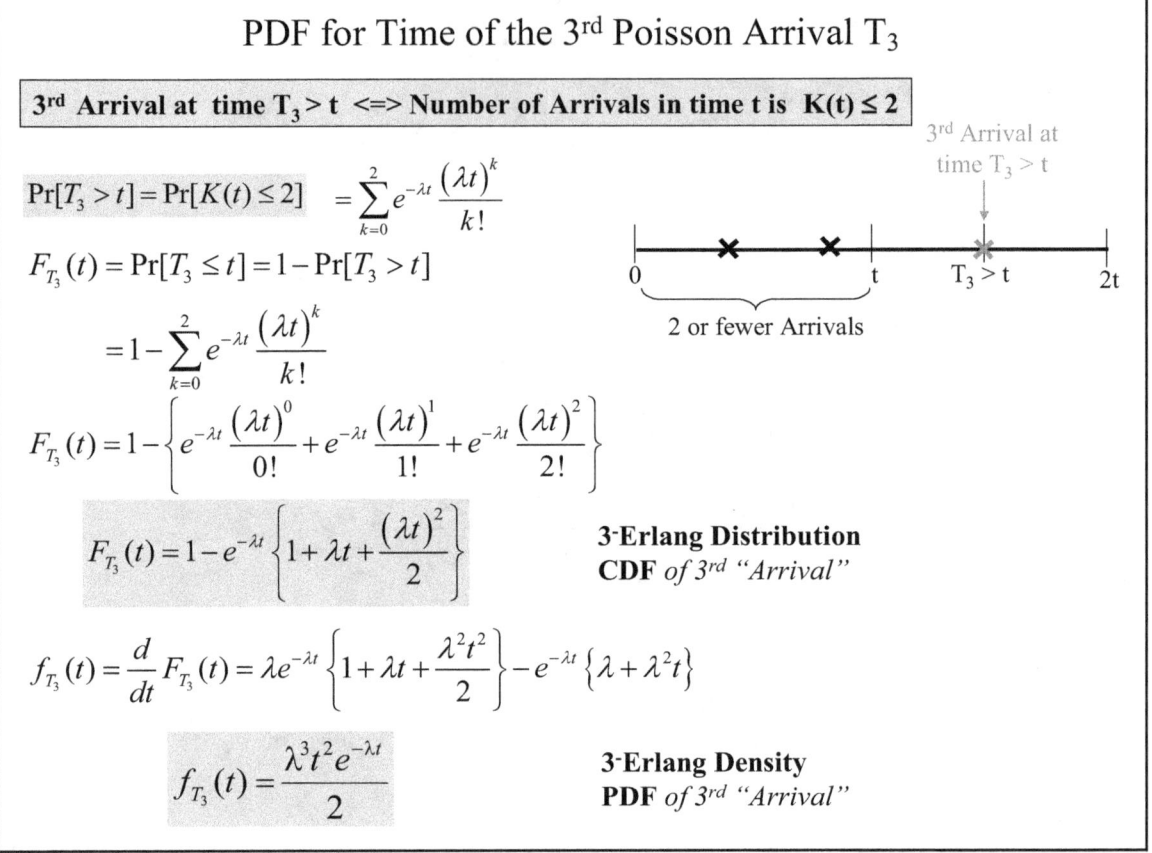

We have previously stated that the "time of the arrival" is an important quantity for many physical observations, but up to this point we have only described the Poisson process in terms of the RV "K(t)" giving the number of arrivals (successes) in a fixed interval of time. Note K is a function of the time "t" in the sense that the number of arrivals K depends upon the length of the time interval from 0 to t.

Let us introduce a new continuous RV T_3 representing the time of the 3rd arrival, and consider the "time-line" of Poisson arrival events indicated by "x"s in the figure. The probability that the 3rd arrival occurs at a time $T_3 > t$ (denoted by the (red) "x" outside the interval [0,t]) must be equal to the probability that no more than two arrivals have already occurred $K(t) \leq 2$ in the interval [0,t]. That is, the two probabilities must be the same and we may write $\Pr[T_3 > t] = \Pr[K(t) \leq 2]$; this can be restated in terms of the cumulative distribution functions CDFs $F_{T3}(t)$ and $F_K(K=2)$ for the two RVs as follows:

$$1 - F_{T3}(t) = F_K(2) = p_K(k=0;\lambda, t) + p_K(k=1;\lambda, t) + p_K(k=2;\lambda, t) = e^{-\lambda t}(\lambda t)^0/0! + e^{-\lambda t}(\lambda t)^1/1! + e^{-\lambda t}(\lambda t)^1/1!$$

This expression gives a cumulative distribution called the 3-Erlang distribution shown in the slide; taking its derivative with respect to the continuous time variable t yields the continuous PDF $f_{T3}(t)$ for the 3rd arrival time

$$f_{T3}(t) = \lambda (\lambda t)^2 e^{-\lambda t}/2!$$

The probability that the 3rd arrival of a Poisson process occurs between t and t+dt is thus $dP = f_{T3}(t)\, dt$. We see that the 3-Erlang distribution is intimately related to the underlying Poisson Process and answers questions about the timing of the 3rd arrival; the actual number of Poisson arrivals is given by the discrete Poisson distribution. These two distributions represent *complementary aspects* of the same *Poisson Process*, but they answer entirely different questions. Their relation to one another is encapsulated in the equality of two probabilities $\Pr[T_3 > t] = \Pr[K(t) \leq 2]$ which states "the probability that the **third arrival** T_3 occurs *after a time duration* "t" is equal to the probability that the **number of arrivals** K(t) (during same time interval "t") is ≤ 2".

5.3.1 r-Erlang PDF for Time of the rth Poisson Arrival T_r

r-Erlang PDF for Time of the rth Poisson Arrival T_r

- General Case for RV T_r: *time* of the rth arrival (success)
- CDF:

$$\Pr[T_r \le t] = F_{T_r}(t) = \sum_{k=r}^{\infty} p_K(k;t,\lambda)$$

r^{th} Arrival at time $T_r > t$

0   t $T_r > t$

r-1 or fewer Arrivals

$$\Pr[T_r \le t] = 1 - \sum_{k=0}^{r-1} p_K(k;t,\lambda) = 1 - \sum_{k=0}^{r-1} e^{-\lambda t} \frac{(\lambda t)^k}{k!}$$

r-Erlang Distribution
CDF *of rth "Arrival"*

$$f_{T_r}(t) = \frac{d}{dt} F_{T_r}(t) = \frac{d}{dt} \left\{ 1 - \sum_{k=0}^{r-1} e^{-\lambda t} \frac{(\lambda t)^k}{k!} \right\} = -\sum_{k=0}^{r-1} \frac{\lambda^k}{k!} e^{-\lambda t} \left\{ kt^{k-1} - \lambda t^k \right\}$$

$$= \lambda e^{-\lambda t} \left\{ \underbrace{\sum_{k=0}^{r-1} \frac{(\lambda t)^k}{k!} - \sum_{k=1}^{r-1} \frac{(\lambda t)^{k-1}}{(k-1)!}} \right\} = \lambda e^{-\lambda t} \frac{(\lambda t)^{r-1}}{(r-1)!} = \frac{r}{t} \left[\frac{(\lambda t)^r}{r!} e^{-\lambda t} \right] = \frac{r}{t} p_K(r;t,\lambda)$$

term-by-term cancellation
except for $k=r-1$ in 1st sum

$$f_{T_r}(t) = \frac{\lambda^r t^{r-1}}{(r-1)!} e^{-\lambda t} \quad ; r = 1,2,\cdots ; t \ge 0$$

r-Erlang Density
PDF *of rth "Arrival"*

Just as the 3-Erlang distribution answers questions about the timing of the 3rd arrival, the r-Erlang distribution describes the timing of the rth arrival in terms of the RV T_r (with subscript "r"). The derivation follows exactly the same arguments given for the 3-Erlang distribution presented on the previous slide. The cumulative distribution $F_{Tr}(t)$ for the rth arrival is given in the second boxed equation of this slide and its associated PDF is found by taking the derivative d/dt $[F_{Tr}(t)]$ and yields the result given by the boxed equation at the bottom of this slide. Note that the r-Erlang density has the units of a rate (1/time) because of an *extra factor* of λ in the numerator and clearly must be multiplied by a *dt* to give a dimensionless differential probability. To make this more apparent, the numerator is often re-written as $\lambda \cdot (\lambda \cdot t)^{r-1}$ where the separated out factor of λ provides the (1/time) dimension to multiply the *dt* when setting up the dimensionless probability integral over time *t*.

5.3.2 Derivation of r-Erlang PDF from Negative Binomial PMF

Derivation of r-Erlang PDF from Negative Binomial PMF

Negative Binomial with probability $p = \lambda \, \Delta t$, $t_r = n_r \, \Delta t$ **time of rth arrival**

$$\Delta P(t_r \leq T \leq t_r + \Delta t) = [\text{Neg. Binomial Density} /(\text{unit time})] \cdot \Delta t$$

$$\lim_{\substack{n \to \infty \\ \Delta t \to 0}} b_{neg}(n_r \Delta t; p = \lambda \cdot \Delta t) = \lim_{\substack{n_r \to \infty \\ n_r \Delta t \to t_r \\ \lambda \cdot \Delta t \to p}} \binom{n_r - 1}{r - 1} (\lambda \cdot \Delta t)^{r-1} (1 - \lambda \cdot \Delta t)^{n_r - r} (\lambda \cdot \Delta t)^1$$

$$= \lim_{\substack{n \to \infty \\ n_r \Delta t \to t_r \\ \lambda \cdot \Delta t \to p}} \overbrace{\frac{(n_r - 1) \cdots (n_r - r + 1)}{(r-1)!}}^{\cong (n_r)^{r-1}} \cdot \frac{(\lambda \cdot t_r)^{r-1}}{(n_r)^{r-1}} \cdot \underbrace{\left(1 - \frac{\lambda \cdot t_r}{n_r}\right)^{n_r}}_{\to e^{-\lambda \cdot t_r}} \cdot \underbrace{\left(1 - \frac{\lambda \cdot t_r}{n_r}\right)^{-r}}_{\substack{\to 0 \\ \to 1}} \lambda \cdot \Delta t$$

$$\boxed{f_{T_r}(t_r; \lambda)dt = \frac{(\lambda \cdot t_r)^{r-1}}{(r-1)!} e^{-\lambda \cdot t_r} \lambda \, dt} \qquad \begin{array}{c} t_r = n_r \Delta t \\[4pt] n_r = r, r+1, \cdots, \infty \end{array} \qquad \begin{array}{c} \textit{In the limit time } t_r \\ \textit{is continuous!} \end{array}$$

r-Erlang PDF – continuous density with time variable for the rth arrival $T_r = t$ and rate λ

$$f_{T_r}(t; \lambda) = \frac{(\lambda \cdot t)^{r-1}}{(r-1)!} e^{-\lambda \cdot t} \cdot \lambda \qquad \text{drop subscript 'r' on time 't'}$$

$$\lambda \cdot t = \lim_{\substack{n \to \infty, \Delta t \to 0 \\ n \Delta t \to t}} (\lambda \cdot n \Delta t) = \lim_{\substack{n \to \infty, \Delta t \to 0 \\ \lambda \Delta t \to p}} (n \cdot \lambda \Delta t) = n \cdot p$$

r-Erlang Properties

$$E[T_r] = r \cdot \frac{1}{\lambda} \qquad \text{var}(T_r) = r \cdot \frac{1}{\lambda^2} \qquad \varphi_{T_r}(s) = \left(\frac{\lambda}{\lambda - s}\right)^r$$

We have previously stated that a binomial PMF becomes a Poisson PMF if we identify the single trial probability of success "p" for the binomial with $\lambda \Delta t$, the product of the Poisson rate with the small interval during which the observation "trial" takes place. One would think that the negative binomial PMF, which gives the number of trials n_r required for r-successes, should relate to the r-Erlang probability density $f_{Tr}(t)$ in the same limit. In fact, making the substitutions $p = \lambda \Delta t$ and $t_r = n_r \cdot \Delta t$ in the negative binomial expression $b_{neg}(n_r; p) \to b_{neg}(n_r \cdot \Delta t; \lambda \, \Delta t)$ and taking the combined limits as both $n_r \to \infty$ and $\Delta t \to 0$, this slide shows how we can identify the probability that the rth arrival occurs between t and t +dt as

$$\Pr(t \leq T_r \leq t + dt) = f_{Tr}(t; \lambda) \, dt = \{(\lambda t)^{r-1} / (r-1)!\} \, e^{-\lambda t} \lambda \, dt,$$

Thus, the discrete negative binomial PMF yields the r-Erlang continuous distribution because the limit of the product "$n_r \cdot \Delta t$" yields the continuous time variable "t" representing the time of the rth Poisson arrival. [Compare this derivation with that for the Poisson PDF from the Binomial PMF on Slide#5-3)

5.4 r-Erlang and Gamma PDFs

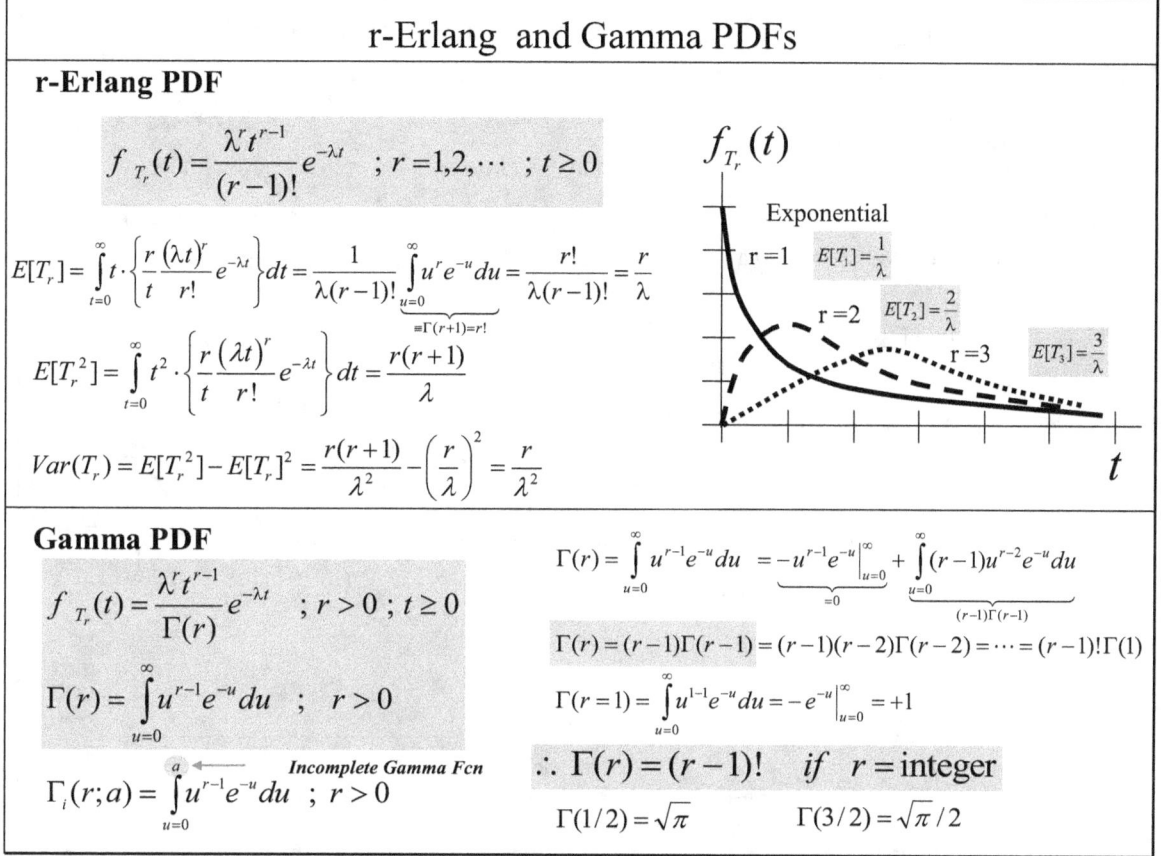

r-Erlang and Gamma PDFs

r-Erlang PDF

$$f_{T_r}(t) = \frac{\lambda^r t^{r-1}}{(r-1)!} e^{-\lambda t} \quad ; r = 1, 2, \cdots \quad ; t \geq 0$$

$$E[T_r] = \int_{t=0}^{\infty} t \cdot \left\{ \frac{r}{t} \frac{(\lambda t)^r}{r!} e^{-\lambda t} \right\} dt = \frac{1}{\lambda(r-1)!} \underbrace{\int_{u=0}^{\infty} u^r e^{-u} du}_{\equiv \Gamma(r+1) = r!} = \frac{r!}{\lambda(r-1)!} = \frac{r}{\lambda}$$

$$E[T_r^2] = \int_{t=0}^{\infty} t^2 \cdot \left\{ \frac{r}{t} \frac{(\lambda t)^r}{r!} e^{-\lambda t} \right\} dt = \frac{r(r+1)}{\lambda}$$

$$Var(T_r) = E[T_r^2] - E[T_r]^2 = \frac{r(r+1)}{\lambda^2} - \left(\frac{r}{\lambda} \right)^2 = \frac{r}{\lambda^2}$$

$f_{T_r}(t)$

Exponential

$r = 1 \quad E[T_1] = \frac{1}{\lambda}$

$r = 2 \quad E[T_2] = \frac{2}{\lambda}$

$r = 3 \quad E[T_3] = \frac{3}{\lambda}$

t

Gamma PDF

$$f_{T_r}(t) = \frac{\lambda^r t^{r-1}}{\Gamma(r)} e^{-\lambda t} \quad ; r > 0 ; t \geq 0$$

$$\Gamma(r) = \int_{u=0}^{\infty} u^{r-1} e^{-u} du \quad ; \quad r > 0$$

Incomplete Gamma Fcn

$$\Gamma_i(r;a) = \int_{u=0}^{a} u^{r-1} e^{-u} du \quad ; \quad r > 0$$

$$\Gamma(r) = \int_{u=0}^{\infty} u^{r-1} e^{-u} du = \underbrace{-u^{r-1} e^{-u} \Big|_{u=0}^{\infty}}_{=0} + \underbrace{\int_{u=0}^{\infty} (r-1) u^{r-2} e^{-u} du}_{(r-1)\Gamma(r-1)}$$

$$\Gamma(r) = (r-1)\Gamma(r-1) = (r-1)(r-2)\Gamma(r-2) = \cdots = (r-1)!\Gamma(1)$$

$$\Gamma(r=1) = \int_{u=0}^{\infty} u^{1-1} e^{-u} du = -e^{-u} \Big|_{u=0}^{\infty} = +1$$

$$\therefore \Gamma(r) = (r-1)! \quad if \quad r = \text{integer}$$

$$\Gamma(1/2) = \sqrt{\pi} \qquad \Gamma(3/2) = \sqrt{\pi}/2$$

1st Erlang density is the well-known Exponential density function $\lambda e^{-\lambda t}$ for the 1st arrival. It has a mean that is simply 1/"rate", *i.e.*, $\mu_{T_1} = 1/\lambda$; this is also known as the "e-folding" time because at time $t = 1/\lambda$ the probability density is reduced by a factor e^{-1}. The variance is $var(T_1) = 1/\lambda^2$ and the curve decays exponentially from a density of λ at time $t=0$, to a density of λ/e at the e-folding time $t = 1/\lambda$ and continues to decay to zero as $t \rightarrow \infty$.

2nd Erlang density is $\lambda^2 t e^{-\lambda t}$; it starts at zero, reaches a maximum where $d/dt(\lambda^2 t e^{-\lambda t})$ vanishes at $t=1/\lambda$ (*i.e.*, at the mean arrive time for the 1st arrival), and then decays to zero almost exponentially as shown in the figure. (Note that this behavior makes good sense logically since the 2nd arrival must come after a wait of $1/\lambda$ for the 1st arrival.) It has a mean that is simply 2/"rate", *i.e.*, $\mu_{T_2} = 2/\lambda$ which is essentially the sum of two exponential wait times, one for the first arrival and another for the second arrival; the variance is $var(T_2) = 2/\lambda^2$ which is twice that for the 1st arrival.

3rd Erlang density is $\frac{1}{2} \lambda^3 t^2 e^{-\lambda t}$ which starts at zero and reaches a maximum where $d/dt(\frac{1}{2} \lambda^3 t^2 e^{-\lambda t})$ vanishes $t=2/\lambda$ *i.e.*, at the mean arrive time for the 2nd arrival and then decays to zero almost exponentially as shown in the figure. (Note that this behavior again makes sense since the 3rd arrival must come after the 2nd arrival which on average occurs after two arrival waits $2/\lambda$). It has a mean that is simply 3/"rate", *i.e.*, $\mu_{T_3} = 3/\lambda$ which is essentially the sum of three exponential time waits once each for the first , second, and third arrivals; the variance is $var(T_3) = 3/\lambda^2$ which is three times that for the 1st arrival. Finally, we write down the Gamma density which is a generalization of the r-Erlang density obtained by replacing $(r-1)!$ with $\Gamma(r)$ making it valid for non-integer values of r. We give the definition of the gamma function as an integral, and note that for integer values of r, the gamma function $\Gamma(r) = (r-1)!$ reduces to the ordinary factorial of one less than its argument $(r-1)$. (See Slide#8-7 for r-Erlang as a sum of exponentials that approach a Gaussian)

5.4.1 r-Erlang Application: Waiting time on a Telephone

---

r- Erlang PDF Application: Waiting time on a Telephone

You are the rth in the queue (say $r = 20$)

The service rate is $\lambda = 2$ **customers/minute** (Poisson Process)

What is probability you will have to wait more than $t = 8$ **minutes**?

T_r = time of rth success => r-Erlang PDF

Sum 20 geometric RVs; means & variances add (See Slide# 8-7)

$$f_{T_r}(t) = \frac{\lambda^r t^{r-1}}{(r-1)!} e^{-\lambda t}$$

$r = 20$; $\lambda = 2 / \min$

$E[T_r] = r / \lambda = 20/2 = 10 \min$

$Var(T_r) = r / \lambda^2 = 20/2^2 = 5 \min^2$

$$P[T_{20} > 8\min] = 1 - P[T_{20} \le 8\min] = 1 - \int_{t=0}^{8} f_{T_r}(t)dt$$

$$= 1 - \int_{t=0}^{8} \frac{(2)^{20} t^{20-1}}{(20-1)!} e^{-2t} dt \underset{u=2t}{\overset{u=16}{=}} 1 - \underbrace{\frac{1}{19!} \int_{u=0}^{u=16} u^{19} e^{-u} du}_{.188} = .812$$

Note: The integral is a fraction of the Gamma function $\Gamma(20)=19!$

Thus 81.2% chance you wait more than 8 minutes if you are 20th in the queue

$E[T_r] = r / \lambda = 20/2 = 10\min$

$Var(T_r) = r / \lambda^2 = 20/2^2 = 5\min^2$

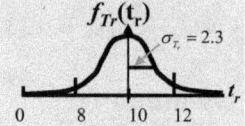

$f_{Tr}(t_r)$, $\sigma_{T_r} = 2.3$, $0 \quad 8 \quad 10 \quad 12 \quad t_r$

Alternately, recalling the relation between time-of-arrival and number-of-arrivals

$$\Pr[T_r > t] = \Pr[K(t) \le r-1] = \sum_{k=0}^{r-1} e^{-\lambda t} \frac{(\lambda t)^k}{k!} \quad ; \quad \lambda \cdot t = (2/\min) \cdot 8\min = 16$$

$$P[T_{20} > 8\min] = \underbrace{P[K(8) \le 19]}_{\substack{no\ more\ than \\ (20-1)\ successes \\ in\ time\ t=8\ min}} = \sum_{k=0}^{20-1} e^{-\lambda t} \frac{(\lambda t)^k}{k!} = \sum_{k=0}^{19} e^{-16} \frac{(16)^k}{k!} = e^{-16}\underbrace{\left(1 + \frac{16}{1}\left(1 + \frac{16}{2}\left(1 + \cdots \left(1 + \frac{16}{19}\right)\cdots\right.\right.\right.}_{=7.217730\times 10^6}\right) = .812$$

t=8 min

$0 \quad$ K< 19 arrivals in 8 min $\quad T_{20} > 8$ min

A classic application of the r-Erlang distribution is to the waiting time for a queue such as in a telephone answering system. It is usually assumed that the occurrence of telephone calls is governed by a Poisson process with a service answering rate of λ customers per minute.

A typical problem asks questions such as " if you are the $r=20^{th}$ customer in the queue, what is the probability that you will have to *wait at least 8 minutes*, given a service rate $\lambda = 2$ customers / min?" Thus, we need to find $\Pr[T_{20} > 8\min] = 1 - \Pr[T_{20} \le 8\min] = 1 - F_{T20}(8)$; this requires the integral of the $r= 20$ Erlang density $\lambda^r t^{r-1} e^{-\lambda t}/(r-1)! = 2^{20} t^{19} e^{-2t}/19!$ between the limits t=0 and t=8 min. Note that the integral is the incomplete gamma function $\Gamma_I(20)$ divided by 19! and so represents a fraction of the complete gamma function $\Gamma_I(20) / \Gamma(20)$. The result $1 - \Gamma_I(20) / \Gamma(20) = 81.2\%$.

Recall the relation between the **timeline and the number of successes,** namely

"the probability that the 20th arrival occurs outside the t=8 minute time interval of the Poisson process is equal to the probability that there are 19 or fewer arrivals inside that same interval"

Thus, for this example the statement translates as

$$\Pr[T_{20} > 8\min] = \Pr[K(8\min) \le 19]$$

and we can compute the sum of the k=0, 1,2,... 19 terms of the Poisson PMF with parameter $a=\lambda t = 2/\min \cdot 8\min = 1$, *i.e.*, sum the terms $p_K(k, a=16) = e^{-16}(16)^k/k!$ which again yields 81.2%

5.4.2 Poisson Process Summary: Number and Timing of Arrivals

Poisson Process Summary: Number and Timing of Arrivals

r-Erlang PDF determines interarrival times

Continuous Distribution

$$u = \lambda t$$

$$f_{T_r}(t)dt = \frac{(\lambda t)^{r-1} e^{-\lambda t}}{(r-1)!} \lambda dt \xrightarrow{} \frac{u^{r-1} e^{-u} du}{\Gamma(r)} = \frac{d\Gamma(r)}{\Gamma(r)}$$

Continuous Time of Arrival "t"

$$\boxed{\Pr[t < T_r < t+dt] = f_{T_r}(t)dt = \frac{d\Gamma(r)}{\Gamma(r)}}$$

Connection Between Two Aspects of Poisson Process

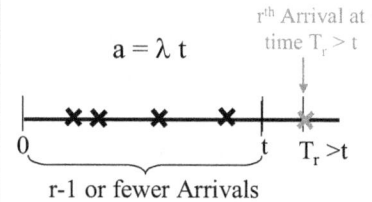

$$a = \lambda t$$

r^{th} Arrival at time $T_r > t$

r-1 or fewer Arrivals

$$\boxed{\Pr[T_r > t] = \Pr[K(t) \le r-1]}$$

Poisson PMF determines number of arrivals

Discrete Distribution

$$p_K(k;t,\lambda) = \frac{(\lambda t)^k}{k!} e^{-(\lambda t)} = \frac{(a^k/k!)}{e^a} = \frac{\Delta_k e^a}{e^a}$$

Discrete Number of Arrivals "k"

$$\boxed{p_K(k;a) = \frac{\Delta_k e^a}{e^a}}$$

$$e^a = \sum_{k=0}^{\infty} \frac{a^k}{k!} \quad ; \quad \Delta_k e^a = (a^k/k!) = k^{th} \text{ term of expansion}$$

We have seen that the Poisson process has two complementary probability densities corresponding to two entirely different characteristics of the process namely, the timeline for arrivals, and the actual number of arrivals. More specifically, we have:

(i) The r-Erlang distribution which gives the probability density for the rth arrival to occur between time t and t+dt, *i.e.*, $f_{Tr}(t)dt = d\Gamma/\Gamma$; integrating this density from t=0 to t yields the cumulative distribution function $F_{Tr}(t) = \Gamma_t(t)/\Gamma$, which is just the ratio of the incomplete gamma function evaluated at time t to the complete gamma function evaluated at t=∞. Thus we have a continuous probability density function (PDF) and a continuous cumulative distribution function (CDF) on the (continuous) time variable t.

(ii) The Poisson distribution which gives the probability mass function of the kth success in a fixed interval of time t with a rate of arrival λ and Poisson parameter a = λt . The probability is given by $p_K(k; a) = [a^k/k!]/e^a = \Delta_k e^a/e^a$ for trials k=0,1,2, ...,∞ ; the individual terms in this discrete probability mass function are easily interpreted as the ratio of a single term in the Taylor expansion of the exponential $[a^k/k!]$ to the exponential e^a itself. The cumulative distribution $F_K(k; a)$ is the ratio of the partial sum over range [0,k] to the completed infinite series e^a. Thus, we have a discrete probability mass function (PMF) and a "quasi-continuous" step-like cumulative distribution function (CDF) on the (discrete) number-of-arrivals variable k.

6 The Gaussian Process

The Gaussian Process

6.1 Gaussian Distribution

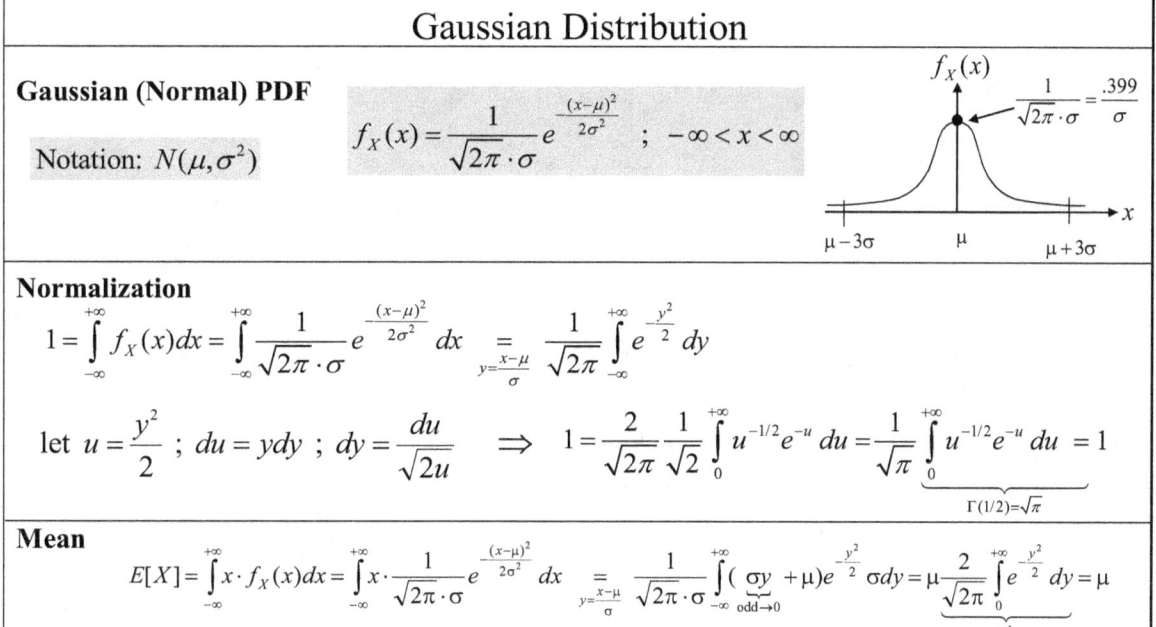

The Gaussian or Normal Process has the familiar "bell-shaped" PDF curve that stretches from $-\infty$ to $+\infty$ symmetrically about its mean value μ, and its "width" is specified by the variance σ^2. A normal random variable X is concisely specified by the notation $X \sim N(\mu, \sigma^2)$ meaning that it has a Normal or Gaussian distribution with mean μ and variance σ^2. Typically distances along the x-axis are measured in units of the standard deviation σ and taken relative to the mean value μ as the origin. The area under the curve between the points $x = \mu \pm 3\sigma$ shown on the plot contains 99.7% of the total probability under the curve and is taken as the standard accuracy or 3-sigma confidence interval.

The normalization integral can be transformed into a gamma function as shown on the slide. The mean and variance are also computed by transforming the integrals into expressions involving the normalization integral.

Both the Gaussian and Poisson processes, derive from the discrete binomial distribution for a large number of Bernoulli trials $n \rightarrow \infty$. However, the Poisson is restricted to *small single trial probability of success* $p \ll 1$, while the Gaussian is valid for *arbitrary* $p \in [0, 1]$. Moreover, the Gaussian extends *symmetrically* between $-\infty$ and $+\infty$, while the Poisson is *asymmetric* and defines a *fixed time interval* "*t*" during which measurements are made. Thus the concept of *interarrival times* makes sense for a Poisson process, but does not for a Gaussian process which measures events over infinite time. Their relationship to the underlying Bernoulli process with probability "p" is different for the two distributions, *viz.*, $\mu_X = np$, $\sigma_X^2 = npq$ for the Gaussian and $\mu_X = np = \lambda t$ and $\sigma_X^2 = npq = np(1-p) \cong np = \lambda t$ for the Poisson. Furthermore, the Poisson process considers the arrival rate λ during the fixed time interval t as its fundamental process descriptor and defines the mean and variance in terms of a dimensionless Poisson parameter defined by their product $a = \lambda t$.

6.1.1 Moment Generating Function for Gaussian

Moment Generating Function for Gaussian

Normal Random Variable $X: N(0,1)$ $\phi_X(s) \equiv E\left[e^{X \cdot s}\right] = \int\limits_{-\infty}^{+\infty} e^{x \cdot s} \cdot \dfrac{e^{-\frac{x^2}{2}}}{\sqrt{2\pi}} \, dx$

"Complete the Square":

$$xs - \frac{x^2}{2} \equiv -\frac{(x-s)^2}{2} + \frac{s^2}{2}$$

$$\phi_X(s) \equiv \int\limits_{-\infty}^{+\infty} \frac{e^{-\frac{(x-s)^2}{2} + \frac{s^2}{2}}}{\sqrt{2\pi}} \, dx = e^{+\frac{s^2}{2}} \cdot \underbrace{\int\limits_{-\infty}^{+\infty} \frac{e^{-\frac{(x-s)^2}{2}}}{\sqrt{2\pi}} \, d(x-s)}_{\text{Gauss Normalization} = 1} = e^{+\frac{s^2}{2}}$$

$$\boxed{\phi_X(s) = e^{+\frac{s^2}{2}}}$$

General Case: $X: N(\mu, \sigma^2)$

Center about mean and normalize: $Y = (X - \mu)/\sigma$ is $N(0,1)$ **Transformed RV:** $X = \mu + \sigma Y$

$$\phi_X(s) \equiv E\left[e^{X \cdot s}\right] = E\left[e^{(\mu + \sigma Y) \cdot s}\right] = e^{\mu s} E\left[e^{Y \cdot (\sigma s)}\right] = e^{\mu s} e^{\frac{(\sigma s)^2}{2}} = e^{\left(\mu s + \frac{\sigma^2 s^2}{2}\right)} \qquad \boxed{\phi_X(s) = e^{\left(\mu s + \frac{\sigma^2 s^2}{2}\right)}}$$

Sum of Two Gaussians is a Gaussian: $X_1: N(\mu_1, \sigma_1^2)$ & $X_2: N(\mu_2, \sigma_2^2)$ $Z = X_1 + X_2$

$$\phi_Z(s) = \phi_{X_1}(s) \cdot \phi_{X_2}(s) = e^{\left(\mu_1 s + \sigma_1^2 \frac{s^2}{2}\right)} \cdot e^{\left(\mu_2 s + \sigma_2^2 \frac{s^2}{2}\right)} = e^{\left((\mu_1 + \mu_2)s + (\sigma_1^2 + \sigma_2^2) \cdot \frac{s^2}{2}\right)} = e^{\left(\mu_Z s + \sigma_Z^2 \frac{s^2}{2}\right)}$$

Means & Variances "Add" $\mu_Z = \mu_1 + \mu_2$; $\sigma_Z^2 = (\sigma_1^2 + \sigma_2^2)$ $\boxed{\phi_Z(s) = e^{\left(\mu_Z s + \sigma_Z^2 \frac{s^2}{2}\right)}}$

The moment generating function for a zero mean unit variance Gaussian denoted N(0,1) is obtained by completing the square, factoring out the positive exponential exp(s^2/2), and showing that the remaining integral is just the normalization integral which evaluates to unity. The moment generating function for the general case of an N(μ, σ^2) Gaussian is obtained using the transformation X = μ + σY and then taking the expectation E[e^{sX}]= E[$e^{s(\mu + \sigma Y)}$] = $e^{s\mu}$ E[$e^{(s \sigma)Y}$] to yield the result in the middle panel.

The PDF for the sum of two independent Gaussians Z=X$_1$+X$_2$ having different means and variances, viz., X$_1$ ~ N(μ_1, σ_1^2) and X$_2$ ~ N(μ_2, σ_2^2), is obtained by multiplying their individual generating functions to obtain the generating function for the sum variable as $\Phi_Z(s) = \Phi_{X1}(s) \cdot \Phi_{X2}(s)$. We find that the generator product expression for $\Phi_Z(s)$ is again that of a Gaussian with mean $\mu_{Z} = \mu_1 + \mu_2$ and variance $\sigma_Z^2 = \sigma_1^2 + \sigma_2^2$, thus proving that the means and variances just add to give the sum variable statistics; or, more concisely, N(μ_1, σ_1^2) + N(μ_2, σ_2^2) = N(μ_1+μ_2, σ_1^2+σ_2^2).

6.2 *Computations with Gaussian*

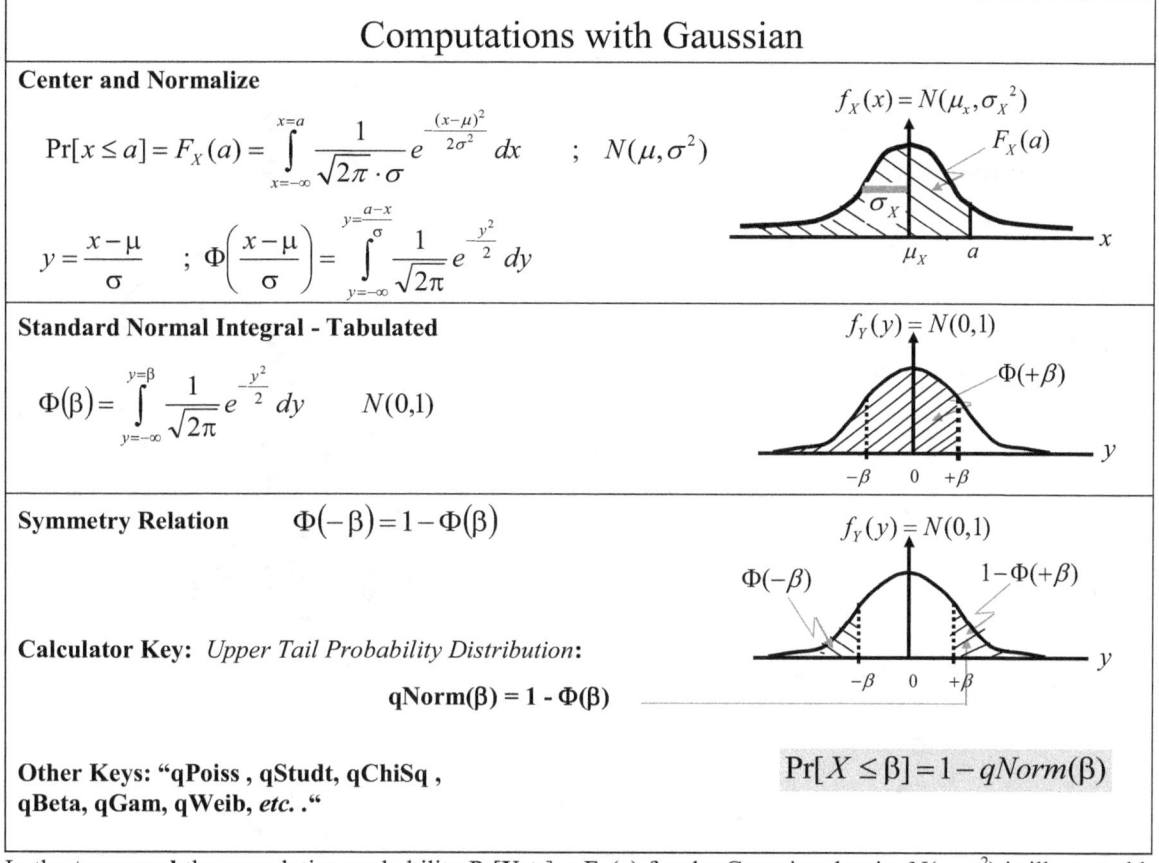

Computations with Gaussian

Center and Normalize

$$\Pr[x \le a] = F_X(a) = \int_{x=-\infty}^{x=a} \frac{1}{\sqrt{2\pi} \cdot \sigma} e^{-\frac{(x-\mu)^2}{2\sigma^2}} \, dx \quad ; \quad N(\mu, \sigma^2)$$

$$f_X(x) = N(\mu_x, \sigma_x^2)$$
$$F_X(a)$$

$$y = \frac{x-\mu}{\sigma} \quad ; \quad \Phi\left(\frac{x-\mu}{\sigma}\right) = \int_{y=-\infty}^{y=\frac{a-x}{\sigma}} \frac{1}{\sqrt{2\pi}} e^{-\frac{y^2}{2}} \, dy$$

Standard Normal Integral - Tabulated

$$\Phi(\beta) = \int_{y=-\infty}^{y=\beta} \frac{1}{\sqrt{2\pi}} e^{-\frac{y^2}{2}} \, dy \qquad N(0,1)$$

$$f_Y(y) = N(0,1)$$
$$\Phi(+\beta)$$

Symmetry Relation $\qquad \Phi(-\beta) = 1 - \Phi(\beta)$

$$f_Y(y) = N(0,1)$$
$$\Phi(-\beta) \qquad 1 - \Phi(+\beta)$$

Calculator Key: *Upper Tail Probability Distribution*:

$$qNorm(\beta) = 1 - \Phi(\beta)$$

$$\Pr[X \le \beta] = 1 - qNorm(\beta)$$

Other Keys: "qPoiss , qStudt, qChiSq , qBeta, qGam, qWeib, *etc.* ."

In the **top panel** the cumulative probability $\Pr[X \le a] = F_X(a)$ for the Gaussian density $N(\mu, \sigma^2)$ is illustrated by the hashed area under the curve from $x=-\infty$ to $x=a$; it is seen that the integral depends upon the values of both μ and σ so each new problem requires a separate numerical integration. To facilitate computations using the Gaussian distribution, the first step is to make the transformation $Y= (X-\mu)/\sigma$ in order to center the RV about its mean and scale it by its standard deviation σ. This new RV "Y" is usually referred to as the "standardized (Gaussian) variable" and designated $Y \sim N(0,1)$, because it is easily shown to have mean 0 and variance 1.

The hashed area under the Gaussian density in the **center panel** represents the probability that the standardized variable Y is less than or equal to β; it is the integral of an $N(0,1)$ density function from $-\infty$ to $+\beta$ and written

$$\Pr[Y \le \beta] = F_Y(Y=\beta) = \Phi(\beta)$$

The cumulative distribution for this standardized RV "Y" is tabulated as $\Phi(\beta)$ and its complement $1-\Phi(\beta)$ also appears as an **Upper Tail Probability** function (UTP) or "q-function" on scientific calculators and in standard computer math libraries. Because of the symmetry of the Gaussian about the origin, the areas in the left and right tail regions must be equal; the **bottom panel** illustrates this equality between the hashed tail areas on the right $1-\Phi(\beta)$ and on the left $\Phi(-\beta)$, and hence the relation

$$\Phi(-\beta) = 1 - \Phi(+\beta);$$

The complementary nature of the calculator tail function qNorm and $\Phi(\beta)$ yields two useful relations

$$qNorm(\beta) = 1 - \Phi(\beta) \qquad \text{and} \qquad \Pr[X \le \beta] = 1 - qNorm(\beta)$$

Note that there are also built-in functions for other distributions such as qPoiss (Poisson), qStudt (Student), qChiSq (Chi Square), qBeta (Beta), qGam (Gamma), qWeib (Weibull), *etc.* .

The Gaussian Process

6.2.1 Normalized Gaussian Computations

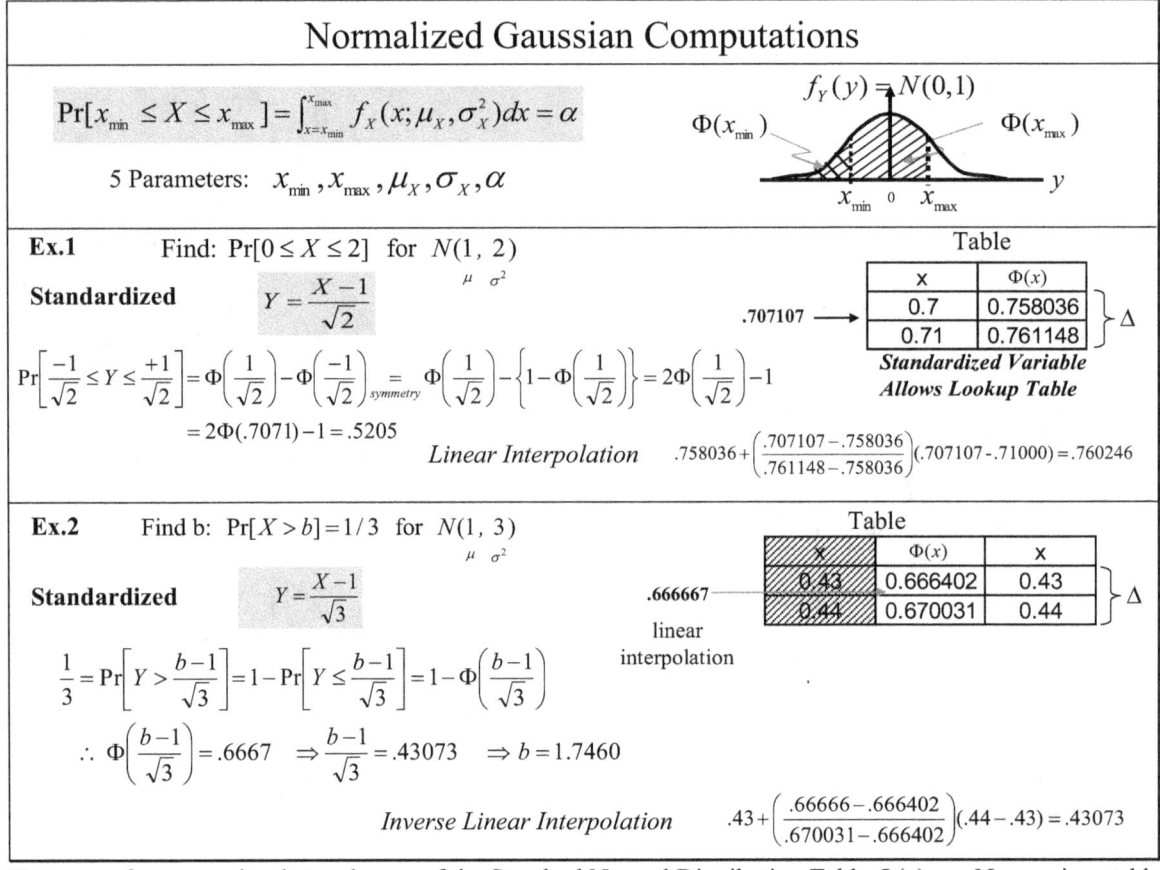

Here are a few examples that make use of the Standard Normal Distribution Table $\Phi(x)$ or qNorm; since tables may not have the particular value specified in the problem, they require linear interpolation between tabulated values as illustrated in these examples. The top panel shows the general "set up" for Gaussian problems which involves the 5 parameters x_{min}, x_{max}, p, μ_X, σ_X^2, and is similar to the discrete case (see Slide#4-6).

Example 1 asks us to find the probability that the RV X is between 0 and 2 for a normal distribution with mean 1 and variance 2. That is, find $Pr[0 \le X \le 2]$ for a N(1,2) distribution. The first step is to recognize that $\mu_X = 1$ and $\sigma_X = (2)^{1/2}$ (not 2) so that the standardized RV is Y= $(X-1)/(2)^{1/2}$; setting X=0 and then X=2 in the transformation gives the corresponding Y-values and transforms $Pr[0 \le X \le 2]$ to

$$Pr[(0-1)/(2)^{1/2} \le Y \le (2-1)/(2)^{1/2}] = \Phi(1/(2)^{1/2}) - \Phi(-1/(2)^{1/2}) = qNorm(-1/(2)^{1/2}) - qNorm(1/(2)^{1/2})$$

The qNorm functions may be input directly on your calculator to give the answer .5205. Otherwise the standard table must be interpolated between the two tabulated values of x= 0.70 and 0.71 to get the value for .707107 to obtain an accurate result.

Example 2 asks us to find value that the RV X must exceed for the probability to be 1/3 given a normal distribution with mean 1 and variance 3. That is, find "b" such that $Pr[X > b] = 1/3$ for a N(1,3) distribution. The first step again is to recognize that $\mu_X = 1$ and $\sigma_X = (3)^{1/2}$ so that the standardized RV is Y=$(X-1)/(3)^{1/2}$; setting X=b in the transformation gives the corresponding Y-value and transforms the expression $Pr[X > b] = 1/3$ to

$$1/3 = Pr[Y > (b-1) / (3)^{1/2}] = 1 - Pr[Y \le (b-1) / (3)^{1/2}] = 1 - \Phi((b-1) / (3)^{1/2})$$

which leads to $\Phi((b-1)/(3)^{1/2}) = .66667$, or equivalently $qNorm((b-1) / (3)^{1/2} = .3333$. Either equation can be solved by trial and error iteration on a calculator. Alternately find table values bracketing .66667 in the $\Phi(x)$ column and (inverse) interpolate to find the argument x to be .43073; then equating it to $(b-1) / (3)^{1/2}$ yields b=1.7460.

6.2.2 Microchip Life Example - (Standardized Φ(x) Table)

Microchip Life Example - (Standardized Φ(x) Table)
Calculators Upper Tail Probability Distributions "q-functions"

Example Microchip Life: $N(\mu=200 \text{ hrs}, \sigma^2)$

If $P(175 \le T \le 225) = 0.9$ as shown in figure

Standardized Variable $Y = \dfrac{T-200}{\sigma}$; $Y: N(0,1)$

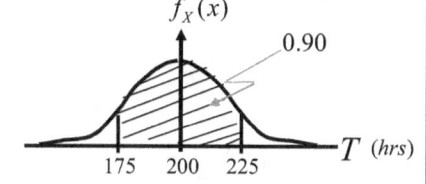

Computational Set-up for Table

$$P\left(\frac{-25}{\sigma} \le Y \le \frac{25}{\sigma}\right) = \Phi\left(\frac{25}{\sigma}\right) - \left\{1 - \Phi\left(\frac{25}{\sigma}\right)\right\}$$

$$= 2\Phi\left(\frac{25}{\sigma}\right) - 1 = 0.9 \implies \Phi\left(\frac{25}{\sigma}\right) = \frac{1.9}{2} = .95$$

Standard Normal Distribution Table

$$\Phi(x) = .95 \implies x = 1.64 + .01\left(\frac{.95 - .949497}{.950529 - .949497}\right) = 1.645$$

Gaussian Table		linear interpolation
x	Φ(x)	
1.64	0.949497	**0.95**
1.65	0.950529	

$$\frac{25}{\sigma} = \Phi^{-1}(0.95) = 1.645$$

$$\implies \sigma = 15.20$$

Calculator "Solve-for"

$$qNormal\left(\frac{25}{\sigma}\right) = 1 - .95 = .05$$

Trial & Error: $\sigma = 15.2$

Here is another example that makes use of the Standard Tables or the qNorm calculator function.

Example 3 considers a microchip whose lifetime is described by a normal distribution with mean of 200 hrs and an unknown variance σ^2. The question asks us to find the value of σ such that the microchips have a lifetime between 175 and 225 hrs with a probability of 90%.

That is, find Pr[175≤T≤ 225] (hashed area under the density curve) given the normal distribution $N(200, \sigma^2)$. The standardized RV is Y= (T-200)/σ and setting T=175 and then T=225 gives the corresponding Y-values and transforms Pr[175≤T ≤ 225] to

Pr[(175-200)/ σ ≤Y ≤ (225-200)/ σ] = Φ(25/ σ) - Φ (-25/ σ)= 2Φ(25/ σ) − 1 =.9

which leads to

Φ(25/ σ) = .95   or   qNorm(25/ σ) = .05

Either equation can be solved by trial and error iteration or by using the standardized table. The table is searched down col#2 (for the probability Φ) to find the two tabulated values .949497 and .950529 that bracket Φ = .95; "inverse interpolation" is then performed to find the corresponding value in col#1 for the argument 25/ σ = Φ⁻¹(.95) =1.645; thus σ = 15.20 is the desired standard deviation. The qNorm functions may be input directly on your calculator and placed in a "solver" application, or simple trial and error iterations will suffice to find an approximate solution σ = 15.20.

6.3 *Binary Communication Channel with Noise*

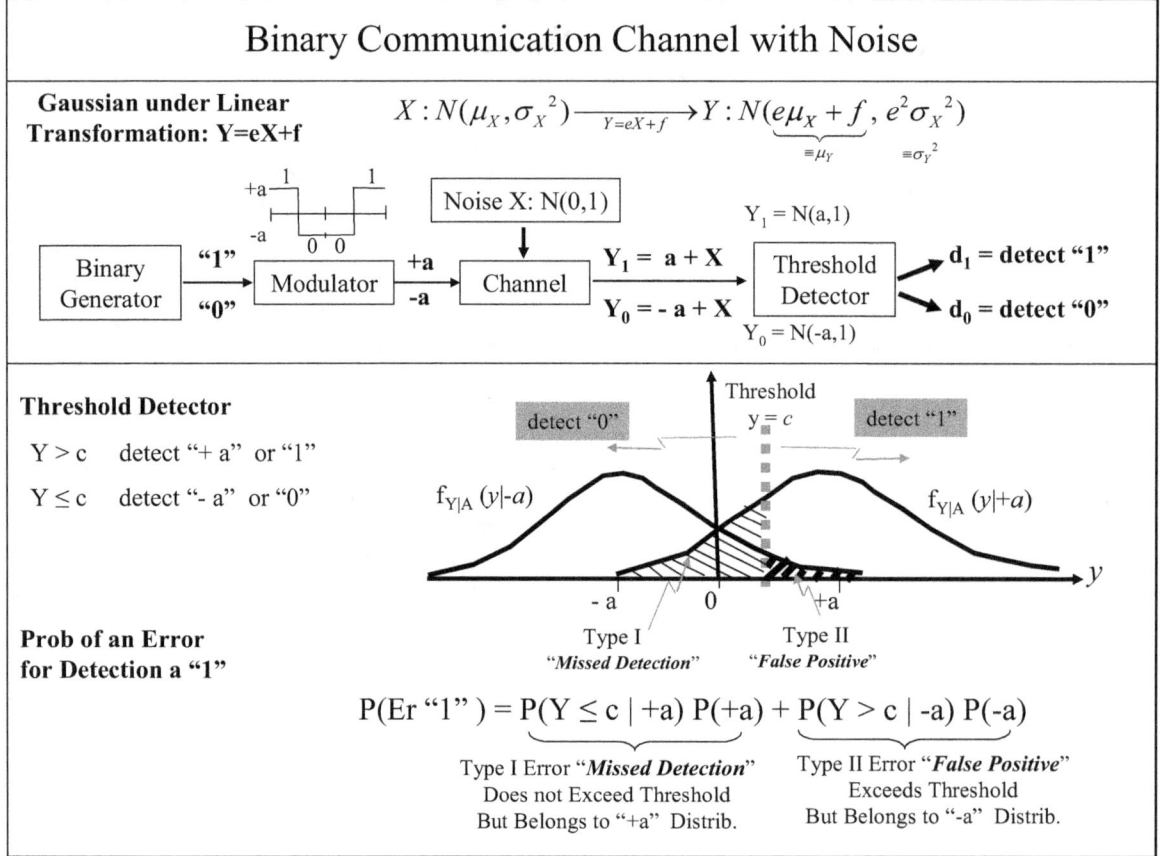

Binary Communication Channel with Noise

Gaussian under Linear Transformation: Y=eX+f

$$X : N(\mu_X, \sigma_X{}^2) \xrightarrow{Y=eX+f} Y : N(\underbrace{e\mu_X + f}_{\equiv \mu_Y}, \underbrace{e^2\sigma_X{}^2}_{\equiv \sigma_Y{}^2})$$

Noise X: N(0,1)

Binary Generator — "1" / "0" — Modulator — +a / -a — Channel

$Y_1 = a + X$ $Y_1 = N(a,1)$
$Y_0 = -a + X$ $Y_0 = N(-a,1)$

Threshold Detector → d_1 = detect "1" / d_0 = detect "0"

Threshold Detector

Y > c   detect "+ a" or "1"

Y ≤ c   detect "- a" or "0"

detect "0" | Threshold $y = c$ | detect "1"

$f_{Y|A}(y|-a)$ $f_{Y|A}(y|+a)$

- a   0   +a

Type I   Type II
"Missed Detection"   *"False Positive"*

Prob of an Error for Detection a "1"

$$P(\text{Er "1"}) = \underbrace{P(Y \le c \mid +a)\, P(+a)}_{} + \underbrace{P(Y > c \mid -a)\, P(-a)}_{}$$

Type I Error *"Missed Detection"*
Does not Exceed Threshold
But Belongs to "+a" Distrib.

Type II Error *"False Positive"*
Exceeds Threshold
But Belongs to "-a" Distrib.

Consider the Binary communication channel depicted in the upper sketch: A binary sequence of "1"s and "0"s is generated and then amplitude modulated by a positive amplitude +a for "1" and –a for "0" as illustrated by the "square wave pulse train" at the modulator. Zero mean unit variance Gaussian noise N(0,1) is added by the "channel" and the (signal + noise) outputs are given by *two distinct* Gaussian RVs: $Y_1 = a + X \sim N(+a, 1)$ and $Y_0 = -a + X \sim N(-a,1)$ about two different means as shown in the probability density plot. This output is presented to a threshold detector which attempts to detect the original sequence of "1"s and "0"s by setting a threshold y = c (vertical dashed line) and then assigning a "1" for Y-values to its right and a "0" for Y-values to its left.

Considering the detection of a digital "1" we see that two types of error can occur as follows:

Type I Missed Detection: P(Y≤c | +a) The larger hatched area on the left with Y<c which belongs to the digital "1" N(+a,1) curve but is rejected because it does not exceed the threshold "c"

Type II False Positive: P(Y>c | -a) The smaller hatched area on the right with Y>c which belongs to the digital "0" N(-a,1) curve, but is falsely detected as a "1" because it exceeds the threshold "c"

The total probability for an error in detecting a "1" is the sum of each conditional multiplied by its *a priori* as shown in the bottom equation. The total probability for an error in detecting a "0" is written down in an analogous fashion as a sum of conditionals multiplied by their *a priori* s (not shown).

6.3.1 Error Probability for Digital "1" - Detailed Break-out

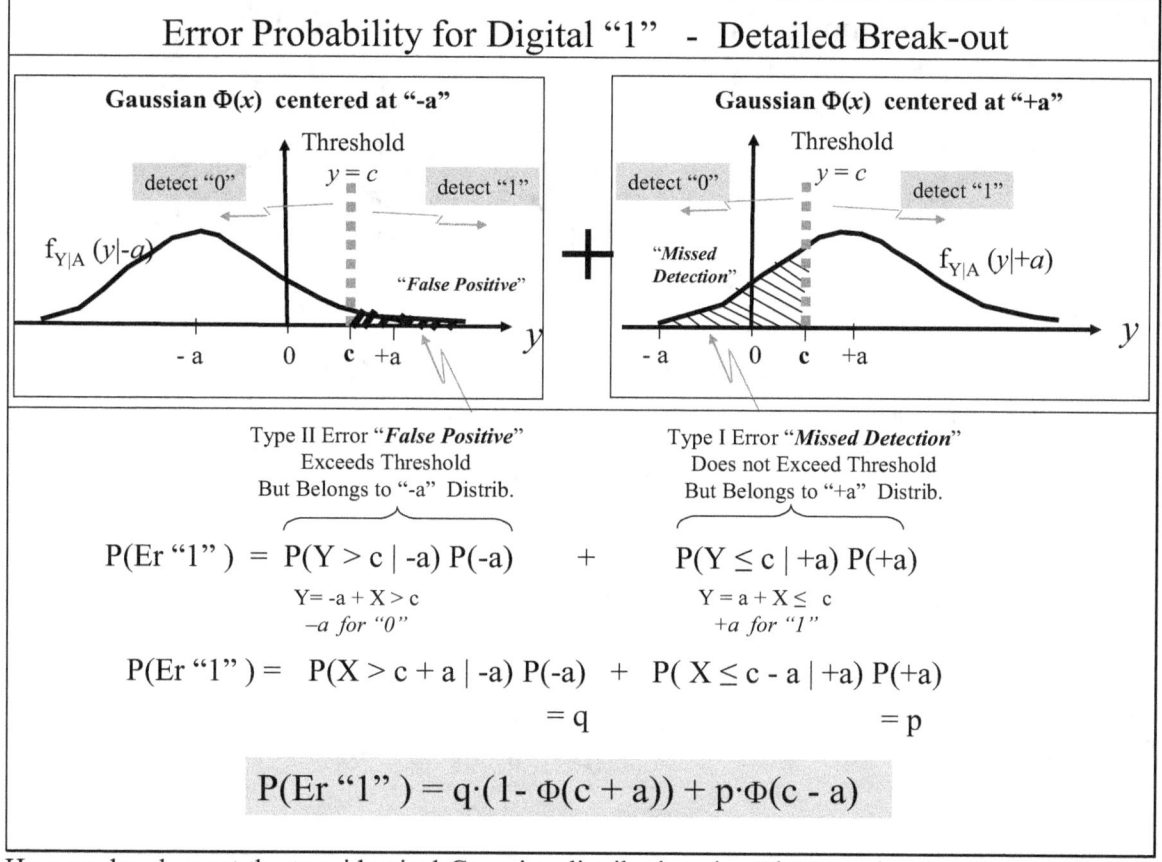

Here we break apart the two identical Gaussian distributions in order to make clear the contributions that each one makes to the total error in detecting a digital "1". It also allows us to more conveniently evaluate the conditional probability terms by expressing them in terms of a standardized $N(0,1)$ Gaussian tabular function $\Phi(y)$.

(i) Left Gaussian for digital "0" has amplitude $-a$, and the signal is $Y_0 = -a + X$. An error occurs when it exceeds the threshold $Y_0 = -a + X > c$ and is falsely detected as a digital "1"; re-expressing this inequality in terms of the standardized $N(0,1)$ noise X, we have $X > c + a$. The conditional probability $P(Y>c|-a)$ is equivalent to $P(Y_0>c)$ because the subscript "0" already specifies correct Gaussian centered at "$-a$". Thus, we may write the *false detection error* for the signal Y_0 in terms of $\Phi(\cdot)$ as

$$P(Y_0>c) = P(X>c+a) = 1-P(X \le c+a) = 1-\Phi(c+a)$$

(ii) Right Gaussian for digital "1" has amplitude $+a$, and the signal is $Y_1 = a + X$. An error occurs when it does not exceed the threshold $Y_1 = a + X \le c$ and becomes a "missed detection" of digital "1"; in terms of the standardized $N(0,1)$ noise X this inequality becomes $X \le c-a$. The conditional probability $P(Y \le c|+a) = P(Y_1 \le c)$ because the subscript "1" already specifies correct Gaussian centered at "$+a$". Thus we may write the *missed detection error* for the signal Y_1 in terms of $\Phi(\cdot)$ as

$$P(Y_1 \le c) = P(X \le c-a) = \Phi(c-a).$$

Assuming a *priori* probabilities for "1"s $P(+a) = p$ and "0"s $P(-a) = q$, the final result is

$$\text{Prob[Err. for "1"]} = (1-\Phi(c+a)) \cdot q + \Phi(c-a) \cdot p \,,$$

where $\Phi(x)$ is the Standard Normal function.

6.3.2 Communication Channel Threshold Trade-offs

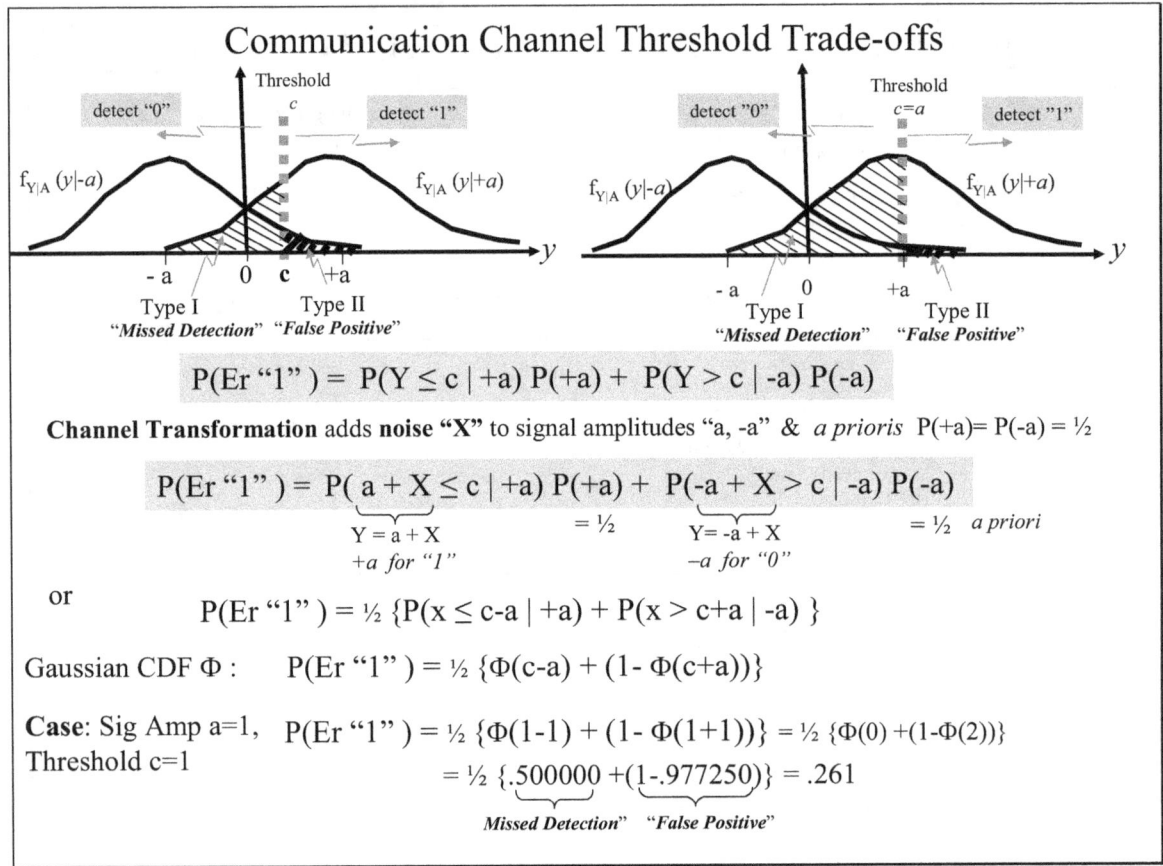

Communication Channel Threshold Trade-offs

$$P(Er\ "1") = P(Y \leq c\ |+a)\ P(+a) + P(Y > c\ |\ -a)\ P(-a)$$

Channel Transformation adds **noise "X"** to signal amplitudes "a, -a" & *a prioris* $P(+a)= P(-a) = \frac{1}{2}$

$$P(Er\ "1") = P(\underbrace{a + X}_{\substack{Y = a + X \\ +a\ for\ "1"}} \leq c\ |+a)\ \underbrace{P(+a)}_{= \frac{1}{2}} + P(\underbrace{-a + X}_{\substack{Y = -a + X \\ -a\ for\ "0"}} > c\ |\ -a)\ \underbrace{P(-a)}_{= \frac{1}{2}\ a\ priori}$$

or

$$P(Er\ "1") = \frac{1}{2}\ \{P(x \leq c-a\ |+a) + P(x > c+a\ |\ -a)\ \}$$

Gaussian CDF Φ : $\quad P(Er\ "1") = \frac{1}{2}\ \{\Phi(c-a) + (1-\Phi(c+a))\}$

Case: Sig Amp a=1, $P(Er\ "1") = \frac{1}{2}\ \{\Phi(1-1) + (1-\Phi(1+1))\} = \frac{1}{2}\ \{\Phi(0) + (1-\Phi(2))\}$
Threshold c=1
$$= \frac{1}{2}\ \{\underbrace{.500000}_{} + \underbrace{(1-.977250)}_{}\} = .261$$
$$\textit{Missed Detection"}\quad \textit{"False Positive"}$$

Continuing with the communication channel, we investigate the effects of changing the threshold value "c" and the modulation amplitude "a". Now, assuming p=q=1/2, the probability for an error in detecting "1" is written down in terms of the Gaussian CDF Φ as

$$P(Er\ "1") = \frac{1}{2}\ \{\Phi\ (c-a) + 1-\Phi(c+a)\}$$

Setting a=1 and c=1 we find P(Er "1") = .261. This choice c=a=1 sets the modulation amplitude to unity and also shifts the threshold c to the mean location y=+a =1 of the "1" distribution. It is clear from the figure that shifting the threshold to the center of the "1" distribution has the effect of "throwing away" half of the "1"s; when this is multiplied by the *a priori* of ½ for a "1" we generate an error of ½.½ = ¼ = .25 for the Type I "missed detections". An additional small error results from the small area under the "0" curve corresponding to Type II "false positives"; this small probability is added to the missed detections to yield the total error of .261.

Clearly trade-offs between separation of the two Gaussians parameterized by the modulation amplitude "a" and the choice of threshold value y=c must be made in order to minimize the total error to "1"s and to "0"s. Given the symmetry of this problem with respect to "1"s and "0"s and identical distributions, the optimal choice for the threshold is clearly in the exact center or at Y=0.

In general, conditions are never this ideal and the introduction of asymmetries such as differences in the *a priori* probabilities p and q for and the "1"s and "0"s and/or differences in the mean and variance of the two Gaussians (or entirely different distributions) would make the trade-off between the various parameters more necessary and more complex.

7 Properties of Common Distributions

7.1 Common Continuous PDFs and Properties

Common Continuous PDFs and Properties

RV Name	PDF	Generating Fcn $\varphi(s)=E[e^{Xs}]$	Mean $\int\limits_{x=-\infty}^{\infty}x\cdot f_X(x)dx$	Variance $\text{var}(X)=E[X^2]-E[X]^2$
Uniform	$f_X(x)=\begin{cases}\dfrac{1}{b-a} & a\le x\le b\\ 0 & Otherwise\end{cases}$	$\dfrac{e^{sb}-e^{sa}}{s(b-a)}$	$\dfrac{a+b}{2}$	$\dfrac{(b-a)^2}{12}$
Exponential	$f_T(t)=\begin{cases}\lambda e^{-\lambda t} & t\ge 0\\ 0 & t<0\end{cases}$ $\lambda>0$	$\dfrac{\lambda}{\lambda-s}$	$\dfrac{1}{\lambda}$ "exponential wait"	$\dfrac{1}{\lambda^2}$
Gamma r-Erlang r = integer	$f_{T_r}(t)=\begin{cases}\dfrac{\lambda e^{-\lambda t}(\lambda t)^{r-1}}{(r-1)!} & t\ge 0\\ 0 & t<0\end{cases}$ $\lambda>0$ Arrival Rate Peaks at $t_{max}=\dfrac{r-1}{\lambda}$	$\left(\dfrac{\lambda}{\lambda-s}\right)^r$	$\dfrac{r}{\lambda}$ For r=3: three "exponential waits" $E[T_3]=\frac{1}{\lambda}+\frac{1}{\lambda}+\frac{1}{\lambda}$	$\dfrac{r}{\lambda^2}$
Normal $N(\mu,\sigma^2)$	$f_X(x)=\dfrac{1}{\sqrt{2\pi}\cdot\sigma}e^{-\frac{(x-\mu)^2}{2\sigma^2}}$ $-\infty<x<\infty$	$e^{\mu s+\frac{(\sigma s)^2}{2}}$	μ	σ^2
Rayleigh	$f_X(x)=a^2xe^{-\frac{a^2x^2}{2}}$ $x>0\ ;\ a>0$	$1+\left(\frac{s}{a}\right)e^{-\frac{(s/a)^2}{2}}\sqrt{\frac{\pi}{2}}$ $\cdot\left[1+erf\left(\frac{(s/a)}{\sqrt{2}}\right)\right]$	$\dfrac{1}{a}\sqrt{\dfrac{\pi}{2}}$	$\dfrac{2-\pi}{2a^2}$

This table gives important properties of some common continuous distributions listed in col#1; across the table are the distribution formula and a typical plot, the generating function, and formulas for the mean and variance.

The Uniform Distribution has a constant magnitude 1/(b-a) over the interval [a,b]; the mean is at the center of the distribution (a+b)/2 and the variance is $(b-a)^2/12$.

The Exponential Distribution decays exponentially with time from an initial probability density λ at t=0. The mean time for an arrival is $E[T]=1/\lambda$ which equals the e-folding time of the exponential; its variance is $1/\lambda^2$. For a Poisson process, the probability that the 1st arrival occurs after time t, $P(T_1>t)$ equals the probability that there have been *no arrivals* up to time t, P(K(t)=0), or mathematically, $\Pr(T_1>t)=\Pr(K(t)=0)$.

The r-Erlang / Gamma Distributions for r>1, all rise from zero to reach a maximum at $(r-1)/\lambda$ and then decay almost exponentially $\sim t^{r-1}e^{-\lambda t}$ to zero. The maximum occurs after a wait of one exponential mean wait time $1/\lambda$ for r=1, two $(1/\lambda)$-waits for r=2, and "r" $(1/\lambda)$-waits for any r. The variance is r times that of the exponential variance $1/\lambda^2$. The Gamma density is a generalization of the r- Erlang density obtained by replacing (r-1)! with $\Gamma(r)$ making it valid for non-integer values of r. For a Poisson process, the probability that the rth arrival occurs after time t, $P(T_r>t)$ equals the probability that there have been *fewer than* r arrivals up to time t, $P(K(t)\le r-1)$, or mathematically, $\Pr(T_r>t)=\Pr(K(t)\le r-1)$.

The Gaussian (Normal) Distribution is the most universal distribution in the sense of that the Central limit theorem requires that sums of many IID RVs approach the Gaussian distribution. The Gaussian is defined for all x ε (-∞, ∞) and is symmetric about the origin (x=μ=0 illustrated) where it peaks at a value of $1/(\sigma\sqrt{2\pi})$.

The Rayleigh Distribution results from the product of two independent Gaussians when expressed in polar coordinates and integrated over the angular coordinate. The probability density is zero at x=0 and peaks at x=1/a before it drops towards zero with a "Gaussian-like" shape for x>0. The Rayleigh and Gaussian are plotted together to emphasize their similarities and differences.

7.2 *Common Discrete PMFs and Properties -1*

Common Discrete PMFs and Properties -1

RV Name	PMF	Mean $E[X] = \sum_{x=0,1} x \cdot p_X(x)$	Variance $\mathrm{var}(X) = E[X^2] - E[X]^2$
Bernoulli 1-Trial X=x succ. "0" or "1" successes	**"Atomic" RV** $p_X(x) = \begin{cases} p & X=1 \text{ (success)} \\ 1-p=q & X=0 \text{ (failure)} \end{cases}$ 	$E[X] = 0 \cdot (1-p) + 1 \cdot p$ $= p$	$E[X^2] = 0^2 \cdot (1-p) + 1^2 \cdot p$ $= p$ $\mathrm{var}(X) = p - p^2 = p(1-p)$ $= pq$
Binomial n - Trials X=x Succ. How many succ "x" in "n" trials ?	**n Independent Bernoulli Trials** $p_X(x) = \binom{n}{x} p^x q^{n-x}$ $x = 0,1,\cdots n$ 	$E[X] = \sum_{x=0}^{n} x \binom{n}{x} p^x q^{n-x}$ $= np$	$\mathrm{var}(X) = npq$
Geometric X=x Trials 1- Success How many trials "x" for "1" succ	**One Sequence** $p_X(x) = \begin{cases} pq^{x-1} & x = 1,2,\cdots \\ 0 & \text{(otherwise)} \end{cases}$ Geom RV = Neg Binom for r=1 succ. 	$E[X] = \sum_{x=1}^{\infty} x \cdot pq^{x-1} = p\frac{d}{dq}\sum_{x=1}^{\infty} q^x$ $= p\frac{d}{dq}\left(\frac{1}{1-q}\right) = \frac{+p}{(1-q)^2} = \frac{1}{p}$ As p decr. Expected num. trials "x" for 1-succ must incr.	$\mathrm{var}(X) = \frac{q}{p^2}$
Negative Binomial X=x Trials r- Successes	**Many Sequences** $p_X(x) = \underbrace{\binom{x-1}{r-1} p^{r-1} q^{x-r}}_{(r-1)\text{succ. in } (x-1)\text{ trials}} \cdot \underbrace{p}_{\substack{\text{succ. on} \\ \text{next trial}}}$ $x = r, (r+1), (r+2), \cdots \infty$ 	$E[X] = \sum_{x=r}^{\infty} x \cdot \binom{x-1}{r-1} p^r q^{x-r} = \frac{r}{p}$ As p decr. Expected num. trials "x" for r-succ must incr.	$\mathrm{var}(X) = r \cdot \frac{q}{p^2}$

This table and one to follow compare some common discrete probability distributions, explore their fundamental properties, and relate them to one another. In the table, a brief description is given under the "RV Name" column followed by the PMF formula and plot in col#2; formulas for the mean and variance are shown in the last two columns.

The Bernoulli RV X answers the question "what is the result of a single Bernoulli trial?" It takes on only two values, namely "1"=Success with probability p and "0"=Fail with probability q=1-p.

The Binomial RV "X" answers the question "how many successes X, are there in n Bernoulli trials?" It takes on values corresponding to the number of successes "X" in "n" independent Bernoulli trials; the sum variable, $X=X_1+ X_2+ \ldots +X_n$, gives the #successes by adding a 1 to the sum for each success only. The PMF $^nC_x\, p^x\, q^{n-x}$, is the binomial term whose coefficient is the #ways to have X=x successes in n trials; the single trial probability of success p enters as the product $p^x(1-p)^{n-x}$. A tree gives a nice visualization by explicitly showing the #tree-paths, ^nC_x, that yield X=x successes at the tree output.

The Geometric RV X answers the question "how many Bernoulli trials X for 1 success?" It takes on values *from 1 to ∞* and is the result of n-1 *failed Bernoulli trials* followed by one success; the sum of n Bernoulli RVs, $X=X_1+ X_2+ \ldots +X_n$ gives just "1" success and the tree has but a single path with X= x trials and only 1-success yields the geometric PMF $q^{x-1}p^1$.

The Negative Binomial RV X answers the question "how many Bernoulli trials X for r-successes?" It takes on values *from r to ∞* and is the *sum of* r *geometric RVs* N_k with probability $p^{r-1}q^{nk-r}p^1$ that each count the #trials $N_k=n_k$ for its one success; the sum of r geometric RVs, $X=N_1+ N_2+ \ldots + N_r$ gives "r" successes and the tree has $^{x-1}C_{r-1}$ paths for X=x-1 trials yielding (r-1)-successes, followed by one final success to yield the negative binomial PMF $^{x-1}C_{r-1}\, p^{r-1}q^{x-r}\cdot p^1 = {}^{x-1}C_{r-1}\, p^r q^{x-r}$ with x = r, r+1, ... ∞.

7.2.1 Common Discrete PMFs and Properties-2

Common Discrete PMFs and Properties-2

RV Name	PMF	Mean $E[X] = \sum_{x=0,1} x \cdot p_X(x)$	Variance $var(X) = E[X^2] - E[X]^2$
Hyper-geometric *X=x -succ* *N= fixed pop* *m= tagged* *n=test sampl* *w/o rplcemt*	$p_X(x) = \begin{cases} \dfrac{\overset{\text{x from}}{\overset{\text{"m-marked"}}{\binom{m}{x}}} \overset{\text{(n-x) from}}{\overset{\text{"(N-m)= unmarked"}}{\binom{N-m}{n-x}}}}{\binom{N}{n}} ; & x_{min} \le x \le x_{max} \\ \\ 0 ; & Otherwise \end{cases}$ $m,n \in [1,N]$; $\max(0, N-m-n) \le x \le \min(m,n)$ PMF Derives from Binomial Identity $n \le m \le N$	$E[X] = n \cdot \dfrac{m}{N} = n \cdot p$ where $p = m/N$ is the "initial" probability of drawing a marked item	$var(X) = \dfrac{(N-n)}{(N-1)} \cdot n \cdot \dfrac{m}{N} \cdot \dfrac{(N-m)}{N}$ $var(X) = \dfrac{(N-n)}{(N-1)} \cdot n \cdot p \cdot q$
	$\binom{N}{n} = \binom{m+(N-m)}{n} = \binom{m}{0}\binom{N-m}{n} + \binom{m}{1}\binom{N-m}{n-1} + \cdots + \boxed{\binom{m}{x}\binom{N-m}{n-x}} + \cdots + \binom{m}{n}\binom{N-m}{0}$		
Poisson *#Trials* $\to\infty$ *X=x Succ.*	$p_X(x) = \begin{cases} \dfrac{(a^x / x!)}{e^a} & x = 0,1,2,\cdots \infty \\ \\ 0 & Otherwise \end{cases}$ Limit of Binomial $a = \lim_{\substack{n\to\infty \\ p\to 0}} (n \cdot p) = \lambda \cdot t$ = (aver. arrival rate)*time	$E[X] = a$	$var(X) = a$
Zeta(Zipf) *#Trials* $\to\infty$ *X=x Succ.*	**Riemann Zeta** $\zeta(s) = \sum_{x=1}^{\infty} 1/x^s$ $\boxed{s>1}$ $p_X(x;s) = \begin{cases} \dfrac{1/x^s}{\zeta(s)} = \dfrac{"\zeta - term"}{\zeta(s)} & x = 1,2,\cdots \\ \\ 0 & (otherwise) \end{cases}$	$E[X;s] = \dfrac{\zeta(s-1)}{\zeta(s)}$ $E[X;3.5] = \dfrac{\zeta(2.5)}{\zeta(3.5)} = 1.191$	$Var(X;s) = \dfrac{\zeta(s-2)}{\zeta(s)} - \left(\dfrac{\zeta(s-1)}{\zeta(s)}\right)^2$ $Var(X;3.5) = \dfrac{\zeta(1.5)}{\zeta(3.5)} - \left(\dfrac{\zeta(2.5)}{\zeta(3.5)}\right)^2$ $= .856$

Common Discrete PMFs given here are: Hypergeometric, Poisson, and Riemann Zeta (or Zipf).
The Hypergeometric RV "X" answers the question "how many successes (*defectives*) x are found in n test samples randomly selected from a production run containing m defective and N-m working items?" The Bernoulli trials here are **dependent** because the selections are made *without replacement*; the distribution is best understood in terms of the binomial identity (Discrete-Slide#2-22)
$$^NC_n = {}^mC_0 {}^{N-m}C_n + \ldots + {}^mC_x {}^{N-m}C_{n-x} + \ldots + {}^mC_m {}^{N-m}C_{n-m}$$
Dividing by ^NC_n yields a sum of terms equal to unity and thus defines the distribution term $^mC_x {}^{N-m}C_{n-x} / {}^NC_n$ where x only takes on values in the restricted range x=[x_{min}, x_{max}], with x_{min}=N-n-m and x_{max}= min(n,m). The table gives the mean as μ_X = np and the variance as var(X) = npq [(N-n)/(N-1)] where p = m/N is the 1st draw probability. Note that for large N these statistics are equivalent to those for a binomial b(x;n,p); but because the underlying Bernoulli trials are not independent the single trial probability of success p changes and there is a correction factor [(N-n)/(N-1)] for the Hypergeometric distribution.
The Poisson RV "X" answers the question "how many successes x in n trials for n very large?" This is also discussed as one aspect of the Poisson process on Slide#7-2 where it is related to the continuous r-Erlang distribution. It is important to understand that the Poisson distribution represents a limiting behavior of the binomial PMF as n$\to \infty$ and that it is given by the ratio of a single Taylor expansion term of e^a divided by e^a. viz., $p_X(x)$={ $a^x/ x!$} / e^a, for x=0,1,2,3,... The product a =$\lambda \cdot t$ defines the so-called Poisson parameter, where λ is the "rate of success" and t is the time interval over which data is taken. The Poisson RV has many applications in physics and engineering.
The Riemann Zeta RV "X" has applications to language processing and prime number theory; a few of its properties are given in the table. Note that the exponent must satisfy s >1 in order to avoid the harmonic series whose sum does not converge and therefore cannot represent an actual PMF.

7.2.2 Bernoulli/Binomial Tree Structures

Bernoulli/Binomial Tree Structures

RV Name	PMF	Tree Graph
Bernoulli 1-*Trial* X=x succ. "0" or "1" successes	$p_X(x) = \begin{cases} p & X=1 \text{ (success)} \\ 1-p=q & X=0 \text{ (failure)} \end{cases}$ "Atomic" RV	**(q+p)**
Binomial 2 - *Trials* X=x Succ. How many succ "x" in "2" trials ?	$p_X(x) = \binom{2}{x} p^x q^{2-x}$ $x = 0,1,2$ Independent Bernoulli Trials	**(q+p)²**
	$(q+p)^2 = q^2 + 2pq + p^2$ $= {}^2C_0\, p^0\, q^2 + {}^2C_1\, p^1\, q^1 + {}^2C_2\, p^2\, q^0$	

The RVs of Slide#7-3 are grouped in pairs {Bernoulli, Binomial} and {Geometric, Negative Binomial} for a reason. The sum of many independent Bernoulli trials generates a Binomial distribution and similarly the sum of many independent Geometric trials generates the Negative Binomial distribution. This slide and the next give a graphical construction of trees for these two groups of paired distributions by repeatedly applying the basic tree structure of the underlying Bernoulli or Geometric tree structure to certain output nodes of the previous stage.

In the **first panel** we show the PMF properties for Bernoulli on the left and on the right we display the Bernoulli tree structure where the upper branch q=Pr[Fail] goes to the state X= 0 and the lower branch p = Pr[Success] goes to the state X= 1.

In the **second panel** we show the PMF properties for a simple n=2 Binomial. The corresponding tree structure for this Binomial is obtained by appending a second Bernoulli tree to each output node of the first trial, thus yielding the 4 output states {{FF}, {FS}, {SF}, {SS}} shown in the **third panel**. We see that there are ${}^2C_0$ tree paths leading to {FF} p^0q^2 , ${}^2C_1$ tree paths leading to {FS} p^1q^1 , and ${}^2C_2$ tree paths leading to {SS} p^2q^0, which is precisely as expected from the Binomial PMF for n=2. This can be continued for n=3, 4, ... by repeatedly appending a binary Bernoulli tree to each new node.

Furthermore, we see that this structure for n=2 is represented algebraically by $(q+p)^2$ since a direct expansion gives $1= q^2 + 2q^1p^1 + p^2$; expanding an expression corresponding to n Bernoulli trials $(q+p)^n$ obviously yields the appropriate Binomial expansion for the exponent n. Thus the Binomial is represented by the repetitive tree structure or by the repeated multiplication of the algebraic structure $1=(q+p)$ by itself n-times to obtain $1^n=(q+p)^n$.

7.2.3 Geometric/NegBinomial Tree Structures

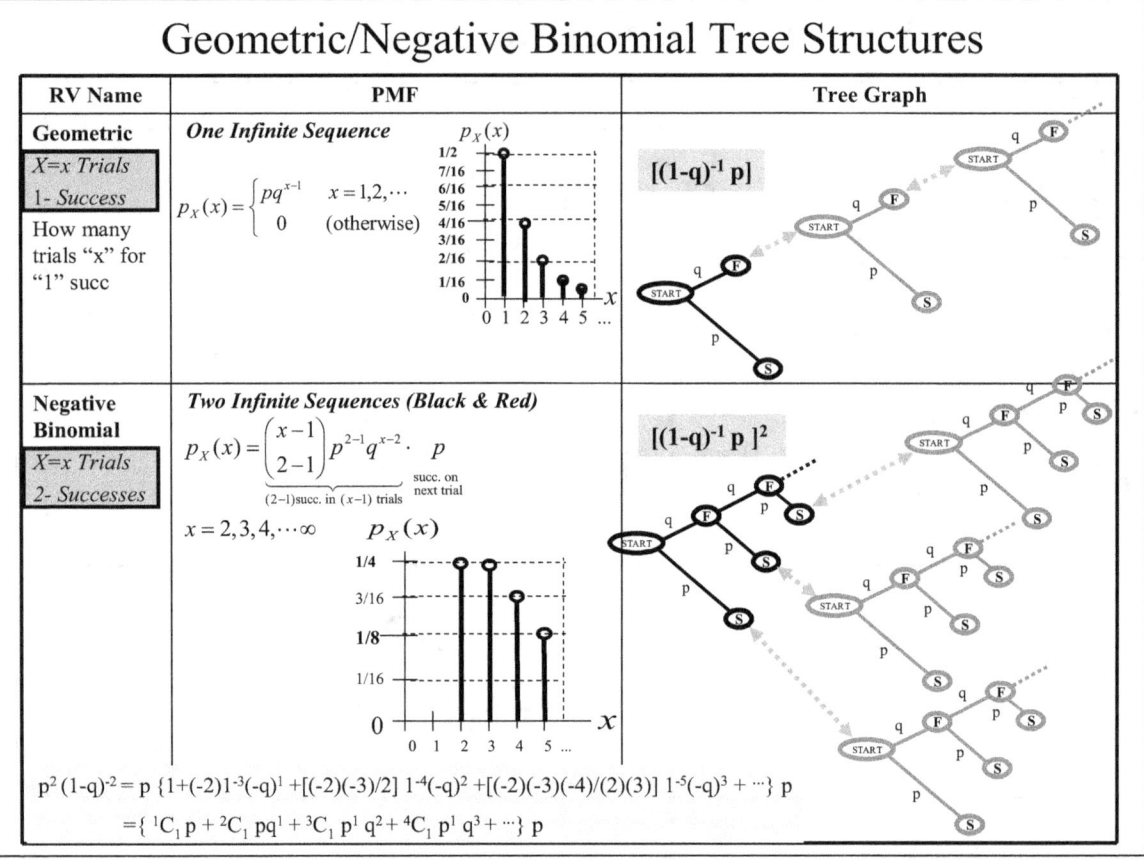

This slide first gives a graphical construction of a Geometric tree from an infinite number of Bernoulli trials and then shows how the Negative Binomial tree is the result of appending a Geometric tree to itself in a manner similar to that of the last slide. In the **first panel** we repeat the PMF properties for Geometric RV. On the right side of this panel we display Geometric tree structure whose branches end in a single success. This tree has a Bernoulli trial appended to *each failure node only* and is constructed from an infinite number of Bernoulli trials. The 1st Bernoulli trial yields X=1 with p=Pr[Success] and this ends the lower branch; its upper branch yields X=0 with q=Pr{Fail}; this failure node spawns a 2nd Bernoulli trial which again leads to X=1 or X=0; this process continues indefinitely. It accurately describes the probabilities for a single success in 1, 2, 3,...,∞ #trials and is algebraically represented by the expression $1=[(1-q)^{-1}p]$ which expands to give an infinite sequence $[1 + q^1 + q^2 + q^3 +....].p$, corresponding to exactly 0, 1, 2, 3,..."failures before a single success."

In the **second panel** we show the PMF properties for an r=2 Negative Binomial; on the right we display the Negative Binomial tree structure obtained by applying the basic Geometric tree only to each S-node (infinite number) corresponding to a 1st success. This leads to a doubly infinite tree structure for the r=2 Negative Binomial representing the number of trials X =x required for r=2 successes. We can verify the first few terms in the Negative binomial expansion given under PMF in the lower panel using the tree.

This process may be extended to r=3, 4, ... successes by repeatedly applying the Geometric tree to each success node. For n=2, direct expansion of the algebraic identity yields

$$1^2 = [(1-q)^{-1}p]^2 = \{\ ^1C_1\,p + {}^2C_1\,pq^1 + {}^3C_1\,p^1\,q^2 + {}^4C_1\,p^1\,q^3 + \cdots\}p,$$

which is in agreement with the n=2 Negative Binomial terms given in the table. In an analogous fashion, expansion of the algebraic generator $1^r = [(1-q)^{-1}p]^r$ for exponent r yields the correct result for the general Negative Binomial. Note the "Negative" modifier to Binomial is a natural designation in view of the $(1-q)^{-1}$ term in its algebraic structure.

8 Approximations and Bounds on Probability Distributions

8.1 Central Limit Theorem (CLT)

Central Limit Theorem (CLT)

General CLT Theorem

Given: RVs $X_1, X_2, X_3, \ldots$ Not necessarily *independent identically distributed* (IID)
Then:

$\quad$ SUM: $\quad Z = X_1 + X_2 + X_3 + \cdots \quad f_Z(z) \xrightarrow[n \to \infty]{} $ Gaussian $\qquad$ *Provided no term is dominant in the sum*

$\quad$ PROD: $Z = X_1 \cdot X_2 \cdot X_3 \cdots$

$\quad$ sum(logs): $\quad \log Z = \log X_1 + \log X_2 + \log X_3 + \cdots$

$\qquad\qquad f_Z(\log z) \xrightarrow[n \to \infty]{} $ Gaussian "log - normal" $\qquad$ *Provided no term is dominant in the product*

Restrictive CLT Theorem

Given: RVs $\{X_i\} \quad$ i=1,2,...,n $\quad$ *independent identically distributed* (IID)

$\quad$ finite means : $E[X_i] < k < \infty$

$\quad$ finite variances : $\mathrm{var}(X_i) = \sigma_{X_i}^2 < \infty$

Then: Sum Variable: $Z = \sum_{i=1}^{n} X_i \qquad$ Normalized Centered Variable $\qquad Y = \dfrac{Z - \mu_Z}{\sigma_Z} = \dfrac{Z - n\mu_X}{\sqrt{n} \cdot \sigma_X}$

$\quad E[Z] = nE[X] = n \cdot \mu_X$

$\quad \mathrm{var}(Z) = n \; \mathrm{var}(X) = n\sigma_X^2 \qquad$ **Standardized Gaussian Distribution** $\qquad f_Y(y) \xrightarrow[n \to \infty]{} \dfrac{1}{\sqrt{2\pi}} e^{-\frac{y^2}{2}}$

It is easily seen that the sum $Z = X_1 + X_2$ of two uniform RVs on [-.5, .5] yields a new RV Z with a "triangular density" over [-1, 1]. (The new range for Z clearly has a minimum value (-.5)+(-.5) = -1 and a maximum value (.5) +(.5) =1.) The convolution can be visualized geometrically as the motion of the X_1 "box density" over the stationary X_2 "box density" and integrating the product of the two in the overlap region. This yields "0" for no overlap, increases linearly to maximum value of "1" at full overlap, and then decreases linearly back down to "0" again when they no longer overlap. This geometrical construct yields the same "triangular density" obtained by formal convolution of the two density functions symbolized by $f_Z(z) = (f_{X_1} * f_{X_2})[z]$. What would happen if we continued this process by convolving the triangular density of the sum RV Z with a 3rd uniform RV X_3, and that result with a 4th uniform RV X_4, *etc.*? If this is repeated indefinitely, the distribution covers the range ($-\infty, +\infty$) and in fact becomes a Gaussian.

The Central Limit Theorem (CLT) justifies and extends the above result for the sum of *independent identically distributed* (IID) **Uniform** RVs to the case of *non-uniform, non-symmetric* RVs (*e.g.*, Exponential), and even to sums of RVs that are not *identical* or *independent*. The **General CLT** stated in the **top panel** is the least restrictive form of the theorem as it does not require the RVs be IID nor does it require they have finite means and variances; the only requirement is that none of the RVs has a **dominant effect** upon their sum. Note the 2nd form recognizes that the product RV $Z = X_1 \cdot X_2 \cdot X_3 \ldots$ is equivalent to the sum RV $\log Z = \log X_1 + \log X_2 + \log X_3 + \ldots$ and results in a "log-Normal" distribution where log(Z) replaces Z as the variable in the standard Gaussian density. The **Restrictive CLT** in the **bottom panel** again allows arbitrary distributions, but is restricted to IID RVs with finite means and variances. In this case, the means and variances simply add to give $\mu_Z = n \cdot \mu_X$ and $\sigma_Z^2 = n \cdot \sigma_X^2$ and the sum RV is $Z \sim N(\mu_Z, \sigma_Z^2)$, a Gaussian in the limit $n \to \infty$. This result for the Standardized Gaussian will be derived using a generating function argument on the next slide.

8.1.1 Restrictive Central Limit Theorem - Proof

Restrictive Central Limit Theorem - Proof

Sum Variable: $Z = \sum_{i=1}^{n} X_i$ $\Rightarrow$ Product of n IID Gen Fcns: $\phi_Z(s) = [\phi_X(s)]^n$

Standardize RV "Z": $Y = \dfrac{Z - \mu_Z}{\sigma_Z}$; $\mu_Z = n\mu_X$; $\sigma_Z = \sqrt{n} \cdot \sigma_X$ $\qquad s \to s/\sigma_Z$

$$\phi_Y(s) = E[e^{sY}] = E\left[e^{s\frac{(Z-\mu_Z)}{\sigma_Z}}\right] = \left[e^{-\frac{\mu_Z s}{\sigma_Z}}\right] E\left[e^{\frac{sZ}{\sigma_Z}}\right] = e^{-\frac{n \cdot \mu_X s}{\sigma_Z}} \cdot \underbrace{\phi_Z(s/\sigma_Z)}_{=[\phi_X(s/\sigma_Z)]^n} = \left[e^{-\frac{\mu_X s}{\sigma_Z}}\right]^n [\phi_X(s/\sigma_Z)]^n$$

Show $= e^{s^2/2}$

Expand both expressions in Taylor Series about s=0 [**Note:** $\phi_X(0) = E[e^{Xs}]_{s=0} = E[e^{X\cdot 0}] = 1$]

$$\phi_Y(s) = \left[e^{-\frac{\mu_X s}{\sigma_Z}}\right]^n \cdot [\phi_X(s/\sigma_Z)]^n = \left[1 - \frac{\mu_X s}{\sigma_Z} + \frac{1}{2}\left(\frac{\mu_X s}{\sigma_Z}\right)^2 + \cdots\right]^n \cdot \left[1 + \underbrace{\frac{d\phi_X}{ds}\Big|_{s=0}}_{E[X]=\mu_X}(s/\sigma_Z) + \underbrace{\frac{d^2\phi_X}{ds^2}\Big|_{s=0}}_{E[X^2]=\mu_X^2 + \sigma_X^2}\frac{(s/\sigma_Z)^2}{2!} + \cdots\right]^n$$

define $\zeta \equiv \dfrac{\mu_X s}{\sigma_Z}$

cancel

$$\phi_Y(s) = \left[1 - \zeta + \frac{1}{2}\zeta^2 + \zeta - \zeta^2 + \frac{1}{2}\zeta^2 + \frac{s^2}{2}\left(\frac{\sigma_X^2}{\sigma_Z^2}\right) + O(s^3)\right]^n = \left[1 + \frac{s^2}{2}\left(\frac{\sigma_X^2}{n \cdot \sigma_X^2}\right) + O(s^3)\right]^n$$

cancel

$$\sigma_Z = \sqrt{n} \cdot \sigma_X$$

$$\varphi_Y(s) = \lim_{\substack{n\to\infty \\ s\to small}} \left(1 + \frac{s^2}{2n}\right)^n = e^{s^2/2}$$

This is moment generating function for a Gaussian N(0,1)!

The **IID property** for the "X"s allows us to write the generator for their sum as the product of the "n" individual generators, *viz.*, $\phi_Z(s) = (\phi_X(s))^n$. Now, forming the standardized RV: $Y = (Z-\mu_Z)/\sigma_Z$, with $\mu_Z = n\mu_X$ and $\sigma_Z = n^{1/2}\sigma_X$, we must show that as $n\to\infty$ the generator $\phi_Y(s) \to \exp(s^2/2)$ corresponding to a normal N(0,1) RV. The details are given in the slide as follows:

(i) By definition the Y-generator is $\phi_Y(s) = E[\exp(s\cdot Y)] = E[\exp(s\cdot(Z-n\mu_X)/\sigma_Z)]$ and this becomes the product of two terms $[\exp(-\mu_X \cdot s/\sigma_Z)]^n \cdot \phi_Z(s/\sigma_Z)$; further, for IID RVs X the Z-generator is the product of "n" X-generators or $[\phi_X(s/\sigma_Z)]^n$ leaving us with product $[\exp(-\mu_X s/\sigma_Z)]^n \cdot [\phi_X(s/\sigma_Z)]^n$.

(ii) These two factors are individually expanded about s=0 up to $O(s^3)$ and then their product is expanded to the same order to yield the 6 terms shown in the next to last equation. Cancelation of terms within the square bracket yields the last equality, an expression raised to the nth power.

(iii) In the limit as $n\to\infty$ the latter expression is equal to "$\exp(s^2/2)$" to $O(s^3)$; this is precisely the generator for an N(0,1) RV, thus proving the theorem.

Note 1: We substituted $\sigma_Z = n^{1/2}\sigma_X$ to make the "n" dependence explicit prior to taking the limit.

Note 2: The Taylor expansion of the generating function $\phi_X(s/\sigma_Z)$ about s=0 starts with the term $\phi_X(s=0) = 1$ and then requires the evaluation of the first two derivatives at s=0, which are the moments of the underlying distribution with mean μ_Z and standard deviation σ_X.

Note 3: Lower case "ϕ" (as in $\phi_X(s/\sigma_Z)$) is the generating function, not the Gaussian Integral Φ.

8.1.2 Central Limit Theorem as Approximation Tool

Central Limit Theorem as Approximation Tool

CLT states: Limiting form of the sum variable Z has a Gaussian distribution

$$\lim_{n \to \infty} Z = \text{Gaussian RV} \quad \sim N(\mu_Z, \sigma_Z^2)$$

The sum variable Z just adds the means and variances of the X_is which yields

$$Z = \sum_{i=1}^{n} X_i \qquad \mu_Z = n\mu_X \quad \sigma_Z = \sqrt{n}\sigma_X$$

For "reasonably" large but finite "n" the sum variable Z is well approximated by the Gaussian Distribution

$$N(\mu_Z, \sigma_Z^2)$$

The CDF is found from Standard Gaussian table

$$\Phi(y) = \Phi\left(\frac{z - \mu_Z}{\sigma_Z}\right) = \Phi\left(\frac{z - n\mu_X}{\sqrt{n} \cdot \sigma_X}\right)$$

1) Sum "n" Independent Identically Distributed (IID) RVs X_i yields new RV "Z_n"

2) Common PDF $f_X(x)$: *Uniform, Poisson, Exponential, Bernoulli, Binomial, Geometric,* etc., ... any continuous or discrete RV.

3) Use Gaussian to Approximate

$$\Pr[z_{min} \leq Z \leq z_{max}] = \int_{\xi = z_{min}}^{z_{max}} f_Z(\xi)d\xi = \frac{1}{\sigma_Z\sqrt{2\pi}} \int_{\xi = z_{min}}^{z_{max}} e^{-(Z - \mu_Z)^2/2\sigma_Z^2}d\xi)$$

4) Centered & Scaled Coordinate Transf.

$$Y = \frac{Z - \mu_Z}{\sigma_Z} \implies \left[\frac{z_{min} - \mu_Z}{\sigma_Z} \leq Y \leq \frac{z_{max} - \mu_Z}{\sigma_Z}\right] \implies \mu_Z = 0; \ \sigma_Z = 1$$

5) Use Gaussian Normal Table $\Phi(x)$

$$\Pr[z_{min} \leq Z \leq z_{max}] \cong \Phi\left(\frac{z_{max} - \mu_Z}{\sigma_Z}\right) - \Phi\left(\frac{z_{min} - \mu_Z}{\sigma_Z}\right)$$

The Central Limit Theorem states that the sum of n random variables yields a Gaussian RV (as n→∞) no matter whether the "X"s being summed are uniform, exponential, r-Erlang, *etc.*. This rather surprising theorem is of course the reason for the common assumption of Gaussian Noise Processes in so many scientific and engineering models. Just start taking sums (averaging) and all the peculiarities of the underlying individual distributions become washed out and the sum approaches a Gaussian with appropriate mean and variance.

Although, we never actually reach the limiting Gaussian form, we often find that after averaging just 4 or so RVs, the effects of the original distribution are sufficiently washed out to make the Gaussian a good approximation to the sum RV. This is especially true if the underlying distribution of the summed RV is symmetric as in a uniform RV (see Slide#8-6), but also yields a reasonably good approximation even for a very asymmetric RV, such as an exponential, for sums of 8 or more (see Slide#8-7). Thus, in most cases, relatively large but finite sums are all that is needed to be able to use a Gaussian approximation with good confidence.

For any finite sum of IID "X"s, the sum variable Z has a mean and variance that are just n times that for an individual X, *i.e.*, $\mu_Z = n \cdot \mu_X$ and $\sigma_Z^2 = n \cdot \sigma_X^2$; moreover, upon standardizing Z *via* the transformation $Y = (Z - \mu_Z)/\sigma_Z$ we can use the standard Gaussian function $\Phi(Y) = \Phi((Z - \mu_Z)/\sigma_Z)$ to evaluate the cumulative distribution. This technique is a convenient tool for approximating answers to questions involving finite sums of RVs and works reasonably well even for sums containing just a few terms. The bottom panel gives a step-by-step procedure to approximate the probability of a sum RV Z over an interval $[z_{min}, z_{max}]$ using a Gaussian Normal distribution Table function $\Phi(y)$.

8.1.3 Computer Simulation of Central Limit Theorem

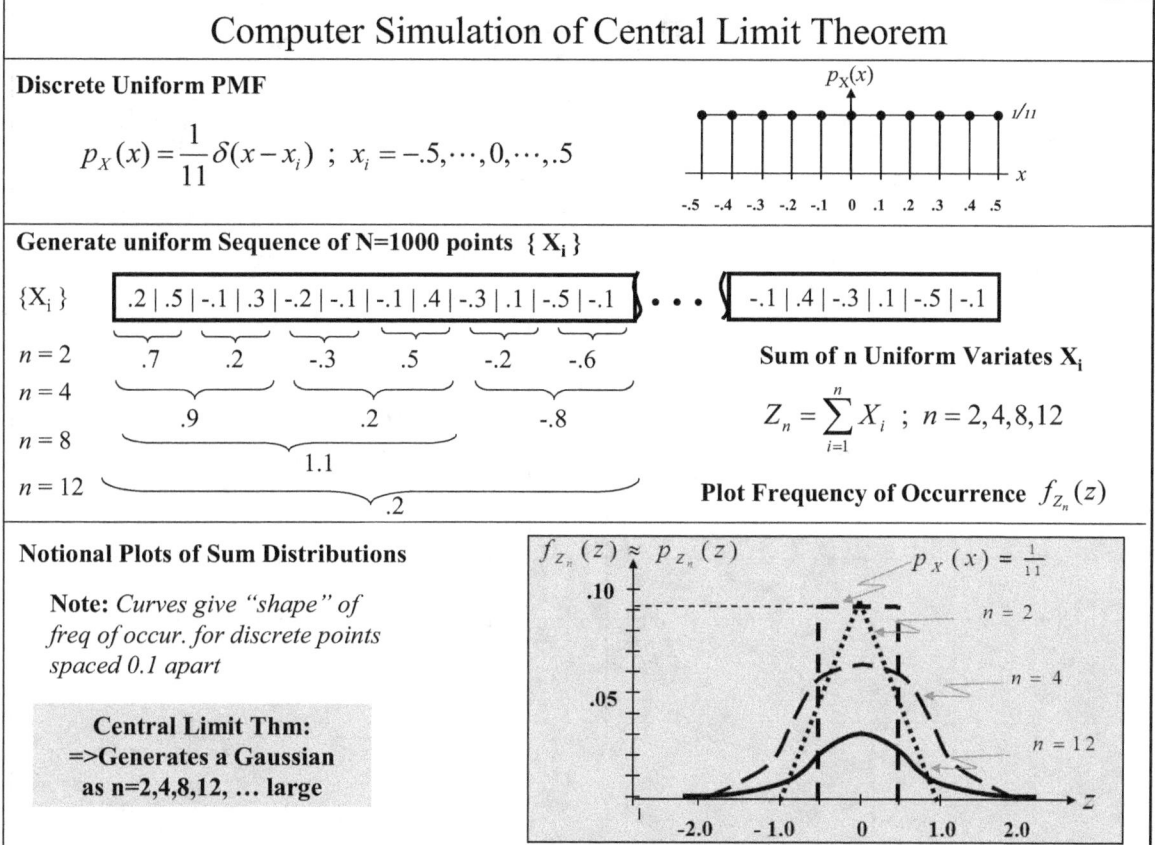

The Discrete Uniform PMF with values at 11 discrete points ranging from x ={-.5, -.4, -.3, -.2,-.1, 0, .1, .2,.3,.4.,.5} can be expressed as a sum of 11 δ-functions with magnitude 1/11 at each of these points as shown in the figure. This discrete PMF is generated by a "sample and hold" transform in the u-x plane on the *continuous* Uniform PDF $f_U(u) = 1/1.1$ with u ε [-.6, +.5]. Thus, integrating this density over the interval u= -.6 to u= -.5 accumulates probability (-.5 - (-.6))/1.1 =1/11 at the *single point* x= -0.5, which is precisely the desired δ-function term $1/11 \cdot \delta(x-(-.5))$ with amplitude 1/11 at x= -.5. Note that the integration on the *u-variable* yields a δ-function term on the *x-variable* (see Slides#2-14, 2-15 for details of the transformation and plots.)

Suppose that a sequence of 1000 numbers from the discrete set {-.5, -.4, -.3, -.2,-.1, 0, .1, .2,.3,.4.,.5} are randomly generated on a computer to create the data run notionally illustrated in the 2nd panel. Now we can create sum random variables Z_n by summing n =2, or n= 4, or n= 8, or n=12 of these samples. According to the CLT, as we increase "n", the resulting frequency distribution of the sum variable "Z_n" must approach a Gaussian. The notional illustration shows what we should expect. The dashed rectangle shows the bounds of the original uniform discrete PMF and the other curves show the march towards a Gaussian. Note that, unlike a Gaussian, all these distributions are zero outside a finite interval determined by the number of variables that are summed. The triangle shape is the sum of two RVs and their sum Z_2 is constrained to the interval [-1, 1], whereas the RV Z_{12} covers the larger interval [-6, 6]. The interval clearly increases as we sum more variables, but only as n→ ∞ does the sum variable fully capture the small Gaussian "tails" for large |x| as required by the CLT. This result can also be thought of in terms of an n-fold convolution of the IID RVs X_k k=1,2,...,n which also spreads out with each new convolution in the sequence. The next two slides show the results of actual MatLab® simulations for sums of uniform and exponential RVs and their approach to a Gaussian. The plots of the zero mean uniform case confirm the notional sketch on this slide, while the *asymmetric* exponential distribution with mean 1/λ *migrates* to higher means as it approaches a Gaussian.

8.1.4 Sum of Uniform RVs Approaches Gaussian

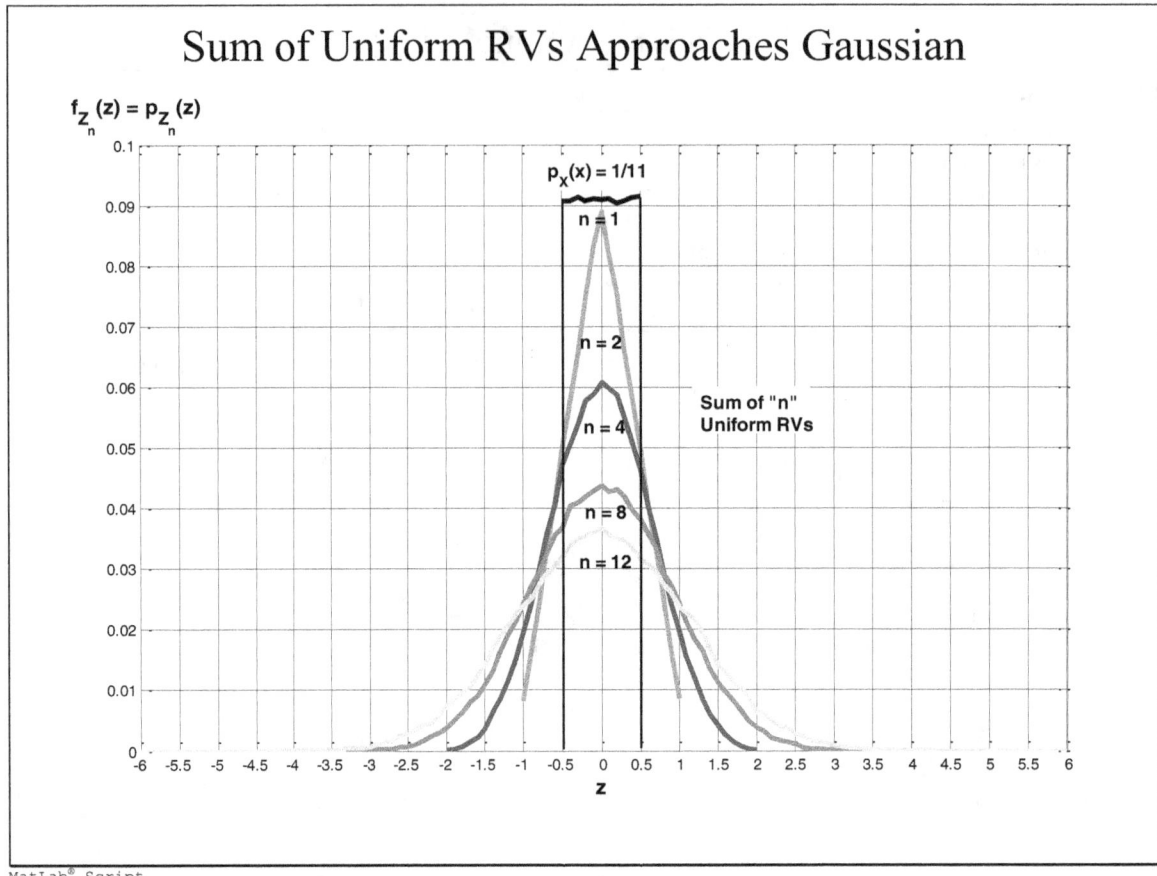

```
MatLab® Script
N=12*8*4*2*1000
xrv=1.1*rand(N,1)-.55;
x2=sum(reshape(xrv,N/2,2),2);
x4=sum(reshape(xrv,N/4,4),2);
x8=sum(reshape(xrv,N/8,8),2);
x12=sum(reshape(xrv,N/12,12),2);
binedge1=[-.55:.1:.55]; bincenter1=[-.5:.1:.5]';
binedge2=[-1.05:.1:1.05];bincenter2=[-1:.1:1]';
binedge4=[-2.05:.1:2.05];bincenter4=[-2:.1:2]';
binedge8=[-4.05:.1:4.05];bincenter8=[-4:.1:4]';
binedge12=[-6.05:.1:6.05];bincenter12=[-6:.1:6]';
[n1]=histc(xrv,binedge1);
[n2]=histc(x2,binedge2);
[n4]=histc(x4,binedge4);
[n8]=histc(x8,binedge8);
[n12]=histc(x12,binedge12);
n1=n1/sum(n1);  n1=n1(1:length(n1)-1);
n2=n2/sum(n2);n2=n2(1:length(n2)-1);
n4=n4/sum(n4);n4=n4(1:length(n4)-1);
n8=n8/sum(n8);n8=n8(1:length(n8)-1);
n12=n12/sum(n12);n12=n12(1:length(n12)-1);
figure(1); plot(bincenter1,n1)
hold on
plot(bincenter2,n2,'r-');plot(bincenter4,n4,'b-');plot(bincenter8,n8,'r-')
plot(bincenter12,n12,'g-');grid on
hold off
```

8.1.5 Sum of Exponential RVs Approaches Gaussian

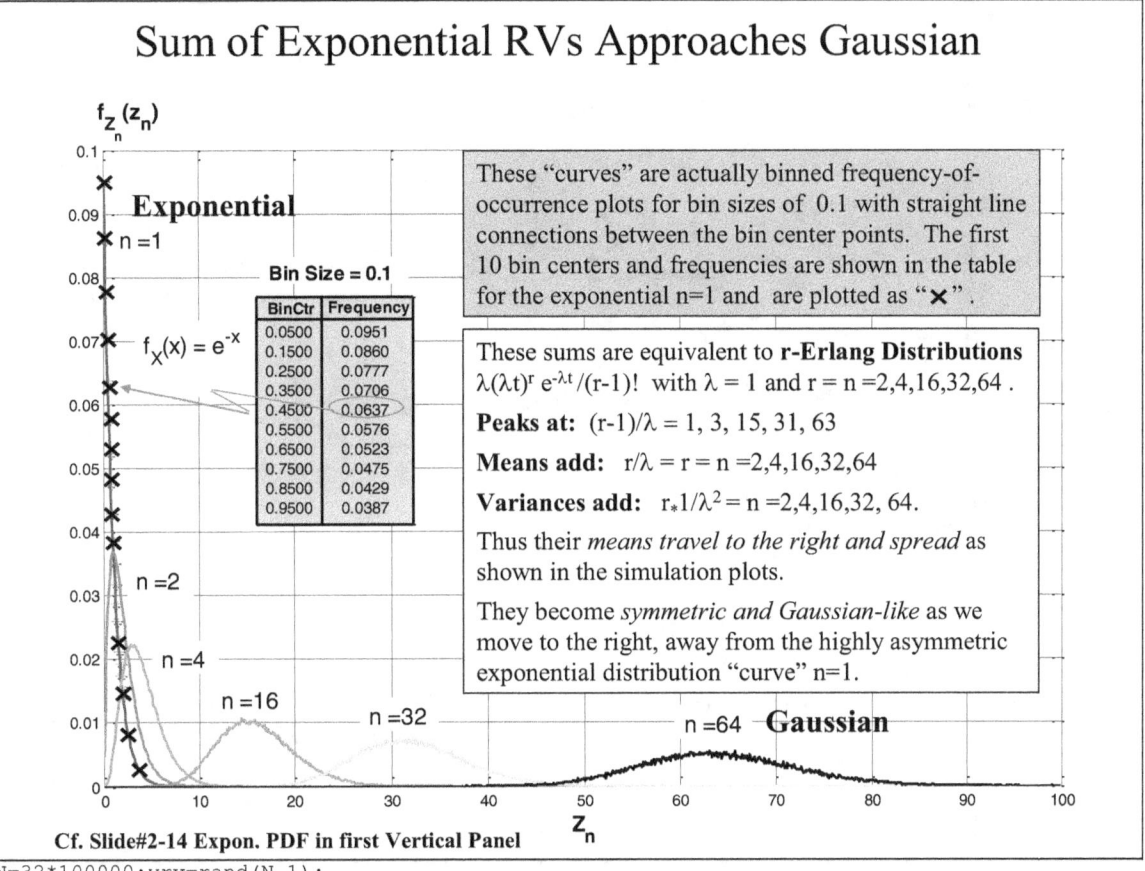

Sum of Exponential RVs Approaches Gaussian

$f_{Z_n}(z_n)$

Exponential n =1

$f_X(x) = e^{-x}$

Bin Size = 0.1

BinCtr	Frequency
0.0500	0.0951
0.1500	0.0860
0.2500	0.0777
0.3500	0.0706
0.4500	0.0637
0.5500	0.0576
0.6500	0.0523
0.7500	0.0475
0.8500	0.0429
0.9500	0.0387

n =2

n =4

n =16

n =32

n =64 **Gaussian**

These "curves" are actually binned frequency-of-occurrence plots for bin sizes of 0.1 with straight line connections between the bin center points. The first 10 bin centers and frequencies are shown in the table for the exponential n=1 and are plotted as "✕".

These sums are equivalent to **r-Erlang Distributions** $\lambda(\lambda t)^r e^{-\lambda t}/(r-1)!$ with $\lambda = 1$ and r = n =2,4,16,32,64 .

Peaks at: $(r-1)/\lambda = 1, 3, 15, 31, 63$

Means add: $r/\lambda = r = n = 2,4,16,32,64$

Variances add: $r*1/\lambda^2 = n = 2,4,16,32, 64$.

Thus their *means travel to the right and spread* as shown in the simulation plots.

They become *symmetric and Gaussian-like* as we move to the right, away from the highly asymmetric exponential distribution "curve" n=1.

Z_n

Cf. Slide#2-14 Expon. PDF in first Vertical Panel

```
N=32*100000;yrv=rand(N,1);
xshift=0; delx=0.1; % shift exponential "x+xshift"
ernge1=0;% -100; ernge2= 100;% bin edge ranges
crnge1=ernge1+delx/2; crnge2=ernge2-delx/2; % bin center ranges
xrv=xshift-log(1-yrv);% exponential RV "X"
x2=sum(reshape(xrv,N/2,2),2);x4=sum(reshape(xrv,N/4,4),2);
x16=sum(reshape(xrv,N/16,16),2);x32=sum(reshape(xrv,N/32,32),2);
x64=sum(reshape(xrv,N/64,64),2); % Sums of 2,4,16,32,64 Exponential RVs
binedge=[ernge1:delx:ernge2];bincenter=[crnge1:delx:crnge2]';% "histc" bin edges
[n1]=histc(xrv,binedge);[n2]=histc(x2,binedge);[n4]=histc(x4,binedge);
[n16]=histc(x16,binedge);[n32]=histc(x32,binedge);[n64]=histc(x64,binedge);
n1=n1/sum(n1); n1=n1(1:length(n1)-1); % frequencies for histogram plot
n2=n2/sum(n2); n2=n2(1:length(n2)-1);
n4=n4/sum(n4); n4=n4(1:length(n4)-1);
n16=n16/sum(n16); n16=n16(1:length(n16)-1);
n32=n32/sum(n32); n32=n32(1:length(n32)-1);
n64=n64/sum(n64); n64=n64(1:length(n64)-1); % "histc" puts out extra bin
figure(1);plot(bincenter,n1);hold
on;plot(bincenter(1:20),n1(1:20),'k*');plot(bincenter,n2,'r-'); plot(bincenter,n4,'b-
');plot(bincenter,n16,'r-');plot(bincenter,n32,'g-')
plot(bincenter,n64,'k-');grid on; hold off
out=zeros(length(n1),7);out(:,1)=bincenter;out(:,2)=n1;out(:,3)=n2;out(:,4)=n4;
out(:,5)=n16;out(:,6)=n32;out(:,7)=n64; % output table for reference
x2=find(out(:,2)==max(out(:,2))); max1=out(x2,2); x5=find(out(:,5)==max(out(:,5)));
center16=out(x5,1); x6=find(out(:,6)==max(out(:,6))); center32=out(x6,1);
x7=find(out(:,7)==max(out(:,7))); center64=out(x7,1); % Histogram Peak Computation
```

8.2 Approximation Examples - Sum of 12 Uniform RVs using Gaussian

Approximation Examples - Sum of 12 Uniform RVs using Gaussian

Probabilities for the "Sum Variable" Z using Gaussian Approximation (CLT)

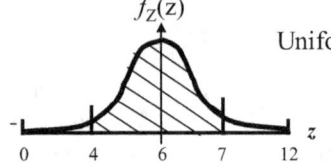

Sum Variable $Z = \sum_{i=1}^{12} X_i$

Find: $\Pr[4 \le Z \le 7]$

Uniform X: $f_X(x) = \begin{cases} 1 & 0 \le x \le 1 \\ 0 & Otherwise \end{cases}$

Note: $z \, \varepsilon \, [0,12]$, but Gaussian $(-\infty, +\infty)$

Uniform Distrib Statistics:

$$\mu_X = E[X] = \int_0^1 1 \cdot x\, dx = \frac{1}{2} \quad ; \quad E[X^2] = \int_0^1 1 \cdot x^2\, dx = \frac{1}{3}; \quad \sigma_X^2 = \frac{1}{3} - \frac{1}{2^2} = \frac{1}{12}$$

Sum Variable Statistics:

$$\therefore \;\; \mu_Z = n\mu_X = 12(.5) = 6 \quad ; \quad \sigma_Z^2 = n \cdot \sigma_X^2 = 12 \cdot \frac{1}{12} = 1$$

Standardized Variable $Y = \dfrac{Z-6}{1}$

$$\Pr\left[\frac{4-6}{1} \le Y \le \frac{7-6}{1} \right] \cong \Phi(1) - \Phi(-2) = \Phi(1) - (1 - \Phi(+2)) = \underbrace{\Phi(1)}_{.841345} + \underbrace{\Phi(+2)}_{.977250} - 1 = .8186$$

The CLT says that the sum of a large number of IID RVs approximates a Gaussian. In fact, the sum of 12 uniform random variables $X_k \sim U[0,1]$ is often deemed accurate enough to represent computer samples of a Gaussian distribution $N(\mu_Z, \sigma_Z^2)$. This example considers such a sum of 12 uniform variables and first computes the mean and variance of a single uniform variable X to be $\mu_X = 1/2$ and $\sigma_X^2 = 1/12$; the mean and variance of the sum variable Z are the sums of the n identical contributions, *i.e.,* $\mu_Z = n \cdot \mu_X = 12(.5) = 6$ and $\sigma_Z^2 = n \cdot \sigma_X^2 = 12(1/12) = 1$ and accordingly the Gaussian RV is $Z \sim N(6,1)$. Using these values, the standardized variable becomes $Y = (Z-6)/1$ and thus the probability within the z-interval [4, 7] translates into a y-interval [(4-6)/1, (7-6)/1] = [-2, 1] and hence the different between the tabulated values $\Phi(1) - \Phi(-2)$ approximates the desired probability as detailed in the last equation. The result $\Pr[4 \le X \le 7] = .8186$ is reasonably accurate because the sum variable Z only omits the *small* Gaussian tails to *the left of 0* and *to the right of 12*, but is otherwise quite accurate in the truncated region [0,12].

The inset figure at the top of the slide shows the shaded area under the $f_Z(z)$ curve which is symmetric about $\mu_Z = 6$, has a variance of $\sigma_Z^2 = 1$, and only has non-zero values in the finite interval [0, 12] unlike a true Gaussian whose tails extend to $\pm\infty$.

8.2.1 Approximate Sum of 10 Uniform RVs using Gaussian

Approximate Sum of 10 Uniform RVs using Gaussian

Sum of 10 Uniform RV => Use Central Limit Thm $\quad Z = \sum_{i=1}^{10} X_i$

$\Pr[Z > 6] = 1 - \Pr[Z \le 6]$ $\qquad\qquad\qquad X_i$ indep unif RV on $[0,1]$

$E[X] = \int_0^1 x\,dx = \frac{1}{2}$

Note: This includes negative values of z "negligible contribution" (See below)

$E[X^2] = \int_0^1 x^2\,dx = \frac{1}{3}$; $Var(X) = \frac{1}{3} - \left(\frac{1}{2}\right)^2 = \frac{1}{12}$

$E[Z] = 10 \cdot \frac{1}{2} = 5$; $Var(Z) = 10 \cdot \frac{1}{12} = \frac{5}{6}$

Standardize: $Y = \dfrac{Z - 5}{\sqrt{5/6}}$

$\Pr[Z \le 6] = \Pr\left[Y \le \dfrac{6-5}{\sqrt{5/6}}\right] \approx \Phi(\sqrt{1.2}) = .863$

$\Pr[Z > 6] = 1 - .863 = .137$

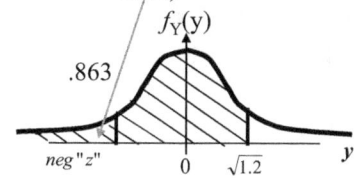

Strictly speaking $\quad 0 \le Z \le 6 \quad$ *i.e.*, Z cannot be negative

$\Pr[0 \le Z \le 6] = \Pr\left[\dfrac{-5}{\sqrt{5/6}} \le Y \le \dfrac{1}{\sqrt{5/6}}\right] = \Phi\left(\sqrt{1.2}\right) - \Phi\left(-5\sqrt{1.2}\right)$

$= \Phi\left(\sqrt{1.2}\right) - \underbrace{\left\{1 - \Phi\left(5\sqrt{1.2}\right)\right\}}_{<10^{-4}} \cong \Phi\left(\sqrt{1.2}\right)$

small contribution subtracted out

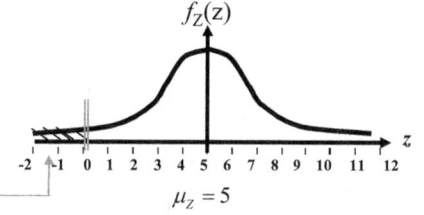

Here the CLT is applied to the sum Z_{10} of 10 uniform RVs X in order to calculate the probability that the sum variable Z_{10} is greater than 6, *i.e.*, $\Pr[Z_{10} > 6]$. The mean and variance of a single uniform variable X are the same as the previous slide $\mu_X = \frac{1}{2}$ and $\sigma_X^2 = 1/12$, but the mean and variance of the sum variable Z_{10} are now $\mu_Z = n\,\mu_X = 10(.5) = 5$ and $\sigma_Z^2 = n\,\sigma_X^2 = 10(1/12) = 5/6$ and correspondingly the Gaussian RV is $Z_{10} \sim N(5, 5/6)$.

The standardized variable is $Y = (Z_{10}-5)/(5/6)^{1/2}$ and the tabulated Φ function can be used to approximate the probability

$$\Pr[Z_{10} \le 6] = \Pr[Y \le 1/(5/6)^{1/2}] = \Phi((1.2)^{1/2}) = .863 \quad \text{or} \quad \Pr[Z > 6] = 1 - .863 = .137 .$$

The hashed area under the standardized Gaussian density $f_Y(y)$ in the **top panel** figure represents the probability computed using $\Phi((1.2)^{1/2})$. A problem arises because this computation clearly has contributions from *negative values* that cannot occur in a sum variable Z_{10} generated from positive "X"s in $[0,1]$. The **lower panel** figure illustrates the "correct" evaluation over only *positive* values of Z_{10}, *viz.*, $\Pr[0 \le Z_{10} \le 6]$; the calculation has a correction term corresponding to negative values of Z_{10} in the hashed tail of the density $f_Z(z)$; this term is $\sim 10^{-4}$ (braced term {} of bottom equation) and can be safely ignored. Thus, the standardized Gaussian calculation in the **top panel** gives a good approximation even though the Φ function evaluation includes points outside the range of the sum variable Z_{10}; this is because contributions of these negative values are in the tail of the Gaussian and have a negligible effect.

8.2.2 Approximate Single Binomial Term with Gaussian

Approximate Single Binomial Term with Gaussian

Sum of n Indep. Bernoulli RVs : $Z = \sum_{i=1}^{n} X_i$

Yields Binomial for Z : $p_Z(z) = b(z\ ;n, p) = \binom{n}{z} p^z q^{n-z}$

Exact Binomial evaluation for $n = 40$, $p = 0.5$:

$$Pr[Z = 20] = \binom{40}{20}(.5)^{20}(.5)^{40-20} = .1254$$

DeMoivre-Laplace Thm

CLT Approx for large $n = 40$, $p = 0.5$:

$$Pr[Z = 20] \cong Pr[19 \leq Z \leq 21]$$
$$\cong Pr[19.5 \leq Z \leq 20.5]$$

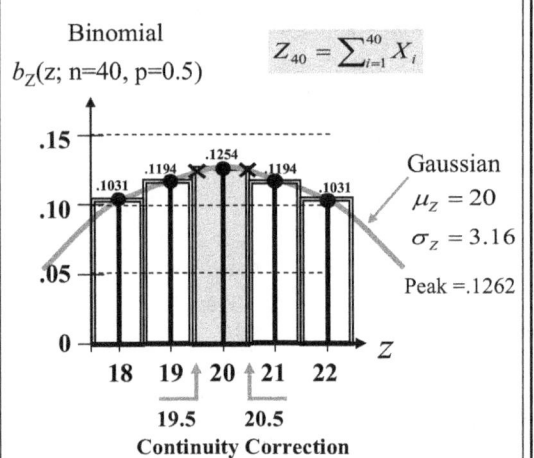

Binomial
$b_Z(z; n=40, p=0.5)$

$Z_{40} = \sum_{i=1}^{40} X_i$

Gaussian
$\mu_Z = 20$
$\sigma_Z = 3.16$
Peak = .1262

Continuity Correction

Mean : $\mu_Z = E[Z] = n \cdot p = 40(.5) = 20$;

Sigma : $\sigma_Z = \sqrt{n \cdot p \cdot q} = \sqrt{40 \cdot (.5) \cdot (.5)} = \sqrt{10} = 3.162$

Standardize : $Y = \dfrac{Z - \mu_Z}{\sigma_Z} = \dfrac{Z - 20}{3.16}$

$$Pr[Z = 20] \cong Pr\left[\frac{19.5 - 20}{3.16} \leq Y \leq \frac{20.5 - 20}{3.16}\right] = \Phi(.1581) - \Phi(-.1581) = qNorm(-.1581) - qNorm(.1581) = .1256$$

Error in Approx $= \dfrac{.1256 - .1254}{.1254} = .0016 = .16\%$

As previously discussed, the sum of n independent Bernoulli RVs X yields the sum RV Z_n which represents the number of successes "z" from "n" draws and is characterized by the binomial PMF b(z;n,p). The binomial is a discrete distribution taking on values for z = 0, 1, 2, ..., n successes only. In this example we set n=40, p=0.5 and compute the probability of exactly 20 successes directly from the binomial PMF as the *single term* $Pr[Z_{40}=20] = {}^{40}C_{20}\ (.5)^{20}(.5)^{40-20} = .1254$.

Alternately, from the point of view of the CLT Theorem, the *sum of 40 Bernoulli variables* defines a sum variable Z_{40} that should be well-approximated by a Gaussian distribution with a mean $\mu_Z = n \cdot p = 40(.5) = 20$ and a variance $\sigma_Z = (n \cdot pq)^{\frac{1}{2}} = (40(.5)(.5))^{\frac{1}{2}} = 3.162$. Transformation to the standardized variable $Y = (Z_{40} - \mu_Z)/\sigma_Z = (Z_{40} - 20)/3.162$ then allows us to evaluate the probability in any interval $[Y_{min}, Y_{max}]$ using the tabulated function $\Phi(y)$ as $Pr[Y_{min} \leq Y \leq Y_{max}] = \Phi(Y_{max}) - \Phi(Y_{min})$.

Thus, in order to estimate the Binomial probability $Pr[Z_{40}=20]$ with a Gaussian, we must specify some interval about $Z_{40}=20$. The stick plot of the Binomial with a Gaussian fit to the discrete values $Z_{40} = 18, 19, 20, 21, 22$ suggests perhaps computing $Pr[19 \leq Z_{40} \leq 21]$, but a little thought shows this to be incorrect. From the figure it is clear that the integral is approximately the sum of two rectangles of width =1 and average height (.1254+.1194)/2=.1224, or ~.245; indeed the exact Gaussian integral yields $\Phi(-1/3.162) - \Phi(1/3.162) = .2482$, a value roughly twice too large as predicted. Clearly, we must integrate over a *single-integer interval*! The DeMoivre-Laplace theorem does precisely that (next slide) by invoking a so-called "continuity correction" and integrating over the surrounding *half-integer interval* on either side [20-.5, 20+.5] to yield a much more accurate approximation (accurate to .16% in this case)

$Pr[Z_{40}=20] = Pr[19.5 \leq Z_{40} \leq 20.5]$

$\qquad = Pr[(19.5-20)/3.16 \leq Y \leq (20.5-20)/3.16\ ] = \Phi(-.1581) - \Phi(.1581) = .1256$.

8.2.3 DeMoivre-Laplace Limit Theorem – Continuity Correction

DeMoivre-Laplace Limit Theorem – Continuity Correction

$$\Pr[a \le Z \le b] = \underbrace{\sum_{z=a}^{b} \binom{n}{z} p^z q^{n-z}}_{\text{Sum Binomial PMF btwn a \& b}} \approx \underbrace{\Phi\left(\frac{b - \mu_Z}{\sigma_Z}\right) - \Phi\left(\frac{a - \mu_Z}{\sigma_Z}\right)}_{\text{Integral of Gaussian btwn a \& b}}$$

a, b integers

$$\approx \Phi\left(\frac{(b+.5) - \mu_Z}{\sigma_Z}\right) - \Phi\left(\frac{(a-.5) - \mu_Z}{\sigma_Z}\right)$$

Continuity
<= Correction
Changes Integr.
limits by 0.5

If $p = q = 1/2$, then binomial is symmetric. If p >> q the binomial becomes **very asymmetric** and will peak closer to 0 than to n; *vice-versa* for q >> p. Thus the CLT approximation using a **symmetric Gaussian** will not be very accurate in these cases. Restrictions on *p* and *q* below attempt to address this issue

Good Approximation by Gaussian Requires: $\quad np > 2\sigma_Z \quad\quad nq > 3\sigma_Z$

Mean : $\mu_Z = E[Z] = n \cdot p = 40(.5) = 20$;

Sigma : $\sigma_Z = \sqrt{n \cdot p \cdot q} = \sqrt{40 \cdot (.5) \cdot (.5)} = \sqrt{10} = 3.16$

$np > 2\sigma_Z \quad np = 20 > 2(3.16) = 6.32 \quad\quad 20 > 6.32 \quad$ **OK for "p"**

$nq > 3\sigma_Z \quad nq = 40(.5) = 20 > 3(3.16) = 9.48 \quad\quad 20 > 9.48 \quad$ **OK for "q"**

In order to use the CLT to "capture" a single point of a discrete distribution, the discussion of the previous slide concluded that we must integrate the Gaussian over one full interval between k-.5 and k+.5 . Thus, when we have an *integer interval* [a, b], integration between a and b does not capture the discrete points "a" and "b" unless we extend the limits of integration between a-.5 to a+.5 to capture point "a" and between b-.5 to b+.5 to capture point "b". This .5 extension of the integration limits results in the DeMoivre-Laplace Limit Theorem displayed in the top equation of this slide; it essentially relates the sum of the n binomial terms Z_n between z=a and z=b to an equivalent expression using the tabulated Gaussian "Φ"s. This so-called "continuity correction" extends the Gaussian integer evaluation limits [a, b] outside that integer interval by 0.5 to give [a-.5, b+.5] as the "corrected" evaluation limits as shown in the second line of the top equation.

Since the Gaussian is a symmetric distribution about its mean, it will provide the best approximation to the Binomial if the value of p is near ~.5 because the Binomial is symmetric for p=.5 . On the other hand, if p (or q) is very small or very large the Binomial is no longer symmetric and the approximation degrades.

It has been found that for the approximation by a Gaussian to be good two conditions must hold, namely, (i) np > 2 σ_Z and (ii) nq > 3σ_Z, where σ_Z =(npq)$^{1/2}$. Applying this to the example on the last slide, we see that both conditions hold.

8.2.4 Stirling Factorial Approximation

Stirling Factorial Approximation

1) Sum n Poisson RVs: Let X_i be IID Poisson RVs with $E[X_i]=1$ and $Var(X_i)=1$

$Z_n = \sum_{i=1}^{n} X_i$ Poisson parameter $a = E[Z_n] = n \cdot 1 = n$

$Var(Z_n) = n \cdot 1 = n$

$\Pr[Z_n = n] = e^{-a} \left. \frac{a^n}{n!} \right|_{a=n} = e^{-n} \frac{n^n}{n!}$ *Single Poisson Term*

2) Use Gaussian to Find same Probablity

CLT with Continuity Correction (DeMoivre) $P[Z_n = n] = \Pr[n-.5 \leq Z_n \leq n+.5]$

Standardize $Y = \frac{Z_n - \mu_{Z_n}}{\sigma_{Z_n}} = \frac{Z_n - n}{\sqrt{n}}$

$\Pr[n-.5 \leq Z_n \leq n+.5] = \Pr\left[\frac{(n-.5)-n}{\sqrt{n}} \leq Y \leq \frac{(n+.5)-n}{\sqrt{n}} \right] = \Phi\left[\frac{1}{2\sqrt{n}} \right] - \Phi\left[\frac{-1}{2\sqrt{n}} \right] = 2\Phi\left[\frac{1}{2\sqrt{n}} \right] - 1$

Approx with Taylor Series $\Phi\left(\frac{1}{2\sqrt{n}} \right) = \Phi(0) + \Phi'(0) \cdot \frac{1}{2\sqrt{n}} + O\left(\left(\frac{1}{2\sqrt{n}} \right)^2 \right) \cong \frac{1}{2} + \frac{1}{\sqrt{2\pi n}} \cdot \frac{1}{2}$

$\Phi(x) = \int_{-\infty}^{x} \frac{e^{-t^2/2}}{\sqrt{2\pi}} dt$; $\Phi(0) = \frac{1}{2}$; $\Phi'(x)|_{x=0} = \frac{d}{dx} \int_{-\infty}^{x} \frac{e^{-t^2/2}}{\sqrt{2\pi}} dt \underset{(Leibnitz)}{=} \left. \frac{d(x)}{dx} \frac{e^{-x^2/2}}{\sqrt{2\pi}} \right|_{x=0} = \frac{1}{\sqrt{2\pi}}$

$\Pr[Z_n = n] \cong 2 \cdot \frac{1}{2}\left(1 + \frac{1}{\sqrt{2\pi n}} \right) - 1 = \frac{1}{\sqrt{2\pi n}}$

Z_n is Poisson "a = n": $\Pr[Z_n = n] = e^{-n} \frac{n^n}{n!}$

Equate

Stirling Factorial Approximation

$e^{-n} \frac{n^n}{n!} = \frac{1}{\sqrt{2\pi n}}$ $\Rightarrow$ $n! = \sqrt{2\pi n} \cdot n^n \cdot e^{-n}$

The Stirling approximation to the factorial $n! = (2\pi n)^{1/2} n^n e^{-n}$ is needed to approximate very large factorials that otherwise would exceed the integer precision of most computers and produce "∞" as the overflow answer. The derivation uses the DeMoivre-Laplace Limit Theorem approximation to evaluate a single Poisson term containing the factorial $n!$. Specifically consider the nth term in a Poisson distribution with parameter $a = n$, *viz.*,

$$p_K(K=n;a=n)=e^{-n} n^n/n! \ .$$

We have previously found that the sum of "n" Poisson RVs each with $a_i=1$ for $i=1,2,\dots n$, yields a sum variable Z_n that is also Poisson, but now with parameter $a=\sum a_i =\sum 1= n$. This allows us to relate the single Poisson term to the probability that the sum variable Z_n takes on the integer value "n" as

$$\Pr[Z_n=n] =p_{Zn}(Z_n=n;a=n)=e^{-n} n^n/n!$$

We now apply the DeMoivre-Laplace Limit Theorem to the sum variable Z_n by first approximating $\Pr[Z_n=n]$ using the continuity correction as $\Pr[n-.5 \leq Z_n \leq n+.5]$; to evaluate this we re-cast it in terms of the standardized RV $Y=(Z_n - \mu_{Zn})/\sigma_{Zn} =(Z_n - n)/ n^{1/2}$. Solving for $Z_n = \sigma_{Zn}Y+\mu_{Zn}$ and substituting the values $\sigma_{Zn} = n^{1/2}\sigma_X = n^{1/2} \cdot 1$ and $\mu_{Zn} = n \cdot \mu_X = n \cdot 1$ yields $Z_n = n^{1/2}Y +n$ and the resulting DeMoivre-Laplace expression for $\Pr[Z_n=n]$ is now expressed in terms of Y to give

$$\Pr[Z_n=n] \approx \Pr[n-.5 \leq Z_n \leq n+.5]=P[-.5/n^{1/2} \leq Y \leq +.5/n^{1/2}] =2\Phi(.5/n^{1/2}) - 1 \ .$$

The remaining part of the derivation is to make a Taylor expansion of $\Phi(.5/n^{1/2})$ about the origin to find $\Pr[Z_n=n] \approx 1/ (2\pi n)^{1/2}$ (details in the dashed box), then equate this expression to the single Poisson term $e^{-n} n^n/n!$, and finally solve for $n!$ to obtain the Stirling result $n!= (2\pi n)^{1/2} n^n e^{-n}$.

8.2.5 Approximate Binomial Sum with Poisson Sum

Approximate Binomial Sum with Poisson Sum

Large # Trials "n"
Small Prob of success "p"

$$\lim_{\substack{n\to\infty \\ p\to 0}} n\cdot p = \lim_{\Delta t\to 0}\left(\frac{t}{\Delta t}\right)\cdot(\lambda\cdot\Delta t) = \lambda t$$

$$b(k;n,p) = \underbrace{\binom{n}{k}p^k q^{n-k}}_{\text{Binomial Term}} \approx \underbrace{\frac{(\lambda t)^k}{k!}e^{-\lambda t}}_{\text{Poisson Term}} \underset{\lambda t\to np}{=} \underbrace{\frac{(np)^k}{k!}e^{-np}}_{\substack{t\to n\Delta t \\ p\to\lambda\Delta t}}$$

$$\left.\vphantom{\int}\right\} \quad \text{Binomial} \Leftrightarrow \text{Poisson}$$

$$np \Leftrightarrow \lambda t$$

Ex 1. Microchips 10% Defective $p=0.1$
("success") $n=10$ selected at random
Find Prob no more than $k=1$ defective

$$\Pr[0,1\,defec.] = \sum_{k=0}^{1}\binom{10}{k}(.1)^k(.9)^{10-k} = \binom{10}{0}(.1)^0(.9)^{10-0} + \binom{10}{1}(.1)^1(.9)^{10-1} = .7361$$

Sum Binomial Terms

$$np \Leftrightarrow \lambda t$$

$$n=10\;;\;p=.1\;;\;\lambda t = np = 10(.1) = 1$$

$$\Pr[0,1\,defec.] = \sum_{k=0}^{1}\frac{(\lambda t)^k}{k!}e^{-\lambda t} = \frac{(1)^0}{0!}e^{-1} + \frac{(1)^1}{1!}e^{-1} = .7358$$

Sum Poisson Terms

Ex 2. Transmit $n=10^4$ bits with Bit Error Rate $p = 10^{-4}$
Find Prob > 4 errors ($k=4$) $n=10^4\;;\;p=10^{-4}\;;\lambda t = np = 10^4(10^{-4}) = 1$

Binomial Sum

$$\Pr[>4Err] = 1 - \Pr[\le 4Err] = 1 - \sum_{k=0}^{4}\underbrace{\binom{10^4}{k}\cdot\left(10^{-4}\right)^k\cdot\underbrace{(1-10^{-4})^{10^4-k}}_{0.9999^{10^4}=.3679}}_{(10^4 10^4\cdots 10^4)10^{-4k}/k!} \cong 1 - 0.36786\cdot\sum_{k=0}^{4}(1/k!) = .00366$$

Poisson Sum

$$\Pr[>4Err] \cong 1 - \sum_{k=0}^{4}\frac{(\lambda t)^k}{k!}e^{-\lambda t} = 1 - \sum_{k=0}^{4}\frac{(1)^k}{k!}e^{-1} = .00366 = 0.4\%$$

Alternate question: Find p such that Prob > 4 errors is less than 1%

For highly asymmetric Binomials with small probability of success $p \ll 1$, a Poisson distribution with parameter $a = \lambda t = np$ gives a better approximation to the Binomial term than a Gaussian distribution evaluation. Two examples are given on this slide:

Example 1: 10 microchips with a defect probability (success) of $p = 0.1$ are selected at random and we are asked to find the probability that no more than $k=1$ is found defective. That probability is found directly by summing the $k=0, 1$ Binomial terms $b_K(k;n,p) = b_K(k;10,.1)$ to yield .7361. The Poisson computation uses the parameter $a = np = 10(.1) = 1$ and sums two Poisson terms $p_K(k;a) = p_K(k;1)$ for $k=0, 1$ to obtain .7358 which compares well with the exact value of .7361.

Example 2: 10^4 bits in a digital transmission with a "bit error rate" given by the probability $p = 10^{-4}$ and we are asked to find the probability of *more than 4 errors* in the transmission. The problem is solved by computing $1 - \Pr[K \le 4\text{ err}]$ so we only have to sum the $k=0,1,2,3,4$ terms of the Binomial distribution and subtract it from unity. The binomial sum evaluation is ***easily performed on a calculator*** or even "by hand" because $[^{10000}C_k](10^{-4})^k(1-10^{-4})^{10000-k} \approx [(10^{+4})^k/k!](10^{-4})^k(.36786) \approx (.36786)/k!$; the sum from $k=0,..,4$ is simply $(.36786)[1+1+1/2+1/6+1/24] = .99629$ and $1-.99629 = .00371$. The Poisson approximation is easily performed by summing 5 terms as shown on the slide and yields $\Pr[K >4\text{ err}] = 1-\Pr[K \le 4\text{ err}] = .00366$. In this case the approximate Binomial and exact Poisson sums yield the same results; however, the binomial coefficient $^{10000}C_k$ for large k will present numerical problems, whereas the Poisson sum generally does not.

Note that the alternate question asking us to instead find a bit error rate "p" such that there are *no more than* 4 errors, is expressed by the equation, $\Pr[K >4\text{ err}] < 1\%$, and therefore requires the solution of a 10,000th degree polynomial equation in "p" ! This is best solved by trial and error iterations.

8.2.6 Approximate Single Binomial Term - Gaussian *versus* Poisson

Approximate Single Binomial Term - Gaussian *versus* Poisson

1) Exact Binomial Evaluation n=40 ; p=0.1 $\Pr[Z=20] = \binom{40}{20}(.1)^{20}(.9)^{40-20} = 1.676 \times 10^{-10}$

Note: Binomial highly asymmetric for p=0.1   *Equivalent to Sum of 40 IID Bernoulli RVs*

2) Gaussian Approximation   Mean : $\mu_Z = E[Z] = n \cdot p = 40(.1) = 4$;

$$Z_{40} = \sum_{i=1}^{40} X_i$$

Sigma : $\sigma_Z = \sqrt{n \cdot p \cdot q} = \sqrt{40 \cdot (.1) \cdot (.9)} = \sqrt{3.6} = 1.9$

Standardize : $Y = \dfrac{Z - \mu_Z}{\sigma_Z} = \dfrac{Z-4}{1.90}$ $\Pr[Z=20] \cong \Pr\left[\dfrac{19.5-4}{1.90} \le Z \le \dfrac{20.5-4}{1.90}\right] = \Phi(8.68) - \Phi(8.16) = 1.655 \times 10^{-16}$

Conditions on p and q for "good" approx.

$n \cdot p > 2\sigma_Z$ $np = 4 > 2(1.90) = 3.8$

$4 > 3.8$ marginal for "p"

$n \cdot q > 3\sigma_Z$ $nq = 40(.9) = 36 > 3(1.9) = 5.7$

$36 > 5.7$ OK for "q"

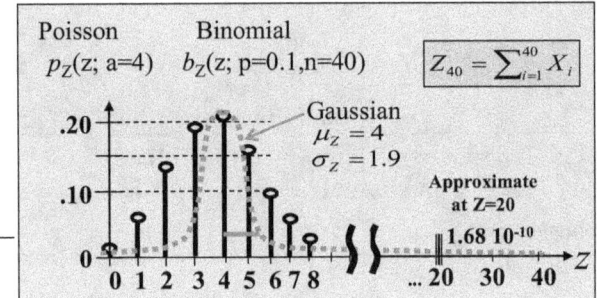

3) Poisson Approximation   n=40 ; p=0.1

$$K = Z_{40} = \sum_{i=1}^{40} X_i \qquad a = \lambda\, t = n\, p = 40(.1) = 4$$

Note: $a = \lambda\, t = \lambda\,(n\,\Delta t) = n\,(\lambda\,\Delta t) = n\,p$

$$\Pr[K=20] = p_K(20, a=4) = \frac{(4)^{20}}{20!} e^{-4} = 8.277 \times 10^{-9}$$

Let us revisit the evaluation of a single Binomial term for n=40, z=20, but now for a small p=0.1 (rather than p=q=.5); the exact evaluation of this Binomial term gives $\Pr[Z=20] = 1.676 \times 10^{-10}$, while the CLT Gaussian approximation in the middle panel gives 1.655×10^{-16} which is 10^{-6} too small. A short calculation given on the slide shows that the two criteria for a good Gaussian approximation namely, (i) $np > 2\,\sigma_Z$ and (ii) $nq > 3\sigma_Z$, are both satisfied, *viz.*, (i) $np = 4 > 2(npq)^{1/2} = 3.8$ and (ii) $nq = 36 > 3(npq)^{1/2} = 5.7$, although (i) is only marginally satisfied. However, because of the small value of p, we expect a Poisson approximation to the Binomial term to give better results; this is indeed the case as we have

$$a = \lambda\, t = \lambda\, n\, \Delta t = n(\lambda\, \Delta t) = n\, p = 40(.1) = 4$$

so that

$$p_K(20, a=4) = e^{-4}\, 4^{20}/20! = 8.277 \; 10^{-9}$$

which is still about 50 times too large. However, it is much closer to the actual value of 1.676×10^{-10} than the Gaussian approximation 1.655×10^{-16} which is 6 orders of magnitude too small.

This improvement is not unexpected since the Poisson distribution is derived from the Binomial on the basis of a small probability p and thus displays the same shape asymmetry as the Binomial. In fact it tracks the Binomial so well that the two stick plots are indistinguishable in the sketch; the evaluation of but a single term means there will be no loss of precision due to subtraction of small numbers. On the other hand, the approximating Gaussian $\sim N(\mu_Z = 4, \sigma_Z^2 = 1.9^2)$ is a narrow *symmetric* distribution about its mean at Z=4; but we must evaluate it using Φ in an interval about Z=20 which is in the vanishingly small tail region shown in the sketch. Thus the Gaussian yields poor results because it does not capture the asymmetric nature of the Binomial and is evaluated in the computationally difficult tail region where it subtracts nearly equal small numbers. The Poisson is somewhat better since it at least follows the asymmetric shape of the Binomial.

8.2.7 Approximate Poisson Sum using Gaussian

Approximate Poisson Sum using Gaussian

Sum of Poisson Terms

K Poisson RV $\quad E[K] = \lambda t = 100 \;;\; Var(K) = 100$

$$P[K \geq 120] = e^{-100} \sum_{k=120}^{\infty} \frac{100^k}{k!} = 1 - e^{-100} \sum_{k=0}^{119} \frac{100^k}{k!} = 1 - 0.97177 = \boxed{.0282}$$

MatLab result

Z_{100} : Sum of 100 IID Poisson RVs "X"

Key: Since the sum of Poissons is Poisson with *Mean and Variance being sum of Poisson parameters,* we can consider the variable K in problem statement to be **the sum of 100 IID Poisson RVs**

$$Z_{100} = \sum_{i=1}^{100} X_i$$

$$E[Z_{100}] = \sum_{i=1}^{100} E[X_i] = 100 \cdot 1$$

$$Var(Z_{100}) = 100 \cdot 1$$

Gaussian Approximation using Z_{100}

Standardize & Approx with Gauss CDF

$$Y = \frac{Z_{100} - 100}{\sqrt{100}}$$

$$Pr[Z_{100} \geq 120] = Pr\left[Y \geq \frac{120 - 100}{10}\right]$$

$$= 1 - \Phi(2) = \boxed{.0228}$$

Note: Continuity correction uses 120 -0.5 = 119.5 and yields .0256

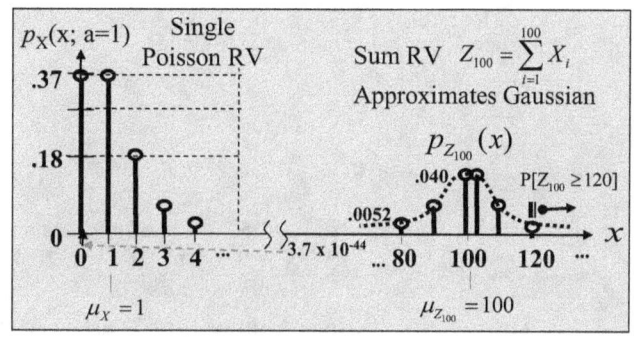

Here is an example in which the sum of 119 terms of a Poisson PMF are replaced by a suitable Gaussian Φ evaluation. The evaluation of the Poisson probability $Pr[K \geq 120]$ is found by taking the complement of the large, but finite sum $Pr[K \leq 119]$. This problem causes an overflow on most calculators because of the 120! evaluation, but is easily done on a computer; using MatLab® the answer is found to be $Pr[K \geq 120] = .0282$. An application of the CLT theorem provides an alternate method that does not have numerical round off issues and is reasoned as follows:

(i) The Poisson RV K with mean $\mu_K = 100$ and variance $\sigma_K^2 = 100$ can be thought of as the sum of 100 IID Poisson RVs each with mean $\mu_K = 1$ and variance $\sigma_K^2 = 1$. This is true because the sum of 2 independent Poisson RVs is again a Poisson RV whose mean and variance are the sum of the two means and two variances (see Slide#5-5.)

(ii) Thus the sum RV Z_{100} has the same mean and variance as K and we can evaluate $Pr[K > 120]$ $= Pr[Z_{100} > 120]$ using a Gaussian Φ evaluation on the standardized RV $Y = (Z_{100}-100)/10$ as $Pr[Y > (120-100)/10] = 1 - \Phi(2) = .0228$. This result is ~ 19% too low compared with the MatLab® answer of .0282 using the direct sum of Poisson terms. Using the continuity correction replaces 120 by 119.5 and evaluates $1 - \Phi(1.95) = .0256$ which is only 9% too low.

This approximation of the sum of *n* Poisson RVs using a Gaussian is yet another example of the Central Limit Theorem; the sketched illustration in this slide should be compared with the plots resulting from the explicit simulation of the sum of *n* exponential RVs shown in Slide#8-7.

8.3 General Bounds on Probability Distribution Tails

General Bounds on Probability Distribution Tails

1) Markov Inequality: given *mean non-neg RV*

If X is RV whose values are "**non-negative**", Then

$$P[X \geq c] \leq \frac{E[X]}{c} = \frac{\mu_X}{c} \quad where \ c > 0$$

Proof:

$$E[X] = \int_{x=0}^{\infty} x \cdot f_X(x)dx = \int_{x=0}^{c} x \cdot \underbrace{f_X(x)}_{\geq 0}dx + \int_{x=c}^{\infty} x \cdot f_X(x)dx$$

$$\geq \int_{\substack{x=c \\ \geq c}}^{\infty} x \cdot f_X(x)dx \geq c \cdot \int_{x=c}^{\infty} f_X(x)dx = c \cdot P[x \geq c]$$

$$\therefore P[x \geq c] \leq \frac{E[X]}{c} \quad Q.E.D.$$

$$P[X \geq r \cdot \mu_X] \leq \frac{1}{r} \quad where \ \boxed{c = r \cdot \mu_X}$$

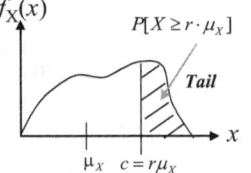

2nd Form says Prob X is greater than r times its mean is less than "1/r"

2) Chebyshev Inequality: given *mean & variance, pos/neg RV*

If X is RV with mean μ_X and variance σ_X^2, Then

Apply Markov Inequality : $X \rightarrow (X - \mu_X)^2$ & $c \rightarrow k^2$

$$P\left[\underbrace{(X - \mu_X)^2 \geq k^2}_{Event \ E_1}\right] \leq \frac{E\left[(X - \mu_X)^2\right]}{k^2} = \frac{\sigma_X^2}{k^2}$$

Equivalent Events : $E_1 = \{(X - \mu_X)^2 \geq k^2\}$ & $E_2 = \{|X - \mu_X| \geq k\}$

$$\therefore P\left[\underbrace{|X - \mu_X| \geq k}_{Event \ E_2}\right] \leq \frac{\sigma_X^2}{k^2} \quad \text{is true}$$

$$P\left[|X - \mu_X| \geq k\right] \leq \frac{\sigma_X^2}{k^2} \quad where \ k > 0$$

$$P\left[|X - \mu_X| \geq r\sigma_X\right] \leq \frac{1}{r^2} \quad where \ \boxed{k = r\sigma_X}$$

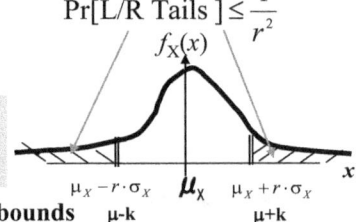

$$\Pr[L/R \ Tails] \leq \frac{1}{r^2}$$

2nd Form says Prob X differs in abs value from its mean by r std deviations is less " $1/r^2$ "

Note: Neither 1) or 2) are very "tight" bounds

The Markov and Chebyshev Inequalities are theoretical bounds on the amount of cumulative probability that **resides in the tail(s)** of **any distribution**, even if only limited information is known about its **mean** μ_X and/or its **variance** σ_X^2.

1) The Markov Inequality considers a RV X with known mean μ_X and assumes that the density $f_X(x)$ is non-zero only for positive values of X as illustrated in the upper sketch. The 1st form of the inequality states that the probability of being in the "tail region (hatched) beyond X=c" has an upper bound equal to its mean μ_X divided by "c" , *i.e.*, $\Pr[X \geq c] \leq \mu_X /c$. The 2nd more intuitive form results from the substitution c = r μ_X and states that the probability that X is greater than r times its mean is less than or equal to 1/r. This means that tail region starting at $2\mu_X$ contains $\leq 1/2$ the total probability, the tail region starting at $3\mu_X$ contains $\leq 1/3$ the total probability, *etc.* .

2) The Chebyshev Inequality considers a RV X on $(-\infty, +\infty)$ with known mean μ_X and variance σ_X^2 as illustrated in the lower sketch. The 1st form of the inequality states that the probability of being in the "two tail regions (hatched) outside the interval $[(\mu_X - k), (\mu_X + k)]$" is less than or equal to its variance σ_X^2 divided by "k^2", *i.e.*, $\Pr[||X - \mu_X| \geq k] \leq \sigma_X^2 / k^2$. The 2nd more intuitive form results from the substitution k = $r\sigma_X$ and states that the probability that X exceeds its mean μ_X in absolute value by more than r-sigmas is $1/r^2$. This means that left and right tail regions beyond $\pm 2\sigma_X$ contains no more than $1/2^2 = 1/4$ the total probability, and likewise the tail regions beyond $\pm 3\sigma_X$ contains no more than $1/3^2 = 1/9$ the total probability.

Proofs: The **Markov inequality** is easily proved by (i) using the definition of E[X], (ii)breaking the resulting integral up into a main part [0,c] and a tail part [c, ∞) (iii) recognizing that the two integrals can only make positive contributions (since x ≥ 0 and $f_X(X) \geq 0$), (iv) dropping the 1st integral so that E[X] is greater than the 2nd integral, and (v) minimizing the remaining integral by taking its smallest value x=c outside the integral to leave the Markov inequality $\Pr[X \geq c] \leq E[X] /c$.

The **Chebyshev inequality** is proved by letting $X \rightarrow (X - \mu_X)^2$ and $c \rightarrow k^2$ in the Markov inequality to yield $\Pr[(X - \mu_X)^2 \geq k^2] \leq E[(X - \mu_X)^2] / k^2$; specifying the tail regions now in terms of the equivalent absolute value inequality $| X - \mu_X | \geq k$ suffices to give the Markov Inequality as $\Pr[| X - \mu_X | \geq k] \leq E[(X - \mu_X)^2] / k^2 = \sigma_X^2 / k^2$.

8.3.1 Examples Using Markov and Chebyshev Bounds

Examples Using Markov and Chebyshev Bounds

Markov

Prob "value" of RV X exceeds "r" times its mean is 1/r

$$P[X \geq r\mu_X] \leq \frac{1}{r}$$

or

$$P[X \geq c] \leq \frac{E[X]}{c} = \frac{\mu_X}{c}$$

Chebyshev

Prob "deviation" of RV X exceeds "r" times its std dev r σ_X is $1/r^2$

$$P\big[|X - \mu_X| \geq r\sigma_X\big] \leq \frac{1}{r^2}$$

or

$$P\big[|X - \mu_X| \geq k\big] \leq \frac{\sigma_X^2}{k^2}$$

Examples:

Kindergarten Class mean height = 42" Find bound on Prob of a student being taller than 63"

$\mu_X = 42 \quad r \cdot 42 = 63 \Rightarrow r = 1.5 \, ; \; \Pr[X \geq 1.5 \cdot 42] \leq 1/1.5 = 66.7\%$

Directly: $\mu_X / c = 42/63 = 2/3$

Note that for r =1 the bound is "1" or 100%; Thus useful bounds require r >1

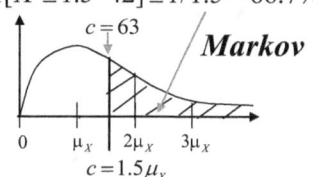

Factory production

a) Given **mean =50**, find bound on Prob production exceeds 75, *i.e.*, Prob[X>75] $\quad P[X \geq 75] \leq \dfrac{E[X]}{c} = \dfrac{50}{75} = .667 \quad$ *Markov*

Note an **upper bound**: at most 66.7%

b) Also given **variance = 25** , find bound on Prob production **between 40 and 60** $\quad P\big[|X - 50| \geq 10\big] \leq \dfrac{25}{10^2} = .25 \;$ ***Chebyshev***

$$\Rightarrow \; 1 - P\big[|X - 50| \geq 10\big] \geq 1 - .25 = .75$$

75% is a **lower bound**

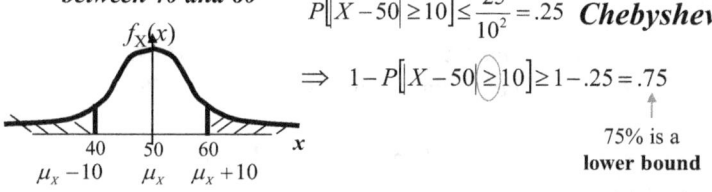

Alternately: $10 = 2\sigma_X \Rightarrow r=2$ thus $1/r^2 = 1/2^2 = .25$

Here are two examples of the application of the Markov and Chebyshev Bounds. The two forms for each are stated on the LHS of the slide for reference purposes. The decision to use one or the other of these bounds depends upon what type of information we have about the distribution. Thus if the RV X takes on only positive values and we only know its mean, μ_X , then we must use the Markov bound. On the other hand, if the RV X takes on both positive and negative values and we know the mean, μ_X , and variance, σ_X^2, then we can use the Chebyshev bound. If in the latter case the RV X takes on only positive values, then we could use either Chebyshev or Markov bounds, but we would choose Chebyshev over Markov because it uses more of the information and hence will always be a tighter upper bound. Neither of these bounds is very tight because the information about the distribution is very limited; knowing the actual distribution always yields the best bounds.

The mean height in a Kindergarten Class is $\mu_X = 42$" and we are asked "what is the probability of a student being taller than 63?" Short of knowing the actual distribution, the best we can do is use the Markov inequality to find an upper bound Pr[X>63] < 42/63=.67 or 67%. This is also easily computed if we realize that the tail is the region beyond 63"= 1.5(42") so r=1.5 and the answer is 1/1.5 =2/3=.67.

2) The factory production has a mean output μ_X = 50 units and we are asked
(a) "what is the probability of a 75 unit output?" This again involves a positive quantity X the number of units and we choose the Markov bound for 1.5(50) = 75 units so again r=1.5 and the resulting probability is 67% .
(b) If we are also given the variance of the production σ_X^2 = 25 the additional information allows us to use the Chebyshev bound to find the probability in the tails on either side of the mean of 50. Thus, if we find the probability in the 2-sigma tails (r=2) to left of 50-10 and to the right of 50+10 as Pr[Tails] $\leq 1/2^2$ = 25%. Hence the production within the bounds [40, 60] is the complementary probability
Pr[40 $\leq$ X $\leq$ 60] =1-Pr[Tails] $\geq$ 1-.25 = .75 or at least 75%

8.3.2 Knowledge of Distribution Gives Tight Bounds Compared to Chebyshev

---

Knowledge of Distribution Gives Tight Bounds Compared to Chebyshev

$$P[|X - \mu_X| \geq r\sigma_X] \leq \frac{1}{r^2}$$

Prob "deviation" of RV
X exceeds "r" times its
std dev r σ_X is $1/r^2$

Chebyshev Bound: Prob X is in the L/R tails "3-sigma outside of mean"

$$P[|X - \mu_X| \geq 3\sigma_X] \leq \frac{1}{3^2} = .1111$$

Pr[L/R tails] $\leq \dfrac{1}{3^2}$

$f_X(x)$

Distribution not specified; general bound

3-σ Prob
99.73%

$\mu_X - 3 \cdot \sigma_X$ μ_X $\mu_X + 3 \cdot \sigma_X$ x

For a Gaussian we Calculate Exactly

$$P\left[\frac{|X - \mu_X|}{\sigma_X} \geq 3 \right] = 2 \cdot qNorm(3) = .0027$$

$$= 1 - \underbrace{[\Phi(3) - \Phi(-3)]}_{\text{3-sigma Gaussian Prob=99.73\%}} = 1 - [2\Phi(3) - 1] = 2[1 - \Phi(3)] = .0027$$

Note1: *You cannot expect a very tight bound when you do not have much information about the distribution*

Note2: *Recall Gauss Probability within 3-σ of mean is .9973 Probability in the Tails beyond 3-σ is 1-.9973 =.0027*

The figure illustrates the Chebyshev bound which states that an upper bound for the probability contained within the Left and Right tails located 3-sigma from the mean is $1/3^2 = 1/9 = .1111$. Now suppose that the distribution is known to be Gaussian so we can compute the tail probability exactly by first transforming to a standardized variable Y=(X- μ_X)/σ_X and then using a Gaussian table (or calculator upper tail probability function) to find Pr[Y>3] =1 – [Φ(3)– Φ(-3)]=2[1– Φ(3)] =.0027, which is 40 times smaller than the Chebyshev upper bound .1111 . The Chebyshev upper bound is some 40 times larger than it needs to be; this "generous upper bound" is to be expected since the result must be true for all distributions. Recall that the derivation required virtually no information about the unknown distribution except that it has a finite mean and variance. The Chebyshev result is indeed an upper bound, but is often useless for any practical purposes. Still, when nothing else is known about the distribution, it does give information that may be useful in some circumstances.

8.3.3 Another Chebyshev Bound Example

Another Chebyshev Bound Example

For a Uniform Distribution we Calculate Exactly

$$f_X(x) = \begin{cases} 1/10 & x \in [-5,5] \\ 0 & else \end{cases}$$

$$\mu_X = E[X] = 0$$

$$\sigma_X^2 = E[X^2] = \int_{-5}^{5} \frac{1}{10} \cdot x^2 dx = \frac{25}{3} = 8.333$$

$$P[|X - 0| \ge 3] = \int_{3}^{5} \frac{1}{10} dx + \int_{-5}^{-3} \frac{1}{10} dx = \frac{4}{10} = 0.4$$

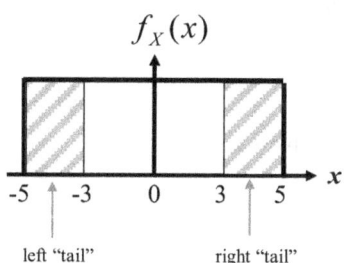

$f_X(x)$

left "tail" right "tail"

Chebyshev Bound: mean=0; variance =25/3

Compare

$$P[|X - 0| \ge 3] \le \frac{\sigma_X^2}{k^2} = \frac{(25/3)}{3^2} = .9259$$

Alternately, $\quad P[|X - \mu_X| \ge r\sigma_X] \le \frac{1}{r^2}$

$$3 = r\sigma = r\sqrt{\frac{25}{3}} \Rightarrow r = \frac{3\sqrt{3}}{5} \qquad \frac{1}{r^2} = \left(\frac{5}{3\sqrt{3}}\right)^2 = \frac{25}{27} = .9259$$

Here is another example that compares the Chebyshev upper bound with the value computed using the actual distribution. If we do not know the distribution but compute (from a data set) or are otherwise given that its mean is $\mu = 0$ and its variance is $\sigma^2 = 25/3$, then we can immediately write down the Chebyshev bound for the tail probability as $Pr[|X-0| \ge 3] \le (25/3)/3^2 = .9259$. This tells us the maximum amount of probability in the tails of the distribution is less than 92.6%; this is an upper bound and the actual amount could be much less. On the other hand, if we in fact know that the data came from a uniform distribution $f_X(x) = 1/10$ on the interval [-5, 5], then we can calculate the actual probability contained in the left and right "tails" [-5,-3] and [3,5] (hashed regions in the figure) by direct integration of the uniform density and find $Pr[|X- 0| \ge 3] = .4$. This is less than half the very generous Chebyshev bound of .9259, so the information in this case pretty useless.

8.4 *Sample Mean (Average) Estimator*

> ### Sample Mean (Average) Estimator
> ### Number Angle Meas. for 0.5'Accuracy with 99% Probability
>
> Angle Measurements
> are IID RVs
> $$\Theta_i = \theta_{true} + E_i \;\; ; \;\; i = 1, 2, \cdots, n$$
>
> E_i = Random Error
> $\theta_{true} = 20^o 00'$ deg$-$min
>
> Errors E_i IID RVs : $E[E_i] = 0$; $Var(E_i) = 3.0 \, \text{min}^2$
>
> Measurements Θ_i : Expected Value = $E[\Theta_i] = E[\theta_{true} + E_i] = \theta_{true} + \underbrace{E[E_i]}_{=0} = \theta_{true}$
>
> $$Var(\Theta_i) = Var(\theta_{true} + E_i) = 0 + Var(E_i) = 3.0 \, \text{min}^2$$
>
> Estimator: Sample Mean RV : $\overline{\Theta}_n = \dfrac{1}{n} \displaystyle\sum_{i=1}^{n} \Theta_i$ (Average of Measurements)
>
> ---
>
> *Determine how many measurements "n" to insure with 99% certainty that the estimate is within 0.5' of the true value.* $\Pr\left[\left|\overline{\Theta}_n - \theta_{true}\right| < 0.5'\right] = .99$
>
> ---
>
> *Statistics of Estimator: Mean & Variance*
>
> $$E[\overline{\Theta}_n] = E\left[\frac{1}{n}\sum_{i=1}^{n}\Theta_i\right] = \frac{1}{n}\sum_{i=1}^{n}\underbrace{E[\Theta_i]}_{=\theta_{true}} = \frac{1}{n}\cdot n \cdot \theta_{true} = \theta_{true} \Rightarrow \overline{\Theta}_n \;\; \textbf{"Unbiased Estimator"}$$
>
> $$Var(\overline{\Theta}_n) = Var\left(\frac{1}{n}\sum_{i=1}^{n}\Theta_i\right) = \frac{1}{n^2}\cdot\sum_{i=1}^{n}\underbrace{Var(\Theta_i)}_{=3.0\,\text{min}^2} = \frac{1}{n^2}\cdot n \cdot Var(E_i) = \frac{3.0\,\text{min}^2}{n} \quad \textbf{"Variance Reduction"}$$
>
> $$\sigma_{\overline{\Theta}_n} = \sqrt{\frac{3}{n}} \;\text{min}$$

Because the measurement of a physical quantity such as an angle usually has random noise E_i added to it, the measured quantity will not equal the "true" value θ_{true} but will be in fact be a new RV $\Theta_i = \theta_{true} + E_i$ obtained by adding the random noise to it.   If each indexed noise contribution E_i is a member of a set of IID RVs with zero mean $\mu_E = 0$ and variance $\sigma_E^2 = 3 \, \text{min}^2$, then the expected value or mean of a single measurement will be $E[\Theta_i] = \mu_{\Theta i} = \theta_{true}$ and its variance will be $\sigma_{\Theta i}^2 = 3 \, \text{min}^2$.

Usually a series of measurements is taken and averaged in order to cancel the noise components of the individual measurements; this average of n IID RVs Θ_i is also a RV called the "sample mean" shown as the first boxed equation labeled "Estimator". The sample mean has an expected value (mean of the sample average) equal to $E[\Theta_n^{av}] = \theta_{true}$ and a variance $var(\Theta_n^{av}) = (3 \, \text{min}^2)/n$ that is reduced by a factor of $1/n$ as detailed on the slide.

The fact that the expected value of the estimator is equal to the true value θ_{true} makes it an "unbiased estimator" and the reduction in variance by the factor $1/n$ is the key reason for averaging many measurements in the first place, *i.e.*, to obtain a more accurate mean value and a tighter estimate of the variance. An important question is how many measurements "n" are needed to attain some specific goal. This is determined by specifying a tolerance for deviation from the true mean value (say 0.5′ sec of arc) and a percentage (99%) of the time for it to be satisfied. This is easily expressed by the simple (Type A) equation (analogous to the Chebyshev inequality) ,

$$\Pr[|\Theta_n^{av} - \theta_{true}| < 0.5'] = .99$$

The answer to this question "how many measurements?" is computed on the next slide, first assuming a zero mean Gaussian noise with variance 3 min^2 (or N(0, 3)) to give n = 80, and then using the Gauss-Markov bound for an unspecified distribution to give n=1200. Clearly knowledge of the noise distribution is quite beneficial in planning experimental observations with given accuracy goals.

8.4.1 Two Estimates for Number of Samples "n" (Gauss and Chebyshev)

Two Estimates for Number of Samples "n" (Gauss and Chebyshev)

1) Gaussian Standardize $Y = \dfrac{\overline{\Theta}_n - \mu_{\overline{\Theta}_n}}{\sigma_{\overline{\Theta}_n}} = \dfrac{\overline{\Theta}_n - \theta_{true}}{\sqrt{3/n}} = \sqrt{n/3}\,(\overline{\Theta}_n - \theta_{true})$

$$\Pr[-0.5' \le (\overline{\Theta}_n - \theta_{true}) \le +0.5'] = \Pr\left[-0.5\sqrt{\frac{n}{3}} \le Y \le +0.5\sqrt{\frac{n}{3}}\right] = .99$$

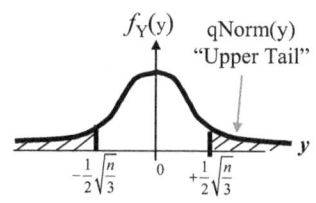

$f_Y(y)$ qNorm(y) "Upper Tail"

$$= \Phi\left(+0.5\sqrt{\frac{n}{3}}\right) - \Phi\left(-0.5\sqrt{\frac{n}{3}}\right) = 2 \cdot \Phi\left(0.5\sqrt{\frac{n}{3}}\right) - 1$$

$$\Phi\left(0.5\sqrt{\frac{n}{3}}\right) = \frac{1+.99}{2} = .995 \quad \Rightarrow \quad 2.576 = 0.5\sqrt{\frac{n}{3}} \quad \Longrightarrow \quad \boxed{n = 79.6 = 80 \text{ meas.}}$$

---

2) Chebyshev Estimate Given Mean and Variance

$$\Pr[|\overline{\Theta}_n - \theta_{true}| \le 0.5] = 1 - \Pr[|\overline{\Theta}_n - \theta_{true}| > 0.5]$$

$$= 1 - \frac{\sigma^2_{\overline{\Theta}_n}}{(0.5)^2} = 1 - \frac{(3/n)}{(0.5)^2} = 1 - \frac{12}{n} = .99$$

$$\boxed{n = \frac{12}{1-.99} = 1200 \text{ meas.}}$$

Knowledge of the distribution (Gaussian) reduces the estimated number of measurements substantially from

	1200	=>	80
	Arbitrary		Gaussian

The number of measurements needed to provide a specified accuracy (.5′) for a given percentage of the time (99%) is easily calculated if we first define the standardized RV: $Y=(\Theta_n{}^{av} - \theta_{true}) / (3/n)^{1/2}$. The Y equation transforms the deviation limits $-.5' \le (\Theta_n{}^{av} - \theta_{true}) \le .5'$ to $-.5'(n/3)^{1/2} \le Y \le .5'(n/3)^{1/2}$, which then leads to the probability bound condition $P[-.5'(n/3)^{1/2} \le Y \le .5'(n/3)^{1/2}] \le .99$ as shown on the slide. The function qNorm$(.5(n/3)^{1/2})$ represents the upper tail illustrated in the inset figure; subtracting both tails from unity leads to the condition qNorm$(.5(n/3)^{1/2}) = .005$, whose solution is n=80 measurements. (The equivalent expression using Φ on the slide is just the complement or $\Phi(.5(n/3)^{1/2}) = 1-.995 = .005$). Alternately given only the mean and variance we may apply the Chebyshev estimate directly to find the tail probability bound $\Pr[|\Theta_n{}^{av} - \theta_{true}| > .5'] = \sigma_{\Theta av}{}^2 / (.5)^2 = (3/n)/.5^2$, and upon setting this equal to $1-.99$ leads to the result n=1200 measurements. Thus if the distribution is known to be Gaussian, then we need make only 80 measurements to attain our goal; however, if we have no idea what the distribution is, then Chebyshev says we need no more than 1200 measurements. It clearly pays to know the distribution you are working with.

8.4.2 Chebyshev Bound Leads to Weak Law of Large Numbers

Chebyshev Bound Leads to Weak Law of Large Numbers

$$\lim_{n\to\infty} \Pr\left\{ \left| \underbrace{\frac{1}{n}\sum_{i=1}^{n} X_i}_{\equiv \bar{X}_n} - \underbrace{E[\bar{X}_n]}_{\equiv \mu_{\bar{X}_n}} \right| \geq \varepsilon \right\} = 0$$

Probability that "sample mean" differs from the "true mean" by an amount ε approaches zero as $n\to\infty$

$f_{\bar{X}_n}(\bar{x}_n)$

$\sigma_X/\sqrt{n}$

$\bar{x}_n$

$\bar{\mu}_X - \varepsilon$ $\bar{\mu}_X$ $\bar{\mu}_X + \varepsilon$

sample mean   true mean $\varepsilon > 0$

The "sample mean" $\bar{X}_n$ is a **Random Variable**

$\Pr[\text{L/R tails}] \underset{n\to\infty}{\longrightarrow} 0$

Proof: $\bar{X}_n = \frac{1}{n}\sum_{i=1}^{n} X_i$ Indep Iden. Distrib X_i with "common" μ_X & σ_X

Compute Mean and Variance of the Sample Mean (Average) RV: $\bar{X}_n$

$$\boxed{\mu_{\bar{X}_n}} \equiv E[\bar{X}_n] = E\left[\frac{1}{n}\sum_{i=1}^{n} X_i\right] = \frac{1}{n}\sum_{i=1}^{n} E[X_i] = \frac{n\mu_X}{n} = \boxed{\mu_X}$$ *sample mean equals mean of distribution*

$$\boxed{\sigma_{\bar{X}_n}^2} \equiv Var(\bar{X}_n) = \left(\frac{1}{n}\right)^2 \sum_{i=1}^{n} Var(X_i) = \left(\frac{1}{n}\right)^2 n \cdot \underbrace{Var(X)}_{=\sigma_X^2} = \boxed{\frac{\sigma_X^2}{n}}$$ *Note: Averaging "reduces" variance by factor of "n"*

Apply Chebyshev Upper Bound and take limit

$$\lim_{n\to\infty} P\left[\left|\bar{X}_n - \mu_{\bar{X}_n}\right| \geq \varepsilon\right] = \lim_{n\to\infty} P\left[\left|\bar{X}_n - \mu_X\right| \geq \varepsilon\right] \leq \lim_{n\to\infty}\frac{\sigma_{\bar{X}_n}^2}{\varepsilon^2} = \lim_{n\to\infty}\frac{(\sigma_X^2/n)}{\varepsilon^2} = 0$$

shown to be equal

The *sample mean* X_n^{bar} is defined to be the average of n independent identically distributed (IID) RVs X_i i=1,2,...,n, and is itself a random variable. The statistics of this RV are easily computed and we find the mean $E[X_n^{bar}] = \mu_X$, the common mean of all the X_i; similarly the variance of X_n^{bar} is that of the underlying distribution σ_X^2 *reduced* n, *viz.*, $var(X_n^{bar}) = \sigma_X^2/n$. If we now apply the Chebyshev inequality to the deviation of the sample mean $|X_n^{bar} - E[X_n^{bar}]|$ we immediately have

$$\Pr[\,|X_n^{bar} - E[X_n^{bar}]| \geq \varepsilon\,] \leq \sigma_{X_n}^{bar\,2}/\varepsilon^2$$

as a bound on the total probability contained in the two tails of the distribution one to the left of $\mu_X - \varepsilon$ and the other to the right of $\mu_X + \varepsilon$. Substituting the values for the mean and variance of X_n^{bar} into the above Chebyshev inequality and taking the limit as $n\to\infty$ yields

$$\lim_{n\to\infty} \Pr[\,|X_n^{bar} - \mu_X| \geq \varepsilon\,] \leq \lim_{n\to\infty}(1/n)\sigma_X^2/\varepsilon^2 = 0$$

This limit states the intuitive notion that as the number of variables summed approaches ∞, the probability in the tails of the distribution approaches zero. The result is known as the "Weak Law of Large Numbers" and is stated more precisely in the following form:

"The probability that the sample mean X_n^{bar} deviates from the true mean μ_X by more than a small value ε, approaches zero in the limit that the number of terms being averaged approaches infinity."

This (weak) law of large numbers supports the notion that (in a probabilistic sense) sample averaging provides a good estimate for any physical measurement subject to random noise. This means that the number of times that the deviation exceeds the ε-value becomes smaller with larger n and hence the probability of the resulting estimate being correct is close to unity. This decrease of probability in the tails is illustrated by the dotted curve which squeezes more and more of the probability into a higher and narrower peak as n increases because of the reduction in variance by 1/n.

8.5 Two Laws of Large Numbers – Experimental "Law of Averages"

Two Laws of Large Numbers – Experimental "Law of Averages"

1) Weak Law of Large Numbers

The sum variable Z is a Random Variable $\quad Z = \sum_{i=1}^{n} X_i$

Compute the "average" of n samples of X dividing Z by n (Sample Mean RV) $\quad \bar{Z}_n = \dfrac{Z}{n}$

$$E[\bar{Z}_n] = E\left[\frac{Z}{n}\right] = \frac{1}{n} E\left[\sum_{i=1}^{n} X_i\right] = \frac{1}{n} \cdot n \cdot E[X] = \mu_X$$

$f_{\bar{Z}_n}(\bar{z}_n)$

Increased n narrows the distribution

$\sigma_X / \sqrt{n}$

$\bar{z}_n$

$\mu_X - \varepsilon \qquad \mu_X \qquad \mu_X + \varepsilon$

Pr[L/R tails] → 0

The Weak Theorem states: *The experimental average z_n exceeds the underlying mean μ_x by an amount ε in each tail of the distribution $f_{\bar{z}_n}(z_n)$ an infinite number of times! However, as the number of trials n approaches infinity, the probability of this happening approaches zero because the distribution "sharpens" with $\sigma_n \to 0$*

"Sample Mean" $\bar{z}_n$ probabilistically approaches μ_X

$$\lim_{n \to \infty} \Pr\left\{\left|\bar{Z}_n - \mu_X\right| \geq \varepsilon\right\} = 0$$

Note: $\bar{Z}_n$ is an Average on n samples $= \dfrac{\sum_{i=1}^{n} X_i}{n}$

while $E[\bar{Z}_n]$ is a statistical Expectation of that RV based on $f_{\bar{z}}(z_n)$, i.e., $\mu_{\bar{Z}_n} = \mu_X$ (see figure)

2) Strong Law of Large Numbers $\quad \lim_{n \to \infty} \Pr\left[\bar{Z}_n = \mu_X\right] = 1$

$$f_{\bar{Z}_n}(\bar{z}_n) \underset{n \to \infty}{=} \delta(\bar{z}_n - \mu_X)$$

The Strong Theorem states: that $\lim_{n \to \infty} \bar{Z}_n = \mu_X$ with a probability =1 => Certainty! *That is, as n approaches infinity, the Expt'l average z_n exceeds the underlying mean μ_x by ε only a finite number of times!!*

μ_X $\qquad \bar{z}_n$

There are a number of limit theorems and bounds in probability theory; we have already discussed the central limit theorem (CLT) which gives the Gaussian as the limiting distribution for summed random variables, and have previously discussed the Markov and Chebyshev limits on the size of the tail probabilities.

The *Weak Law of Large Numbers* has just been shown to be a consequence of the Chebyshev bound applied to the deviation of the sample mean X_n^{ave} from the true mean μ_X. From an experimental point of view this theorem states that averaging data over many samples n *reduces the variance* of the RV X_n^{ave} by a factor of 1/n giving more confidence that the estimated value is close to the true but un-measurable value μ_X. The sketch of the probability density function for X_n^{ave} about its mean $\mu_X^{ave} = \mu_X$ displays the left and right tails under which the Chebyshev inequality computes the probability. It is this tail probability that vanishes as n increases and the PDF variance decreases so its peak becomes more pronounced. The formal statement is repeated for reference purposes:

"The probability that the sample mean X_n^{ave} deviates from the true mean μ_X by more than a small value ε, approaches zero in the limit that the number of terms being averaged approaches infinity."

The **"Strong Law of Large Numbers"** makes a stronger statement about the approach of the sample mean X_n^{ave} to the true mean μ_X by stating that **all the probability** is concentrated at one point and therefore the PDF approaches a Dirac delta function located at the true mean $\delta(X_n^{ave} - \mu_X)$ in the limit as the number of samples being averaged n approaches infinity.

8.5.1 Strong Law Example: Experimental "Frequency-of-Occurrence"

Strong Law Example: Experimental "Frequency-of-Occurrence"

Limiting Frequencies-of-Occurrence for Heads & Tails are Equal

Mis-interpretation: **#Heads = #Tails**

Freq of occurrence are equal only
in the limit as n $\rightarrow \infty$:
$$\lim_{n\to\infty}\left(\frac{\#Heads}{n}\right)=\lim_{n\to\infty}\left(\frac{\#Tails}{n}\right)$$

Ex.: Consider 1st 10 flips of fair coin are "heads";
and that afterwards the "H" and "T" are evenly split;
$$N_H=10+\frac{n-10}{2}\quad;\quad N_T=\frac{n-10}{2}$$

Clearly the numbers of heads
and tails are never equal
$$f_H=\frac{10+\frac{n-10}{2}}{n}=\frac{n+10}{2n}\quad;\quad f_T=\frac{\frac{n-10}{2}}{n}=\frac{n-10}{2n}$$

However Limiting frequencies are
guaranteed to be equal **with certainty**
$$\lim_{n\to\infty} f_H = \lim_{n\to\infty} f_T = \frac{1}{2}$$

The repeated trials associated with coin flips is an example that can be interpreted in terms of the Strong Law of Large Numbers. It is clear that every flip is independent and so for a fair coin, the probability that any flip will produce a head H is 1/2 *independent of how many Heads or Tails have previously been flipped.*

The interpretation of the Strong Law of Large Numbers $\lim_{n\to\infty} \Pr[Z_n^{ave}=\mu_X]=1$ is as follows:

The average number of Head occurrences f_H is computed as the sum of successes (Heads) divided by the number of trials n (Heads + Tails). Although this number f_H (so-computed) may never be exactly equal to the expected value of the "underlying Bernoulli distribution" $\mu_X=E[X]=p=E[Heads]$, the Strong Law states that "in the limit as $n\to\infty$, the frequency of occurrence f_H equals the mean of the underlying distribution "p = 1/2" (for a fair coin) with complete certainty. Generally, in the case of an unfair coin (p $\neq$1/2) the frequency f_H approaches p (whatever its value) with complete certainty.

8.5.2 Strong Law and Expected Outcomes for a Large # of Coin Flips

Strong Law and Expected Outcomes for a Large # of Coin Flips

X_i = **Bernoulli RV** *indicating* the *success or failure* of the ith trial of an *Experimental Coin Flip*

$$f_{X_i}(x_i) = \begin{cases} p & X_i = 1 & \text{success Heads (H)} \\ 1-p & X_i = 0 & \text{failure Tails (T)} \end{cases}$$

$$\mu_{X_i} = E[X_i] = p \qquad Var(X_i) = pq$$

Mean of Underlying Distribution True mean

$$\mu_{X_i} = E[X_i] = p$$

Sum RV represents # Succ. in n trials (counts the # of "1"s)

$$Z_n = \sum_{i=1}^{n} X_i \qquad E[Z_n] = E\left[\sum_{i=1}^{n} X_i\right] = \sum_{i=1}^{n} E[X_i] = n \cdot p$$

Sample Mean RV

$$\bar{Z}_n = \frac{\sum_{i=1}^{n} X_i}{n} \qquad E[\bar{Z}_n] = E\left[\frac{\sum_{i=1}^{n} X_i}{n}\right] = \frac{1}{n}\sum_{i=1}^{n} E[X_i] = \frac{n \cdot p}{n} = p$$

Expected Value

$$E[\bar{Z}_n] = p = \mu_X$$

Strong Law Guarantees that as $n \to \infty$, a "single realization" satisfies $\bar{Z}_n = \mu_X$ with *"certainty"*

In words, as $n \to \infty$, a single realization of the "sample mean" RV equals the mean of the underlying Random Process *with probability "1"*

$$\lim_{n \to \infty} \Pr[\bar{Z}_n = \mu_X] = 1$$

" with certainty"

Strong Law Statement

The Experimental "freq-of-occur for Heads, f_n is *one realization or sample* of the RV $\bar{Z}_n = \bar{z}_n$ whose expected value is p ; *i.e.,* $f_n = \bar{z}_n$ As we increase the number of fair coin flips without limit the value of f_n will always approach the probability p of underlying distribution" $\Pr[Heads] = \lim_{n \to \infty} f_n = p$

Experimental Interpretation

If we sum n IID Bernoulli random variables, the resulting mean and variance of sum variable are found by just adding up the n independent contributions to give n·p and n·(pq) respectively. Subsequently dividing by the "n" results in a new RV called the sample mean or average and it is easily demonstrated that its expected value is $E[Z_n^{ave}] = p = \mu_X$ (that of the underlying Bernoulli RV). The Strong Law of Large Numbers states that in the limit $n \to \infty$ any realization of the sample mean RV Z_n^{ave} (obtained by averaging n Bernoulli trials) will equal μ_X with probability 1 (certainty) . The frequency-of-occurrence for Heads f_n is one realization of the "Sample Mean RV" Z_n^{ave} giving the number of heads in n coin flips. As we increase the number of coin flips without limit, the value f_n approaches p the probability for heads in the underlying Bernoulli distribution.

Thus simulating fair coin flips on a computer with larger and larger numbers of trials gives equal frequencies-of-occurrence for heads and for tails $f_n(H) = f_n(T) \to 1/2$. Similarly for any simulated distribution, as the number of samples generated increases the frequency-of-occurrence "histogram" will approach the values of the underlying distribution.

9 The Bivariate Gaussian Distribution

9.1 Matrix Formulation of the Bivariate Gaussian

Formulation of the Bivariate Gaussian

Two N(0,1) Indep. Gaussians X_1, X_2
$$f_{X_1 X_2}(x_1, x_2) = \frac{1}{\sqrt{2\pi}} e^{-x_1^2/2} \cdot \frac{1}{\sqrt{2\pi}} e^{-x_2^2/2}$$

Vector Notation $X = (X_1, X_2)$
$$f_{\vec{X}}(\vec{x}) = \frac{1}{(\sqrt{2\pi})^2} e^{-\vec{x}^T \vec{x}/2}$$

$$\vec{x} = \begin{bmatrix} x_1 \\ x_2 \end{bmatrix} \quad \text{Col. Vector}$$

$$\vec{x}^T = [x_1, x_2] \quad \text{Row. Vector}$$

Mean
$$\vec{\mu}_X \equiv E[\vec{X}] = \begin{bmatrix} E[X_1] \\ E[X_2] \end{bmatrix} = \begin{bmatrix} \mu_{X_1} \\ \mu_{X_2} \end{bmatrix}$$

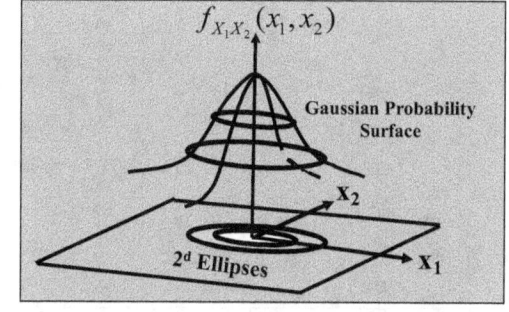

$f_{X_1 X_2}(x_1, x_2)$

Gaussian Probability Surface

2^d Ellipses

X_1 X_2

Covariance $K_{XX} = E[(X - \mu_X) \cdot (X - \mu_X)^T]$

(Drop Vector "Arrows")

$$\begin{pmatrix} \text{Cov.} \\ \text{Matrix} \end{pmatrix} = (\text{Col. Vector}) \times (\text{Row. Vector})$$

$$K_{XX} = E\left[\begin{pmatrix} X_1 - \mu_{X_1} \\ X_2 - \mu_{X_2} \end{pmatrix} \cdot (X_1 - \mu_{X_1} \quad X_2 - \mu_{X_2}) \right] = \begin{bmatrix} E[(X_1 - \mu_{X_1})^2] & E[(X_1 - \mu_{X_1})(X_2 - \mu_{X_2})] \\ E[(X_2 - \mu_{X_2})(X_1 - \mu_{X_1})] & E[(X_2 - \mu_{X_2})^2] \end{bmatrix}$$

The importance of the Gaussian is well established by the Central Limit Theorem and its multidimensional extension is a cornerstone of many scientific and engineering analyses. We initially consider a Bivariate Gaussian in some detail and then generalize those results to a multidimensional Gaussian and develop the important Gauss-Markov Theorem.

The product of two N(0,1) Gaussian densities $f_{X1}(x_1)$ $f_{X2}(x_2)$ yields a joint Gaussian density $f_{X1X2}(x_1, x_2)$ which can be expressed in the vector form $f_{X1X2}(x_1, x_2) = (1/(2\pi)^{1/2})^2 \exp(-\mathbf{x}^T\mathbf{x}/2)$, where the column vector $\mathbf{x} = [x_1, x_2]^T$.

The mean is a vector obtained in the usual manner by taking the expectation $\mu_x = E[\mathbf{X}]$ and the covariance is defined by the expectation of the "outer product" $E[(\mathbf{X}-\mu_X)(\mathbf{X}-\mu_X)^T]$ which forms a 2×2 matrix with the variances of X_1 and X_2 on the diagonal and the covariances on the off the diagonal. Since we started with two independent Gaussian RVs N(0,1) with zero mean and unit variance we expect the mean vector to be zero $\mu_X = [\mu_{X1}, \mu_{X2}]^T = [0,0]^T$ and the covariance to be a diagonal matrix with "1"s on the diagonal. On the next slide we shall explicitly show that this is the case.

A typical Bivariate Gaussian density function is illustrated in the figure as a surface above the x_1-x_2 plane. In general the "level surfaces" obtained by cuts parallel to the x_1-x_2 plane are ellipses centered about the mean values as origin and vertical cuts have a typical Gaussian shape. For the case of two independent N(0,1) Gaussians (with the same variance =1) the level surfaces are circles centered on the origin and the vertical cuts are all N(0,1) Gaussians; in other words the bivariate density is a surface of rotation about the vertical axis.

9.1.1 Computing Covariance Matrix for Independent RVs

Computing Covariance Matrix for Independent RVs

$$K_{11} = E[(X_1 - \mu_{X_1})(X_1 - \mu_{X_1})] = \int dx_1 \int dx_2 f_{X_1 X_2}(x_1, x_2) \cdot (x_1 - \mu_{X_1})^2$$

independence

$$= \int dx_1 \int dx_2 f_{X_1}(x_1) \cdot f_{X_2}(x_2) \cdot (x_1 - \mu_{X_1})^2$$

$$= \int dx_1 f_{X_1}(x_1) \cdot (x_1 - \mu_{X_1})^2 \int dx_2 f_{X_2}(x_2) = \sigma_{X_1}^2$$

$$K_{22} = E[(X_2 - \mu_{X_2})(X_2 - \mu_{X_2})] = \sigma_{X_2}^2$$

Diagonal Covariance Matrix for

***Uncorrelated* Random Variables**

$$K_{XX} = \begin{bmatrix} \sigma_{X_1}^2 & 0 \\ 0 & \sigma_{X_2}^2 \end{bmatrix}$$

$$K_{21} = E[(X_2 - \mu_{X_2})(X_1 - \mu_{X_1})] = \int dx_1 \int dx_2 f_{X_1 X_2}(x_1, x_2) \cdot (x_2 - \mu_{X_2})(x_1 - \mu_{X_1})$$

independence

$$= \int dx_1 f_{X_1}(x_1)(x_1 - \mu_{X_1}) \int dx_2 \cdot f_{X_2}(x_2)(x_2 - \mu_{X_2})$$

$$= \underbrace{\left\{ E[X_1] - \mu_{X_1} \right\}}_{=0} \cdot \underbrace{\left\{ E[X_2] - \mu_{X_2} \right\}}_{=0} = 0$$

$$K_{12} = E[(X_2 - \mu_{X_2})(X_1 - \mu_{X_1})] = 0$$

We introduced the Bivariate Gaussian distribution for the case of two independent N(0,1) Gaussians (zero mean and unit variance) and arrived at an unevaluated form for the covariance matrix; here we consider a more general case of two independent Gaussians with *different means and variances, i.e.,* $N(\mu_{X1}, \sigma_{X1}^2)$ and $N(\mu_{X2}, \sigma_{X2}^2)$ and evaluate the diagonal and off-diagonal terms of the covariance matrix K_{XX} by explicitly computing the expectations.

K_{11} is calculated by integrating the term $(X_1 - \mu_{X1})^2$ over the joint density. Because the Gaussians are independent the integrals separate into the product of a Gaussian normalization integral and an integral over dx_1 which yields $var(X_1) = \sigma_{X1}^2$; similarly for K_{22} we find $var(X_2) = \sigma_{X2}^2$.

K_{21} is calculated by integrating the term $(X_2 - \mu_{X2}) \cdot (X_1 - \mu_{X1})$ over the joint density. Because the Gaussians are independent the integrals separate into the product of two integrals which both vanish because they reduce to $E[X_1] - \mu_{X1} = \mu_{X1} - \mu_{X1} = 0$ and $E[X_2] - \mu_{X2} = \mu_{X2} - \mu_{X2} = 0$; similarly for the symmetric term K_{12} we find $cov(X_1 X_2) = 0$.

The resulting diagonal covariance matrix (shown in the grey boxed equation) is for two independent random variables with variances σ_{X1}^2 and σ_{X2}^2. Because their covariance is zero the two independent RVs X_1 and X_2 are said to be "uncorrelated."

9.1.2 Transformation Y = AX+b and General Bivariate Normal Distribution

Transformation $Y = AX+b$ and General Bivariate Normal Distribution

		Mean	Covariance
X a bivariate normal (indep comp) N(0,1)	$\mu_X = E[\vec{X}] = 0$ $K_{XX} = E[X \cdot X^T] = I$	$\mu_x = \begin{bmatrix} 0 \\ 0 \end{bmatrix}$	$K_{XX} = \begin{bmatrix} 1 & 0 \\ 0 & 1 \end{bmatrix}$
Linear Xform to Y	$Y = AX + b$	$\mu_Y = b = \begin{bmatrix} b_1 \\ b_2 \end{bmatrix}$	$K_{YY} = AA^T$

Computation μ_Y $\quad \mu_Y = E[Y] = E[A \cdot X + b] = A \cdot E[X] + b = b$

Computation K_{YY} $\quad K_{YY} = E\left[(Y-\mu_Y)(Y-\mu_Y)^T \right] = E\left[(\underset{=AX+b}{\underbrace{Y}} \overset{=0}{-} b)(Y-b)^T \right] = E[AX(AX)^T] = E[A(XX^T)A^T] = A \underset{=I}{\underbrace{E[XX^T]}} A^T = AA^T$

Determinant K_{YY} $\quad \det K_{YY} = \det A \cdot \det A^T = (\det A)^2 \quad \Rightarrow \quad \det A = \sqrt{\det K_{YY}}$

A is Jacobian: $\det\left\{ \dfrac{\partial y_i}{\partial x_j} \right\} = \det\{A_{ij}\} \Rightarrow J\binom{y}{x} = \det(A) = \sqrt{\det K_{YY}}$

$$\left(A^{-1} \right)^T \left(A^{-1} \right) = \left(AA^T \right)^{-1} = K_{YY}^{-1}$$

New Prob Density $\quad f_Y(y) = \dfrac{f_X(x)}{\left| J\begin{pmatrix} y_1 & y_2 \\ x_1 & x_2 \end{pmatrix} \right|} = \dfrac{\frac{1}{2\pi} e^{-\frac{1}{2}\left[\left(A^{-1}(y-b)\right)^T \cdot A^{-1}(y-b) \right]}}{\sqrt{\det K_{YY}}}$

General Bivariate Normal Distribution $\quad f_Y(y) = \dfrac{\frac{1}{2\pi} e^{-\frac{1}{2}\left[(y-\mu_y)^T K_{YY}^{-1}(y-\mu_y) \right]}}{\sqrt{\det K_{YY}}}$ ***(No Longer Independent Components or zero means & unit variances)***

We introduced the Bivariate Gaussian distribution for the case of two independent N(0,1) Gaussians (means=0 and variances=1) and arrived at a zero mean vector μ_X and a diagonal covariance matrix K_{XX} =diag(1,1) corresponding to a pair of uncorrelated Gaussian RVs as displayed in the *first row* of the table. The *second row* of the table shows the effects of making a linear transformation of variables **Y=AX+b** from the **X**=[X$_1$, X$_2$]^T coordinates to the new **Y**=[Y$_1$,Y$_2$]^T coordinates, where A is the 2×2 transformation matrix, and **b** =[b$_1$,b$_2$]^T is the displacement vector of the **Y** coordinates relative to **X** =[0,0]^T. We see that the new mean vector is no longer zero but rather μ_Y = **b** and the new covariance K_{YY} =AA^T no longer has unit variances along the diagonal and now has *non-zero off-diagonal elements* as well. The fact that this linear transformation yields non-zero off-diagonal elements in the covariance matrix means that the new RVs Y$_1$ and Y$_2$ are no longer uncorrelated.

The computations supporting these table entries are straightforward. The new mean is obtained by taking the expectation E[**Y**]= E[A**X**+ **b**] and uses the fact that the original mean E[**X**] is zero to give μ_Y = E[**Y**]= **b** . Substituting this value **b** for μ_Y in the covariance expression K_{YY} = E[(Y-b)(Y-b)^T] yields K_{YY} = E[(A**X**)(A**X**)^T] = A E[**XX**^{T}] A^T =A·I·A^T=A·A^T since E[**XX**^{T}] =K$_{XX}$ = I (*i.e.*, the identity matrix diag(1,1)).

In order to find the new bivariate density f$_{Y1,Y2}$(y$_1$,y$_2$) we need to divide f$_{X1,X2}$(x$_1$,x$_2$) by the Jacobian determinant J(**Y,X**) and replace **X** by A^{-1}(**Y-b**). This Jacobian is found by differentiating the transformation **Y=AX+b** to find J=det[∂**Y** / ∂**X**] = det(A); this result is easily verified by writing out the two component equations explicitly and differentiating y$_1$ and y$_2$ with respect to x$_1$ and x$_2$ to obtain the partials ∂y$_i$ / ∂x$_j$ = a$_{ij}$ and then taking the determinant to find the Jacobian. Taking the det(K$_{YY}$) =det(AA^T) =det(A)·det(A^T) and using the fact that the determinant det(A) = det(A^T), we find that detA = det (K$_{YY}$)$^{\frac{1}{2}}$. Finally substituting this and **X** = A^{-1}(**Y-b**) yields the general Bivariate Normal Distribution f$_Y$(**y**) given in the grey boxed equation at the bottom of the slide. Be careful to note that the inverse K$_{YY}$$^{-1}$ occurs in the exponential quadratic form and that the matrix K$_{YY}$ occurs in the denominator det(K$_{YY}$)$^{\frac{1}{2}}$; also observe the "shorthand" **vector** notation for the bivariate density f$_Y$(**y**) in place of the more explicit f$_{Y1,Y2}$(y$_1$,y$_2$). Note that our result K$_{YY}$ = AA^T for a bivariate zero mean Gaussian under the vector transformation **Y=AX** is analogous to the result σ_Y^2 = a^2· σ_X^2 = a^2·1= a^2 for a single N(0,1) Gaussian RV under the scalar transformation Y = aX.

The Bivariate Gaussian Distribution

9.1.3 Covariance Matrix Properties

Covariance Matrix Properties		
New Covariance	$K_{YY} = \begin{bmatrix} E[(Y_1 - \mu_{Y_1})^2] & E[(Y_1 - \mu_{Y_1})(Y_2 - \mu_{Y_2})] \\ E[(Y_2 - \mu_{Y_2})(Y_1 - \mu_{Y_1})] & E[(Y_2 - \mu_{Y_2})^2] \end{bmatrix} = \begin{bmatrix} \sigma_{Y_1}^2 & \rho\sigma_{Y_1}\sigma_{Y_2} \\ \rho\sigma_{Y_1}\sigma_{Y_2} & \sigma_{Y_2}^2 \end{bmatrix}$		
Correlation Coefficient	$\rho \equiv \dfrac{cov(Y_1, Y_2)}{\sigma_{Y_1}\sigma_{Y_2}} = \dfrac{E[(Y_1 - \mu_{Y_1})(Y_2 - \mu_{Y_2})]}{\sqrt{E[(Y_1 - \mu_{Y_1})^2]} \cdot \sqrt{E[(Y_2 - \mu_{Y_2})^2]}} = \dfrac{cov(Y_1, Y_2)}{\sqrt{Var(Y_1)Var(Y_2)}}$		
K_{YY} is Real & Symmetric	$(K_{YY})^T = (AA^T)^T = (A^T)^T A^T = AA^T = K_{YY}$		
K_{YY} is Positive Semi-Definite	$Q = a^T K_{YY} a \geq 0$ for any real vector $a = [a_1, a_2]$		
Proof:	$Q = a^T K_{YY} a = a^T E\left[(Y - \mu_Y) \cdot (Y - \mu_Y)^T\right] a = E\left[a^T(Y - \mu_Y) \cdot (Y - \mu_Y)^T a\right]$ $= E\left[\underbrace{\left((Y - \mu_Y)^T a\right)^T \cdot (Y - \mu_Y)^T a}_{\equiv Z}\right] = E[Z^T Z] = E[|Z|^2] \geq 0$		
Det(K_{YY}) is Positive Semi-Definite $\Rightarrow \det(K_{YY}) \geq 0$ **Correlation Bounds** $\det(K_{YY}) = \det\begin{bmatrix} \sigma_{Y_1}^2 & \rho\sigma_{Y_1}\sigma_{Y_2} \\ \rho\sigma_{Y_1}\sigma_{Y_2} & \sigma_{Y_2}^2 \end{bmatrix} = \sigma_{Y_1}^2\sigma_{Y_2}^2(1 - \rho^2) \geq 0 \quad\Rightarrow\quad -1 \leq \rho \leq +1$ **Note:** The determinant is singular for $\rho = \pm 1$		

The grey boxed equation at the top of the slide displays the general covariance matrix K_{YY} in terms of the expected values defining variances on the diagonal and covariances off the diagonal for the Gaussian vector $\mathbf{Y} = [Y_1, Y_2]^T$. The second equality gives a more useful form in terms of standard deviations σ_{Y1} and σ_{Y2} and the correlation coefficient ρ; the basic definition for ρ as the covariance $cov(Y_1, Y_2)$ divided by the product of $\sigma_{Y1} \sigma_{Y2}$ as well as two equivalent expressions are given directly below covariance matrix properties.

The covariance matrix $K_{YY} = AA^T$ is obviously a real matrix and its symmetry is verified from its definition AA^T by simply taking the transpose as detailed on the slide. Another key property is that K_{YY} is positive semi-definite, which means that the quadratic form $Q = \mathbf{a}^T K_{YY} \mathbf{a} \geq 0$ (positive or zero) for any real vector $\mathbf{a} = [a_1, a_2]^T$. This fact is easily proved by writing out Q explicitly using the definition of K_{YY} to find $Q = \mathbf{a}^T E[(\mathbf{Y}-\mu_Y)(\mathbf{Y}-\mu_Y)^T]\mathbf{a}$, then bringing the constant vector $\mathbf{a}$ inside the expectation, and finally re-expressing it as an inner product of two vectors, thereby yielding a *sum of squares* that must be positive for all non-zero vectors $\mathbf{a}$. This property of K_{YY} implies that its determinant det $K_{YY} \geq 0$, which in turn bounds the correlation coefficient to be between -1 (complete negative correlation) and +1 (complete positive correlation). We note that the determinant K_{YY} becomes singular for $\rho = \pm 1$ corresponding to degenerate cases of the quadratic form in which the ellipses collapse to straight lines at ± 45 degrees in Y_1, Y_2 coordinates. This corresponds to complete correlation or anti-correlation between the random variables Y_1, and Y_2.

9.1.4 Inverse of Covariance Matrix K_{YY}^{-1} and Transformed Gaussian

---

Inverse of Covariance Matrix K_{YY}^{-1} and Transformed Gaussian

Inverse of 2 x 2 Matrix: 1) swap diagonals, 2) negate off-diagonals 3) divide by determinant

$$K_{YY} = \begin{bmatrix} \sigma_{Y_1}^2 & \rho\sigma_{Y_1}\sigma_{Y_2} \\ \rho\sigma_{Y_1}\sigma_{Y_2} & \sigma_{Y_2}^2 \end{bmatrix} \Rightarrow K_{YY}^{-1} = \frac{1}{\sigma_{Y_1}^2\sigma_{Y_2}^2(1-\rho^2)} \cdot \begin{bmatrix} \sigma_{Y_2}^2 & -\rho\sigma_{Y_1}\sigma_{Y_2} \\ -\rho\sigma_{Y_1}\sigma_{Y_2} & \sigma_{Y_1}^2 \end{bmatrix}$$

$$K_{YY}^{-1} = \begin{bmatrix} \dfrac{1}{\sigma_{Y_1}^2(1-\rho^2)} & \dfrac{-\rho}{\sigma_{Y_1}\sigma_{Y_2}(1-\rho^2)} \\ \dfrac{-\rho}{\sigma_{Y_1}\sigma_{Y_2}(1-\rho^2)} & \dfrac{1}{\sigma_{Y_2}^2(1-\rho^2)} \end{bmatrix}$$

Quadratic Form in Gaussian exponent

$$q(y_1,y_2) = [y_1-\mu_1, y_2-\mu_2] \cdot \underbrace{\begin{bmatrix} \dfrac{1}{\sigma_{Y_1}^2(1-\rho^2)} & \dfrac{-\rho}{\sigma_{Y_1}\sigma_{Y_2}(1-\rho^2)} \\ \dfrac{-\rho}{\sigma_{Y_1}\sigma_{Y_2}(1-\rho^2)} & \dfrac{1}{\sigma_{Y_2}^2(1-\rho^2)} \end{bmatrix}}_{=K_{YY}^{-1}} \cdot \begin{bmatrix} y_1-\mu_1 \\ y_2-\mu_2 \end{bmatrix}$$

Transformed Gaussian

$$f_Y(y) = \frac{1}{2\pi\sigma_{Y_1}\sigma_{Y_2}\sqrt{(1-\rho^2)}} e^{-\frac{1}{2}q(y_1,y_2)}$$

Express in Normalized &Centered Coordinates

$$\tilde{y}_1 = \frac{y_1-\mu_{Y_1}}{\sigma_{Y_1}} \; ; \; \tilde{y}_2 = \frac{y_2-\mu_{Y_2}}{\sigma_{Y_2}}$$

$$q(\tilde{y}_1,\tilde{y}_2) = \frac{1}{(1-\rho^2)}\left[\tilde{y}_1^2 - 2\rho\tilde{y}_1\tilde{y}_2 + \tilde{y}_2^2\right]$$

$$q(y_1,y_2) = \frac{1}{(1-\rho^2)}\left[\frac{(y_1-\mu_1)^2}{\sigma_{Y_1}^2} - \frac{2\rho(y_1-\mu_1)(y_2-\mu_2)}{\sigma_{Y_1}\sigma_{Y_2}} + \frac{(y_2-\mu_2)^2}{\sigma_{Y_2}^2}\right]$$

$$dP = \frac{1}{2\pi\cdot\sqrt{1-\rho^2}}\cdot e^{\frac{1}{2}q(\tilde{y}_1,\tilde{y}_2)}d\tilde{y}_1 d\tilde{y}_2$$

The quadratic form in the exponential of general Bivariate Normal Distribution contains the inverse matrix K_{YY}^{-1} and the distribution can be made more explicit by computing the inverse in terms of σ_{Y1}, σ_{Y2}, and the correlation coefficient ρ. The inverse of a 2×2 matrix is easily obtained by swapping the diagonals, changing the signs of the off-diagonals and dividing by its determinant. Performing these operations on K_{YY} immediately yields the explicit form for K_{YY}^{-1} given on the slide. The quadratic form in the exponent of the distribution can now be computed explicitly to yield a more useful form for the bivariate density $f_Y(y) = f_{Y1,Y2}(y_1,y_2)$ shown in the two grey boxed equations on the bottom of the slide.

The quadratic form $q(y_1, y_2)$ contains all the information on the shape of the Bivariate Gaussian distribution characterized by its correlation coefficient ρ and its means μ_{Y1}, μ_{Y2}, and sigmas σ_{Y1}, σ_{Y2}. Such a quadratic expression gives the general mathematical representation of an ellipse which determines the magnitude and orientation of the major and minor principal axes in y_1, y_2 coordinate system as will be discussed on the next few slides. For the simple case of zero correlation $\rho = 0$, the cross term vanishes and the quadratic form immediately becomes a sum of squares and hence the Bivariate Gaussian density function factors into a product of two independent 1-dimensional Gaussians. Moreover, in this case the principal axes of the ellipse are along y_1 and y_2 axes and their values are easily found by first setting $y_1 = 0$ to find $y_2 = \sigma_{Y2}$ as the length of one principal axis and then setting $y_2 = 0$ to find $y_1 = \sigma_{Y1}$ as the length of the other; the major axis of the ellipse a is of course the larger sigma-value. This result should be no surprise since setting the correlation $\rho = 0$, in fact corresponds to the product of two independent Gaussians that we started with on Slide#9-2, but now in y_1, y_2 coordinates.

The Bivariate Gaussian Distribution

9.2 Bivariate Gaussian Distribution and Level Surfaces

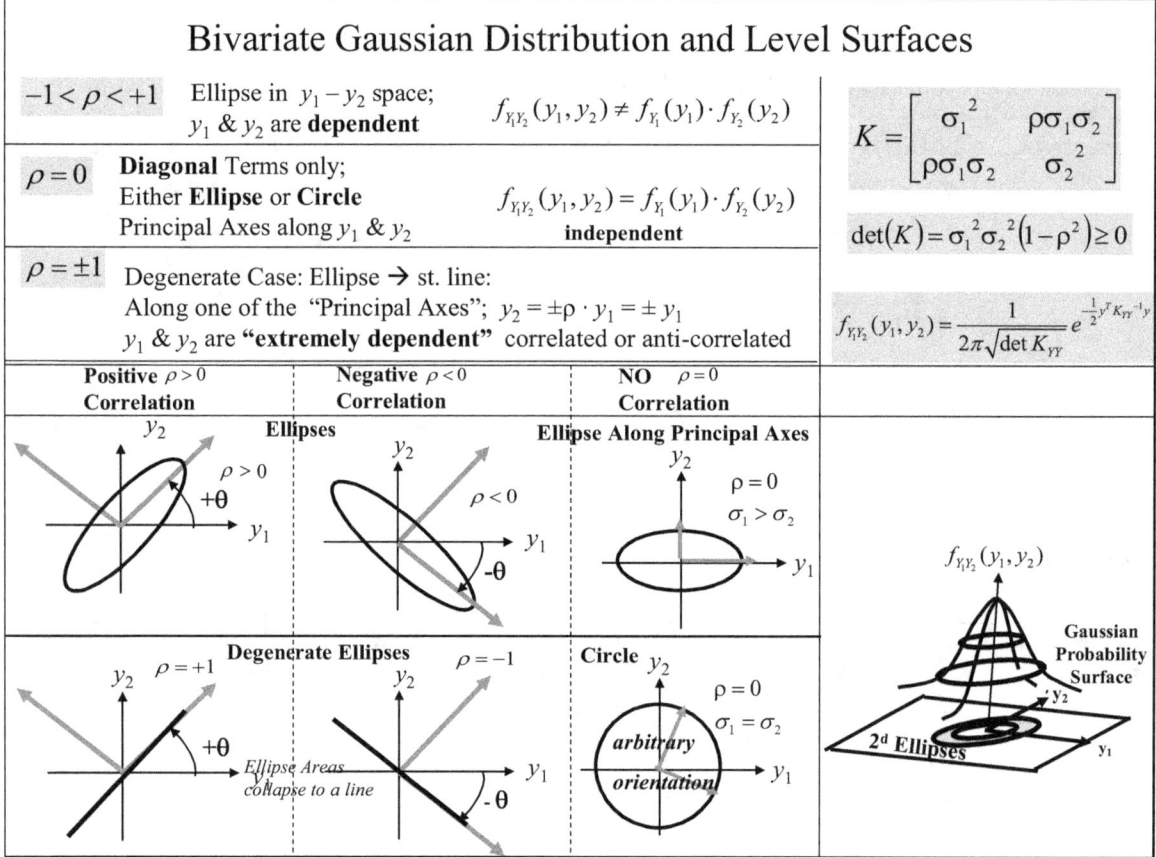

The bivariate density $f_Y(y) = f_{Y1,Y2}(y_1,y_2)$ is completely determined by its mean vector μ_Y and its covariance matrix K_{YY} as given by the boxed equations on the upper right. Consider the Bivariate Gaussian density which is plotted as a 2^d surface relative to its mean vector components μ_{Y1} and μ_{Y2} taken as the origin. The level surfaces represented by cuts parallel to the y_1-y_2 plane are the ellipses given by the quadratic form equation of the last slide. The structure of these ellipses is shown in the tableau consisting of 3 columns for positive, negative, and zero correlation coefficient ρ and by 2 rows corresponding to general (top row) and degenerate cases (bottom row).

The general cases in the top row have unequal sigmas $\sigma_1 > \sigma_2$ and as we go across the row we have an ellipse with positive correlation ($\rho > 0$), one with negative correlation ($\rho < 0$) and an ellipse along its principal axes with no correlation ($\rho = 0$). The (red) arrows show the directions of the principal axes of the ellipse in each case; the zero correlation case on the extreme right has the principal axes coinciding with y_1 and y_2, while the negative correlation case has its principal axes rotated at $-\theta$ to the y_1-axis and the positive correlation case has its principal axes rotated at $+\theta$ to the y_1-axis.

The bottom row illustrates a number of special cases as follows: (i) two degenerate cases for $\rho = +1$ and $\rho = -1$ in which the ellipses in the top row "collapse" to straight lines corresponding to complete correlation or anti-correlation (opposite variations of Y_1 and Y_2) respectively, and (ii) the completely uncorrelated case $\rho = 0$ with equal sigmas ($\sigma_1 = \sigma_2$) for which the ellipse degenerates to a circle with no preferred orientation.

The Bivariate Gaussian Distribution

9.2.1 Rotation to Principal Axes - Diagonalize K_{YY}

Rotation to Principal Axes - Diagonalize K_{YY}

X a Bivariate Gaussian

$$K_{XX} = \begin{bmatrix} \sigma_{X_1}^2 & \rho\sigma_{X_1}\sigma_{X_2} \\ \rho\sigma_{X_1}\sigma_{X_2} & \sigma_{X_2}^2 \end{bmatrix} \; ; \; \mu_X = \begin{bmatrix} 0 \\ 0 \end{bmatrix}$$

Rotation Y=RX $\mu_Y = E[Y] = E[RX] = RE[X] = R\begin{bmatrix} 0 \\ 0 \end{bmatrix} = \begin{bmatrix} 0 \\ 0 \end{bmatrix}$ $K_{YY} = E[YY^T] = E[RX(RX)^T] = RE[XX^T]R^T$

$$K_{YY} = \begin{bmatrix} C\theta & S\theta \\ -S\theta & C\theta \end{bmatrix}\begin{bmatrix} \sigma_{X_1}^2 & \rho\sigma_{X_1}\sigma_{X_2} \\ \rho\sigma_{X_1}\sigma_{X_2} & \sigma_{X_2}^2 \end{bmatrix}\begin{bmatrix} C\theta & -S\theta \\ S\theta & C\theta \end{bmatrix} = \begin{bmatrix} C\theta & S\theta \\ -S\theta & C\theta \end{bmatrix}\begin{bmatrix} \sigma_{X_1}^2 C\theta + \rho\sigma_{X_1}\sigma_{X_2} S\theta & -\sigma_{X_1}^2 S\theta + \rho\sigma_{X_1}\sigma_{X_2} C\theta \\ \rho\sigma_{X_1}\sigma_{X_2} C\theta + \sigma_{X_2}^2 S\theta + & -\rho\sigma_{X_1}\sigma_{X_2} S\theta + \sigma_{X_2}^2 C\theta \end{bmatrix}$$

$$K_{YY} = \begin{bmatrix} \sigma_{X_1}^2 C^2\theta + \sigma_{X_2}^2 S^2\theta + 2\rho\sigma_{X_1}\sigma_{X_2} C\theta S\theta & \boxed{\rho\sigma_{X_1}\sigma_{X_2}(C^2\theta - S^2\theta) + (\sigma_{X_2}^2 - \sigma_{X_1}^2)C\theta S\theta} \\ \rho\sigma_{X_1}\sigma_{X_2}(C^2\theta - S^2\theta) + (\sigma_{X_2}^2 - \sigma_{X_1}^2)C\theta S\theta & \sigma_{X_1}^2 C^2\theta + \sigma_{X_2}^2 S^2\theta - 2\rho\sigma_{X_1}\sigma_{X_2} C\theta S\theta \end{bmatrix}$$

Setting Off-Diagonal Terms to Zero

$$0 = \rho\sigma_{X_1}\sigma_{X_2}\underbrace{(C^2\theta - S^2\theta)}_{=C_{2\theta}} + (\sigma_{X_2}^2 - \sigma_{X_1}^2)\underbrace{C\theta \cdot S\theta}_{=S_{2\theta}/2} \; ; \quad \theta = \frac{1}{2}\arctan\left(\frac{2\rho\sigma_{X_1}\sigma_{X_2}}{\sigma_{X_1}^2 - \sigma_{X_2}^2}\right)$$

Note for $\sigma_{X_1} = \sigma_{X_2}$ $\theta = \frac{1}{2}\arctan(\infty) = \pm\frac{\pi}{4}$ *i.e.*, $\pm 45°$ rotations

$\theta = +45°$
$\rho > 0$

$$K_{YY} = \begin{bmatrix} \sigma_X^2(1+\rho) & 0 \\ 0 & \sigma_X^2(1-\rho) \end{bmatrix} \; ; \; K_{YY}^{-1} = \begin{bmatrix} \frac{1}{\sigma_X^2(1+\rho)} & 0 \\ 0 & \frac{1}{\sigma_X^2(1-\rho)} \end{bmatrix}$$

Ellipse $q(y_1, y_2) = \mathbf{y}^T K_{YY}^{-1}\mathbf{y} = \frac{y_1^2}{\sigma_X^2(1+\rho)} + \frac{y_2^2}{\sigma_X^2(1-\rho)}$

y_1 semi-major axis
$"a" = \sigma_X\sqrt{1+\rho}$

Consider the Bivariate Gaussian with correlated RVs $\mathbf{X} = [X_1, X_2]^T$, zero mean $\mu_X = [0,0]^T$, and covariance matrix K_{XX} as shown with $\rho \neq 0$. Further consider a rotation in the plane through an angle θ given by $\mathbf{Y} = R(\theta)\mathbf{X}$ where $R(\theta)$ is the 2×2 rotation matrix. The rotated mean and covariance are obtained directly from their definitions and are displayed in the top line of the 2nd panel. The zero mean vector is unchanged by the coordinate rotation so we have $\mu_Y = [0,0]^T$ while the new covariance K_{YY} is given by the matrix product $R(\theta)K_{XX}R(\theta)^T$. Explicitly performing the matrix multiplications yields the result for K_{YY} shown on the slide. For this to be a transformation to principal axes the rotated matrix K_{YY} *must be diagonal*; since K_{YY} is symmetric, we only need "zero out" the one circled off-diagonal term and solve for the rotation angle θ.

The resulting expression for the rotation angle θ is given in terms of the correlation coefficient ρ and the sigmas σ_{X1} and σ_{X2} in the shaded box equation in the third panel. For the special case of *equal sigmas* ($\sigma_{X1} = \sigma_{X2}$), the angle of rotation to principal axes is $-45°$ for $\rho < 0$ (top figure) and $+45°$ for $\rho > 0$ (bottom figure) and the new covariance K_{YY} takes on the simple diagonal form shown on the bottom of the slide corresponding to a pair of uncorrelated RVs as desired. The general *arctan* formula for the rotation angle θ and the expressions for the diagonal terms of K_{YY} in terms of θ must be used in all other cases.

9.2.2 Moment Generating Function for Bivariate Gaussian

Moment Generating Function for Bivariate Gaussian

X Bivariate Gaussian, Zero mean, Cov K_{XX}	$\phi_X(t) = E[e^{X^T t}] = \int\int_{-\infty}^{+\infty} e^{x^T t} f_X(\vec{x}) d\vec{x} \quad ; \quad t = \begin{bmatrix} t_1 \\ t_2 \end{bmatrix}$

$$= \int\int_{-\infty}^{+\infty} \frac{1}{2\pi\sigma_{X_1}\sigma_{X_2}} e^{-\frac{1}{2}x^T K_{XX}^{-1} x} \cdot e^{x^T t} d\vec{x} = \int\int_{-\infty}^{+\infty} \frac{1}{2\pi\sigma_{X_1}\sigma_{X_2}} e^{-\frac{1}{2}(x^T K_{XX}^{-1} x - 2x^T t)} d\vec{x}$$

Completing the square in exponent	$x^T K_{XX}^{-1} x - 2x^T t + \underbrace{\{t^T K_{XX} t - t^T K_{XX} t\}}_{\text{"add zero"}} = (x - K_{XX} t)^T K_{XX}^{-1} (x - K_{XX} t) - t^T K_{XX} t$
Mathematical Details	$(x - K_{XX} t)^T K_{XX}^{-1} (x - K_{XX} t) = x^T K_{XX}^{-1} (x - K_{XX} t) - t^T \underbrace{K_{XX}^T K_{XX}^{-1}}_{=I} (x - K_{XX} t)$ $\qquad \underbrace{}_{=K_{XX}}$ $\checkmark\checkmark$

$$= x^T K_{XX}^{-1} x - x^T \underbrace{K_{XX}^{-1} K_{XX}}_{=I} t - t^T (x - K_{XX} t) = x^T K_{XX}^{-1} x - 2x^T t + t^T K_{XX} t$$

Generating Fcn for Zero Mean Bivariate Gaussian	$\phi_X(t) = \left\{ e^{+\frac{1}{2}t^T K_{XX} t} \right\} \underbrace{\int\int_{-\infty}^{+\infty} \frac{1}{2\pi\sigma_{X_1}\sigma_{X_2}} e^{-\frac{1}{2}\{(x-K_{XX}t)^T K_{XX}^{-1}(x-K_{XX}t)\}} d\vec{x}}_{\text{Normalization Integral} = 1} \qquad \boxed{\phi_X(t) = e^{+\frac{1}{2}t^T K_{XX} t}}$

Partials: $\dfrac{\partial \phi_X(t_1, t_2)}{\partial t_1} \quad ; \quad \dfrac{\partial \phi_X(t_1, t_2)}{\partial t_2}$

Transform to Non-zero Mean	$Y = X + \mu_Y \qquad K_{YY} = K_{XX}$

$$\phi_Y(t) = E[e^{Y^T t}] = E[e^{(X+\mu_Y)^T t}] = e^{\mu_Y^T t} E[e^{X^T t}] = e^{\mu_Y^T t} e^{+\frac{1}{2}t^T K_{XX} t} \qquad \boxed{\phi_Y(t) = e^{\mu_Y^T t + \frac{1}{2}t^T K_{YY} t}}$$

The generating function for the Bivariate Gaussian with zero mean and covariance K_{XX} is derived in a manner exactly analogous to the scalar case, *i.e.*, by "completing the square" of a quadratic form, but now involving matrix/vector manipulations. The basic difference is that for the vector random variable $\mathbf{X}^T = [X_1, X_2]$ we must take the expected value of the product $E[\exp(X_1 t_1)\exp(X_2 t_2)] = E[\exp(X_1 t_1 + X_2 t_2)]$; upon defining the column vector $\mathbf{t} = [t_1, t_2]^T$, this may be written as $\phi_X(\mathbf{t}) = E[\exp(\mathbf{X}^T \mathbf{t})]$, where the scalar in the exponential is the result of an inner product.

Performing the expectation requires a 2-dimensional integral over the Bivariate Gaussian density and results in a exponential form involving the two terms $\mathbf{x}^T K_{XX}^{-1} \mathbf{x} + \mathbf{x}^T \mathbf{t}$. Completing the square for this latter expression requires some matrix manipulations shown in the mathematical details panel; the resulting expression for the Bivariate Gaussian has one term that integrates to unity and leaves the quadratic exponential $\phi(\mathbf{t}) = \exp\{+\frac{1}{2}(\mathbf{t}^T K_{XX} \mathbf{t})\}$ as the generating function (note K_{XX}, not its inverse!).

The general Bivariate Gaussian for *non-zero mean* is easily found by substituting $\mathbf{Y} = \mathbf{X} + \mu_X$ into the defining relation $\phi_Y(\mathbf{t}) = E[\exp(\mathbf{Y}^T \mathbf{t})]$ and expanding out the terms as shown in the bottom panel. The resulting generating function is shown in the shaded box equation on the bottom of the slide Note that the transformation $\mathbf{Y} = \mathbf{X} + \mu_Y$ only changes the mean; the covariance matrix is unchanged, $K_{YY} = K_{XX}$. These results should be compared with their scalar versions given in the boxed equation in the top panel of Slide#2-30, *viz.*, $\phi(t) = \exp\{+(\sigma t)^2/2\}$ and $\phi(t) = \exp\{\mu t + (\sigma t)^2/2\}$. It is easily seen that successive mixed partial derivatives evaluated at $t_1 = t_2 = 0$ yield the corresponding moments or covariance K_{YY} matrix, *viz.*,

$$E[Y_1] = \partial/\partial t_1 [\Phi_Y(\mathbf{t})]_{t=0} = \mu_{Y_1} \quad ; \quad E[Y_2] = \partial/\partial t_2 [\Phi_Y(\mathbf{t})]_{t=0} = \mu_{Y_2}$$

$$E[Y_1 Y_1] = \partial^2/\partial t_1^2 [\Phi_Y(\mathbf{t})]_{t=0} = \mu_{Y_1}^2 + \sigma_{Y_1}^2 \quad ; \quad E[Y_2 Y_2] = \partial^2/\partial t_2^2 [\Phi_Y(\mathbf{t})]_{t=0} = \mu_{Y_2}^2 + \sigma_{Y_2}^2$$

$$E[Y_1 Y_2] = \partial^2/\partial t_1 \partial t_2 [\Phi_Y(\mathbf{t})]_{t=0} = \rho\sigma_{Y_1}\sigma_{Y_2} + \mu_{Y_1}\mu_{Y_2}$$

$$Cov[Y_1 Y_2] = E[Y_1 Y_2] - \mu_{Y_1}\mu_{Y_2} = (\rho\sigma_{Y_1}\sigma_{Y_2} + \mu_{Y_1}\mu_{Y_2}) - \mu_{Y_1}\mu_{Y_2} = \rho\sigma_{Y_1}\sigma_{Y_2} = K_{Y_1 Y_2}$$

The Bivariate Gaussian Distribution

9.3 *Ellipses of Concentration as a Measure of Probability*

Ellipses of Concentration as a Measure of Probability

1^D Gaussian Distribution described by
two scalars: mean μ_X & **Var(X)** (intuitive)
Normalized & Centered RV
Standardized Distribution (Tabulation of CDF)

Tabulate Area

$$\Phi(y) = \int_{t=-\infty}^{x} \frac{1}{\sqrt{2\pi}} e^{-t^2/2} dt$$

2^D Gaussian Distributions described by
Vector & Matrix: mean vector $\vec{\mu_X}$ &
Covariance $\mathbf{K_{XX}}$

Vector $\vec{\mu_X}$ and K_{XX} are not very intuitive!

$$f_X(x_1, x_2) = \frac{1}{2\pi\sqrt{\det K_{XX}}} e^{-\frac{1}{2}x^T K_{XX}^{-1} x}$$

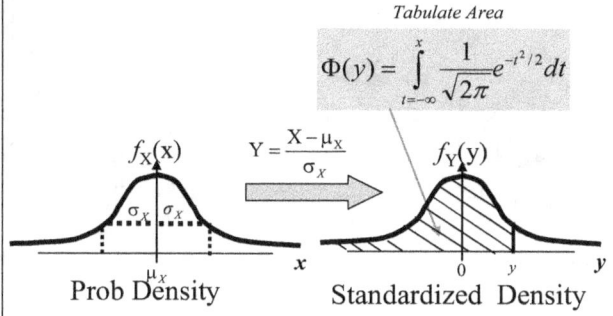

$Y = \dfrac{X - \mu_X}{\sigma_X}$

Prob Density → Standardized Density

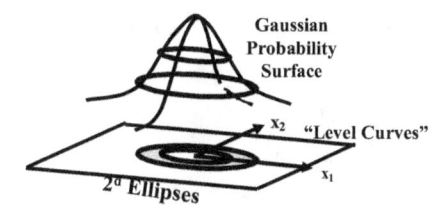

Gaussian Probability Surface

"Level Curves"

2^D Ellipses

"Level curves" of Zero
Mean 2^D Gaussian Surface
with Covariance K_{XX}

$$x^T K_{XX}^{-1} x = \frac{1}{(1-\rho^2)}\left[\frac{x_1^2}{\sigma_{X_1}^2} - \frac{2\rho x_1 x_2}{\sigma_{X_1}\sigma_{X_2}} + \frac{x_2^2}{\sigma_{X_2}^2}\right] = c^2 = const.$$

The 1-dimensional Gaussian distribution is completely described by two scalars the mean μ_X and the variance σ_X^2. The tabulation of a single integral for the cumulative distribution function $F_Y(y)$ shown in the left box is sufficient to characterize all such Gaussians X: $N(\mu_X, \sigma_X^2)$ provided we first transform to the standardized variable $Y = (X - \mu_X) / \sigma_X$. The Gaussian integral representing the probability distribution for the standardized variable Y is the CDF function $F_Y(y) = Pr[Y \leq y]$; it is used so often that it is denoted as the "Normal Integral" $\Phi(x)$.

We would like to extend this concept of a single tabulated integral to describe all 2-dimensional Gaussian distributions; however, as we have seen, the bivariate Gaussian distribution requires more than just the means and variances of two Gaussian RVs as we must also characterize their "co-variation" by specifying the correlation coefficient ρ. Thus we must specify the two elements of the mean vector μ_X and all *three elements* of the (symmetric) covariance matrix K_{XX} in order to completely characterize a Bivariate Gaussian $f_{X1X2}(x_1, x_2)$ given in the right box of the slide.

We have seen that the level "surfaces" (actually curves) of the Gaussian PDF are ellipses centered about the mean vector coordinates μ_{X1} and μ_{X2} and described by quadratic form $x^T K^{-1}_{XX} x$ in the exponent of the PDF. The explicit equation for level curves with zero mean are obtained by setting the quadratic term in the exponential equal to an arbitrary positive constant c^2 as given by the bottom equation of the slide. These ellipses are called ellipses of concentration because the area contained within them measures the concentration of probability for the specific "cut through" the PDF surface. In the next few slides we will show how this leads to a single tabulated function for the Bivariate Gaussian that is analogous to $\Phi(x)$ for the Normal Distribution.

The Bivariate Gaussian Distribution

9.3.1 Area of Covariance Ellipse

Area of Covariance Ellipse

Standard Ellipse $\qquad A = \pi ab$ $\qquad \dfrac{x_1^2}{a^2} + \dfrac{x_2^2}{b^2} = 1$

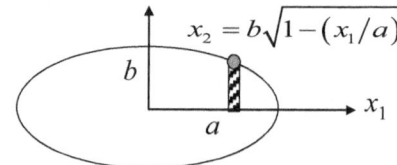

Scale Ellipse
axes by factor "c" $\qquad a' = ca \; ; \; b' = cb$ $\qquad \dfrac{x_1^2}{a^2} + \dfrac{x_2^2}{b^2} = c^2$

Area scales by c^2 $\qquad A' = \pi(ac)(bc) = Ac^2$

Note: If we double (or halve) both axes c=2 (or 1/2) area quadruples (or quarters)

Matrix Formulation of Standard Ellipse

$$x = \begin{bmatrix} x_1 \\ x_2 \end{bmatrix} \qquad K_{XX} = \begin{bmatrix} a^2 & 0 \\ 0 & b^2 \end{bmatrix} \qquad x^T K_{XX}^{-1} x = \begin{bmatrix} x_1, x_2 \end{bmatrix} \begin{bmatrix} 1/a^2 & 0 \\ 0 & 1/b^2 \end{bmatrix} \begin{bmatrix} x_1 \\ x_2 \end{bmatrix} = 1 \quad \Rightarrow \quad \dfrac{x_1^2}{a^2} + \dfrac{x_2^2}{b^2} = 1$$

$$K_{XX}^{-1} = \dfrac{1}{a^2 b^2} \begin{bmatrix} b^2 & 0 \\ 0 & a^2 \end{bmatrix} = \begin{bmatrix} 1/a^2 & 0 \\ 0 & 1/b^2 \end{bmatrix}$$

Note: Since inverse covariance determines the ellipse, the *longer ellipse axis "a>b" has smaller eigenvalue, viz., $1/a^2 < 1/b^2$*

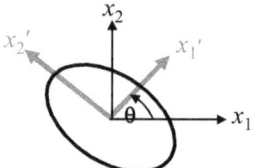

$$\sqrt{\det(K_{XX})} = \sqrt{a^2 b^2} = ab$$

$$A' = \pi(ac)(bc) = \pi \ \underbrace{(ab)}_{=\sqrt{\det K_{XX}}} c^2 = \pi \sqrt{\det(K_{XX})} \, c^2$$

Note: Area scales up or down relative to c=1;

Note: *Since $\det(K_{XX})$ is invariant under coordinate transformations, so too is the Area*

Here we explicitly show that (i) the area of the ellipse whose principal axes {a, b} are multiplied by a number 'c' scales as c^2, (ii) the quadratic form of the type found in the Bivariate Gaussian represents an ellipse, and iii) the area of the ellipse depends upon the square root of the determinant of the covariance matrix $[\det(K_{XX})]^{1/2}$ whose inverse K^{-1}_{XX} defines the ellipse.

(i) The area of an ellipse with principal axes {a,b} is $A = \pi ab$ which can be verified by performing the integration along the vertical-strip in the figure from $x_2=0$ to $x_2=b[1-(x_1/a)^2]^{1/2}$, then integrating over all x_1 from $x_1=0$ to $x_1=a$ and multiplying by 4 for the contributions from all quadrants. The area obviously scales as c^2 because of its dependence upon the product $a \cdot b$.

(ii) Taking a diagonal covariance matrix $\text{diag}(a^2, b^2)$ whose elements are the square of major and minor principal axes of an ellipse {a,b}, then computing its inverse and expanding out the quadratic form $q(x_1', x_2') = \mathbf{x}'^T K^{-1} \mathbf{x}'$ immediately leads to the well-known formula for an ellipse relative to its principal axes. When the ellipse is rotated through a positive angle θ as shown in the figure its representation in the original $x_1 - x_2$ coordinate system will no longer have the simple form of ellipse equation shown on the slide, but now has a cross term $x_1 \cdot x_2$ which we shall see corresponds to the correlation coefficient.

(iii) We note that the product of principal axes "$a \cdot b$" can be expressed as $[\det(K_{XX})]^{1/2}$ and therefore the area of the ellipse scales as $A = \pi [\det(K_{XX})]^{1/2} c^2$ (bottom panel of the slide.)

9.3.2 Bivariate Gaussian – Scaled Ellipses of Concentration

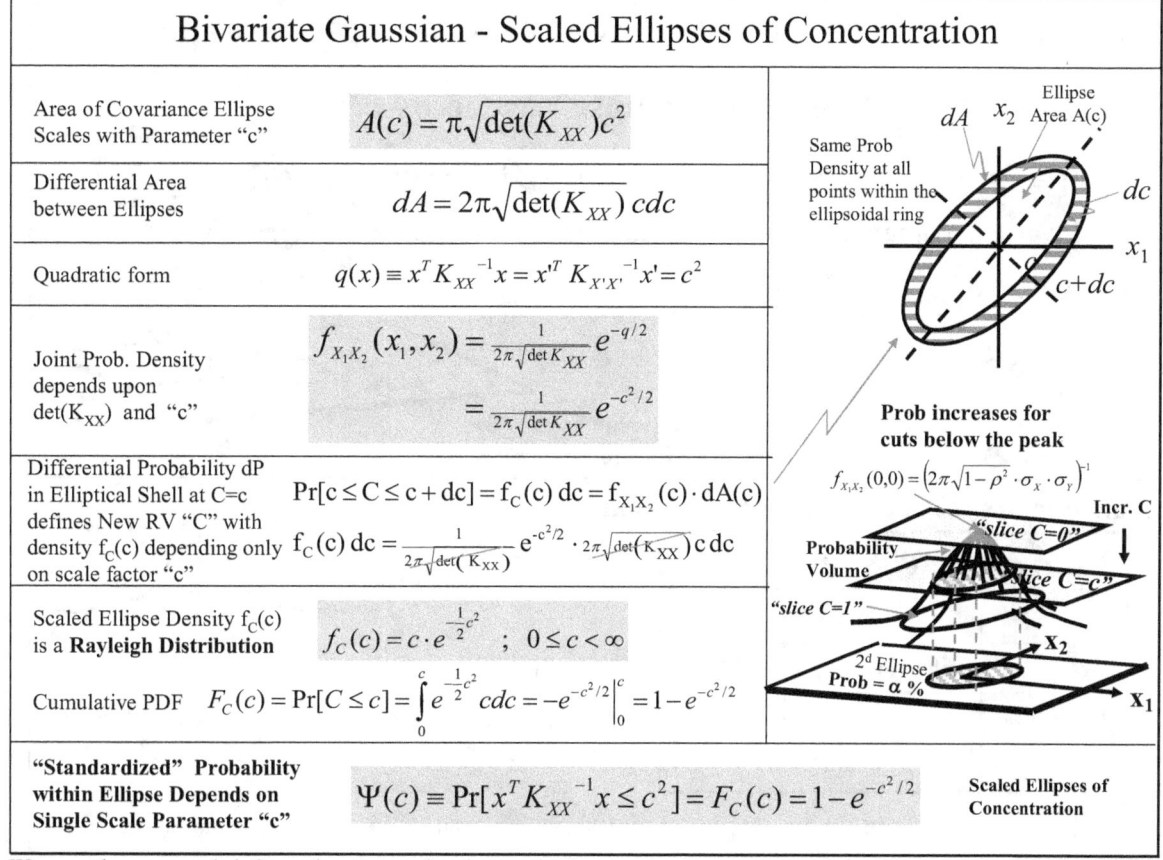

Bivariate Gaussian - Scaled Ellipses of Concentration

Area of Covariance Ellipse Scales with Parameter "c"	$A(c) = \pi\sqrt{\det(K_{XX})}c^2$
Differential Area between Ellipses	$dA = 2\pi\sqrt{\det(K_{XX})}\,c\,dc$
Quadratic form	$q(x) \equiv x^T K_{XX}^{-1} x = x'^T K_{X'X'}^{-1} x' = c^2$
Joint Prob. Density depends upon $\det(K_{XX})$ and "c"	$f_{X_1 X_2}(x_1, x_2) = \frac{1}{2\pi\sqrt{\det K_{XX}}} e^{-q/2}$ $= \frac{1}{2\pi\sqrt{\det K_{XX}}} e^{-c^2/2}$
Differential Probability dP in Elliptical Shell at C=c defines New RV "C" with density $f_C(c)$ depending only on scale factor "c"	$\Pr[c \le C \le c+dc] = f_C(c)\,dc = f_{X_1 X_2}(c) \cdot dA(c)$ $f_C(c)\,dc = \frac{1}{2\pi\sqrt{\det(K_{XX})}} e^{-c^2/2} \cdot 2\pi\sqrt{\det(K_{XX})}\,c\,dc$
Scaled Ellipse Density $f_C(c)$ is a **Rayleigh Distribution**	$f_C(c) = c \cdot e^{-\frac{1}{2}c^2}$; $0 \le c < \infty$
Cumulative PDF	$F_C(c) = \Pr[C \le c] = \int_0^c e^{-\frac{1}{2}c^2} c\,dc = -e^{-c^2/2}\Big|_0^c = 1 - e^{-c^2/2}$
"Standardized" Probability within Ellipse Depends on Single Scale Parameter "c"	$\Psi(c) \equiv \Pr[x^T K_{XX}^{-1} x \le c^2] = F_C(c) = 1 - e^{-c^2/2}$ **Scaled Ellipses of Concentration**

We now have enough information to rewrite the quadratic term in the exponent of the Bivariate Gaussian probability density $f_{X1X2}(x_1,x_2)$ more simply in terms of a single parameter "c" which scales the area of the covariance ellipse. We start by setting the quadratic form in the exponential $q(\mathbf{x}) = c^2$ so the density function $f_{X1X2}(x_1,x_2)$ now becomes a function of "c" and the Jacobian $\det(K_{XX})$ as shown in the middle panel boxed equation.

The top figure shows a "thin" (infinitesimal) elliptical shell between c and c+dc resulting from a pair of cuts defined by a horizontal planes at C=c and C=c+dc (not shown) below the peak of the Bivariate Gaussian probability density surface. The bottom figure shows the cut that is tangent to the top of the joint PDF and has zero area corresponding to C=0; as we proceed downward from the peak, the area of the ellipse C=c increases and so does the area of the thin elliptical shell.

The total area of the thin elliptical shell at C=c is found by taking the differential of the covariance ellipse area, *viz.*, $dA(c)$ $=2\pi[\det(K_{XX})]^{\frac{1}{2}}cdc$; since the probability density $f_C(c)$ is the same at every point within this infinitesimal shell, we compute the total probability as

$$\Pr[c \le C \le c+dc] = f_C(c)\,dc = f_{X1X2}(c) \cdot dA(c) = \{2\pi[\det(K_{XX})]^{\frac{1}{2}}\}^{-1} \exp(-c^2/2) \times 2\pi[\det(K_{XX})]^{\frac{1}{2}} cdc$$

Subsequent cancellation of the terms involving K_{XX} defines the desired one-dimensional density in terms of a single variable c as

$$f_C(c) = \exp(-c^2/2) \text{ for } 0 \le c < \infty$$

This density in fact defines a new "standardized" *scale random variable* "C" that represents the distance downward from the peak of the original PDF $f_{X1X2}(x_1,x_2)$. Integrating this density with respect to "c" yields the cumulative distribution function and defines the desired Standardized Probability integral $\Psi(c)$ for the Bivariate Gaussian distribution as $\Psi(c)$ $\equiv\Pr[\mathbf{x}^T K^{-1}_{XX} \mathbf{x} \le c^2]$ $=\Pr[C \le c]$ $= F_C(c) = 1 - \exp(-c^2/2)$. This is in fact a Rayleigh distribution (Slide#2-18, 9-16). Note that although σ_X and σ_Y determine the shape of the ellipses of concentration; however, they have no affect on the probability between ellipses of concentration which only depend upon their respective scale parameters c_1 and c_2. (See Slide# 9-13 to 9-15 for more discussion)

9.3.3 Gaussian and Bivariate Gaussian Distributions Compared

Gaussian and Bivariate Gaussian Distributions Compared

Probability (x_1, x_2) lies within ellipse "scaled by c":

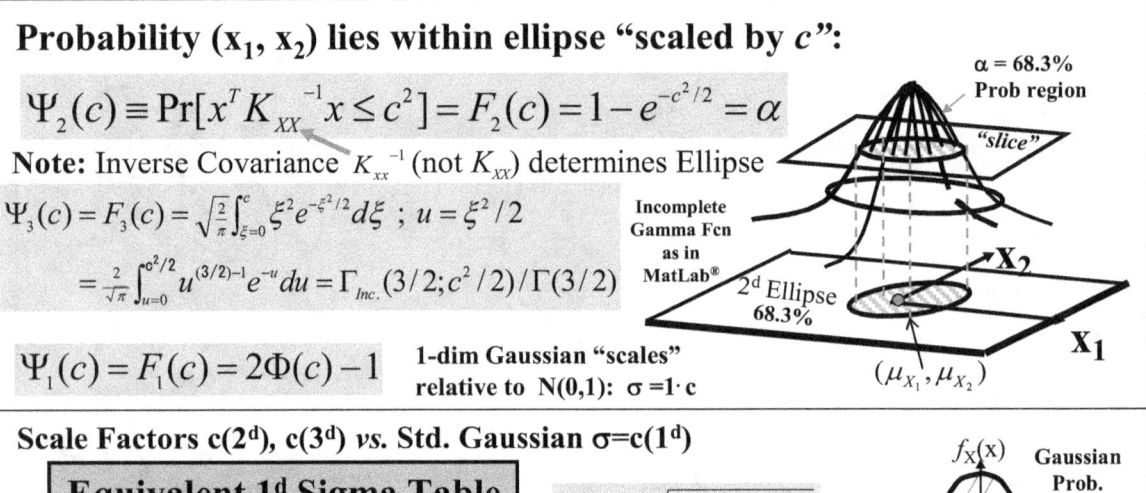

$$\Psi_2(c) \equiv \Pr[x^T K_{xx}^{-1} x \le c^2] = F_2(c) = 1 - e^{-c^2/2} = \alpha$$

Note: Inverse Covariance K_{xx}^{-1} (not K_{xx}) determines Ellipse

$$\Psi_3(c) = F_3(c) = \sqrt{\frac{2}{\pi}} \int_{\xi=0}^{c} \xi^2 e^{-\xi^2/2} d\xi \; ; \; u = \xi^2/2$$

$$= \frac{2}{\sqrt{\pi}} \int_{u=0}^{c^2/2} u^{(3/2)-1} e^{-u} du = \Gamma_{Inc}(3/2; c^2/2)/\Gamma(3/2)$$

Incomplete Gamma Fcn as in MatLab®

$$\Psi_1(c) = F_1(c) = 2\Phi(c) - 1$$

1-dim Gaussian "scales" relative to N(0,1): $\sigma = 1 \cdot c$

α = 68.3% Prob region

"slice"

2^d Ellipse 68.3%

(μ_{x_1}, μ_{x_2})

Scale Factors $c(2^d)$, $c(3^d)$ vs. Std. Gaussian $\sigma = c(1^d)$

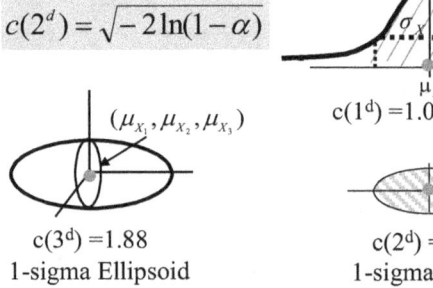

Equivalent 1^d Sigma Table				
σ's	c (1^d)	α (%)	c (2^d)	c (3^d)
1-σ	1.00	68.3	1.52	1.88
2-σ	2.00	95.4	2.48	2.83
3-σ	3.00	99.7	3.41	3.73

$c(2^d) = \sqrt{-2\ln(1-\alpha)}$

$(\mu_{x_1}, \mu_{x_2}, \mu_{x_3})$

$c(3^d) = 1.88$
1-sigma Ellipsoid

$f_X(x)$ Gaussian Prob. Density

$c(1^d) = 1.00 = 68.3\%$

(μ_{x_1}, μ_{x_2})

$c(2^d) = 1.52$
1-sigma Ellipse

$$\Psi_n(c) = \Gamma_{Inc}(n/2; c^2/2)/\Gamma(n/2); \; n = 1,2,3,4,...$$

On the last slide we found that the Bivariate Gaussian cumulative probabilities are described in terms ellipses of concentration specified by the axis scale parameter c. The probability that all the Bivariate Gaussian vectors x=[x1, x2] are within the ellipse of concentration only depends upon the parameter c and is given by the expression Ψ(c) in the boxed equation on this slide. The top figure illustrates a typical horizontal cut at C=c and the projected ellipse in the (x1, x2)-plane that uniquely determines the probability volume in the "cap region" from C=0 at the peak to the cut at C=c as Ψ(c)=1-exp(-c²/2). For c=1.52 this evaluates to 68.3% as indicated in the figure. This CDF is in fact a Rayleigh distribution with "radial distance r" replaced by the ellipse scale parameter "c". Setting the probability *within the ellipse* 1-exp(-c²/2) equal to some specified value α allows us to solve for the corresponding c-value, c = √(-2ln(1-α)). Note that the statement "*probability within the ellipse*" is a bit misleading since we actually mean to say the integral of the density under the "cap" cut off by the ellipse (which is a volume) as shown by the probability region in the top figure. Using this equation, we compute the table values for the Bivariate Gaussian scale factor c corresponding to the Gaussian standard deviations of one-σ (α = .683) , two-σ (α =. 954), and three-σ (α =.997). This may be extended to higher dimensions and we have shown the results for the for a tri-variate Gaussian whose level surfaces are 3ᵈ ellipsoids and whose corresponding CDF for *ellipsoids of concentration* is Ψ3(c) = Γ_Inc(3/2, c²/2) / Γ(3/2). For an n dimensional ellipsoid the result is Ψn(c) = Γ_Inc(n/2, c²/2) / Γ(n/2) for n =1,2,3, ... which yields the specific forms given in the slide for n=1,2,3; for the case n=1 the "ellipsoid" is the line segment from –c to +c. Thus, the 1-σ, 2-σ and 3-σ "standard deviations" of the 1ᵈ Gaussian Normal distribution are compared with the scaling parameters c(2ᵈ) and c(3ᵈ) for ellipses and ellipsoids respectively in the table (bottom panel); extending this tabulation allows us to define standard Bivariate and Trivariate Normal distributions for Ψ2(c) and Ψ3(c) similar to Φ(x) for the Normal Gaussian. The figures illustrate this equivalence by showing the c(2ᵈ) =1.52 cut through the Bivariate Gaussian surface yielding an equivalent "1-σ"ellipse containing α = 68.3% of the probability and then notionally comparing the ellipse area with the "one-σ" area under the standard Gaussian curve. Although we cannot illustrate a 4 dimensional Trivariate Gaussian surface, we can extend these ideas theoretically and take a level surface cut to yield a 3ᵈ concentration ellipsoid and Ψ3(c)=68.3% yields c(3ᵈ) =1.88 given in the last column of the table; c-values for 2-σ and 3-σ as listed as well.

9.3.4 Gaussian Probability Table for Concentration Ellipse Geometry

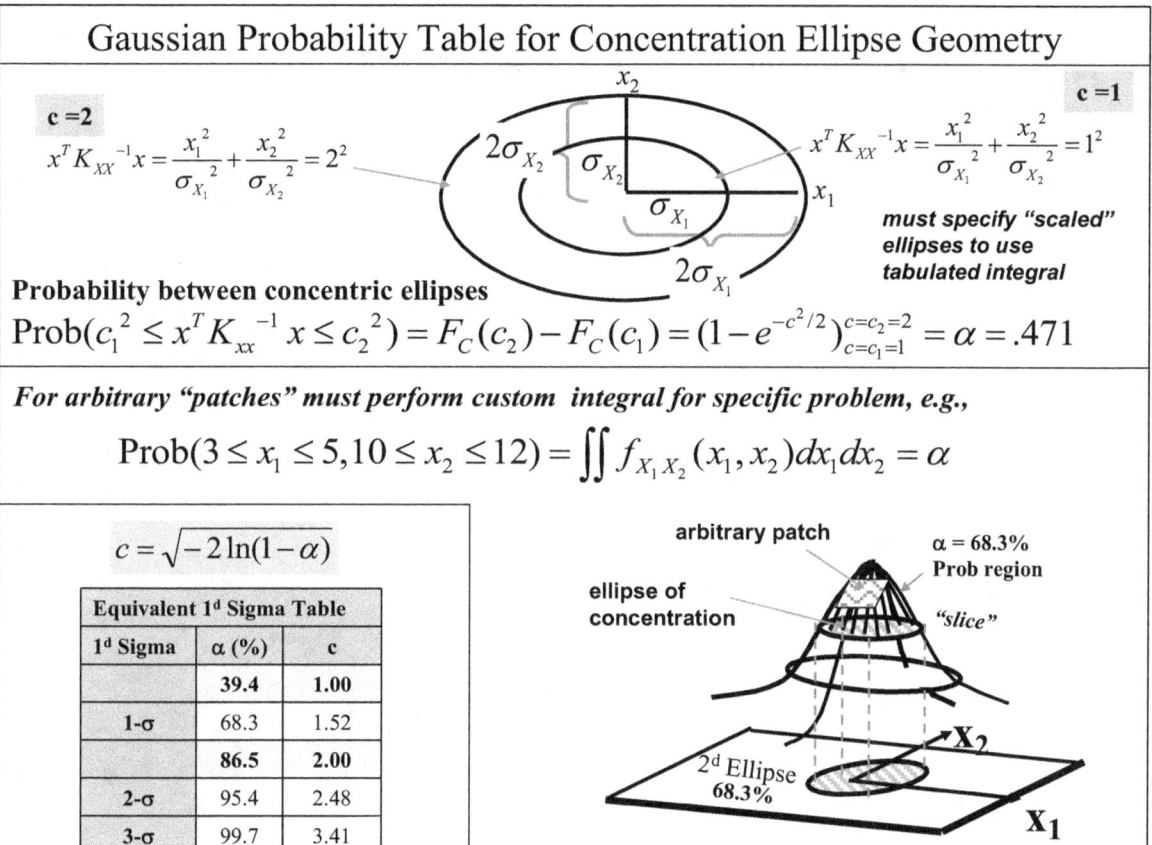

Gaussian Probability Table for Concentration Ellipse Geometry

c =2
$$x^T K_{XX}^{-1} x = \frac{x_1^2}{\sigma_{X_1}^2} + \frac{x_2^2}{\sigma_{X_2}^2} = 2^2$$

c =1
$$x^T K_{XX}^{-1} x = \frac{x_1^2}{\sigma_{X_1}^2} + \frac{x_2^2}{\sigma_{X_2}^2} = 1^2$$

must specify "scaled" ellipses to use tabulated integral

Probability between concentric ellipses

$$\text{Prob}(c_1^2 \le x^T K_{xx}^{-1} x \le c_2^2) = F_C(c_2) - F_C(c_1) = (1 - e^{-c^2/2})_{c=c_1=1}^{c=c_2=2} = \alpha = .471$$

For arbitrary "patches" must perform custom integral for specific problem, e.g.,

$$\text{Prob}(3 \le x_1 \le 5, 10 \le x_2 \le 12) = \iint f_{X_1 X_2}(x_1, x_2) dx_1 dx_2 = \alpha$$

$$c = \sqrt{-2\ln(1-\alpha)}$$

Equivalent 1^d Sigma Table		
1^d Sigma	α (%)	c
	39.4	1.00
1-σ	68.3	1.52
	86.5	2.00
2-σ	95.4	2.48
3-σ	99.7	3.41

The ellipses of concentration table can be used *only when we specify geometric regions corresponding to scaled ellipses* as shown in the top panel. The inner and outer concentric ellipses correspond to c=1 and c=2 respectively with the inner ellipse having length σ_{X1} along the x_1-axis and σ_{X2} along the x_2-axis; the outer ellipse has both axes *scaled by a factor of two* to $2\sigma_{X1}$ and $2\sigma_{X2}$ respectively. The probability in the region between these concentric ellipses is the difference between the two concentration ellipse values $F_C(c) = 1 - e^{-c^2/2} = 1 - e^{-(2)^2/2} = .8646$ and $1 - e^{-(1)^2/2} = .3935$ computed to be .4711 in the top panel equation. Thus we can make *two more table entries* straddling the 1^d Gaussian one-sigma value of 68% corresponding to c= 1.52; explicitly, for c=1 we have entered α = 39.4% and for c=2 we have entered α = 86.5% . Clearly we could generate a large look-up table using *uniformly spaced* values of c by computing the probabilities $\alpha = \Psi(c) = 1 - e^{-c^2/2}$ for each tabulated value of c. This would establish a Bivariate Gaussian Normal table $\Psi(c)$ analogous to the well-known $\Phi(x)$ Normal Probability table.

Note that if we wish to compute the probability for an *arbitrary patch* such as that shown in the 2^d Gaussian probability density we would need to compute the integral corresponding to the volume under that patch as indicated in the bottom figure of the slide. There is no table look up in this case; the table is only available for scaled Gaussian concentration ellipses.

Also note that that although a coordinate rotation *changes the correlation between the RV pair, it does not change the area* of the concentration ellipse. Thus, the probability event region only needs to be specified in the *principal axis coordinate system* and then is computed directly using the formula $\Psi(c)$ or a table constructed from it.

The Bivariate Gaussian Distribution

9.3.5 Computed Error Ellipses over a Sensor Field of Regard

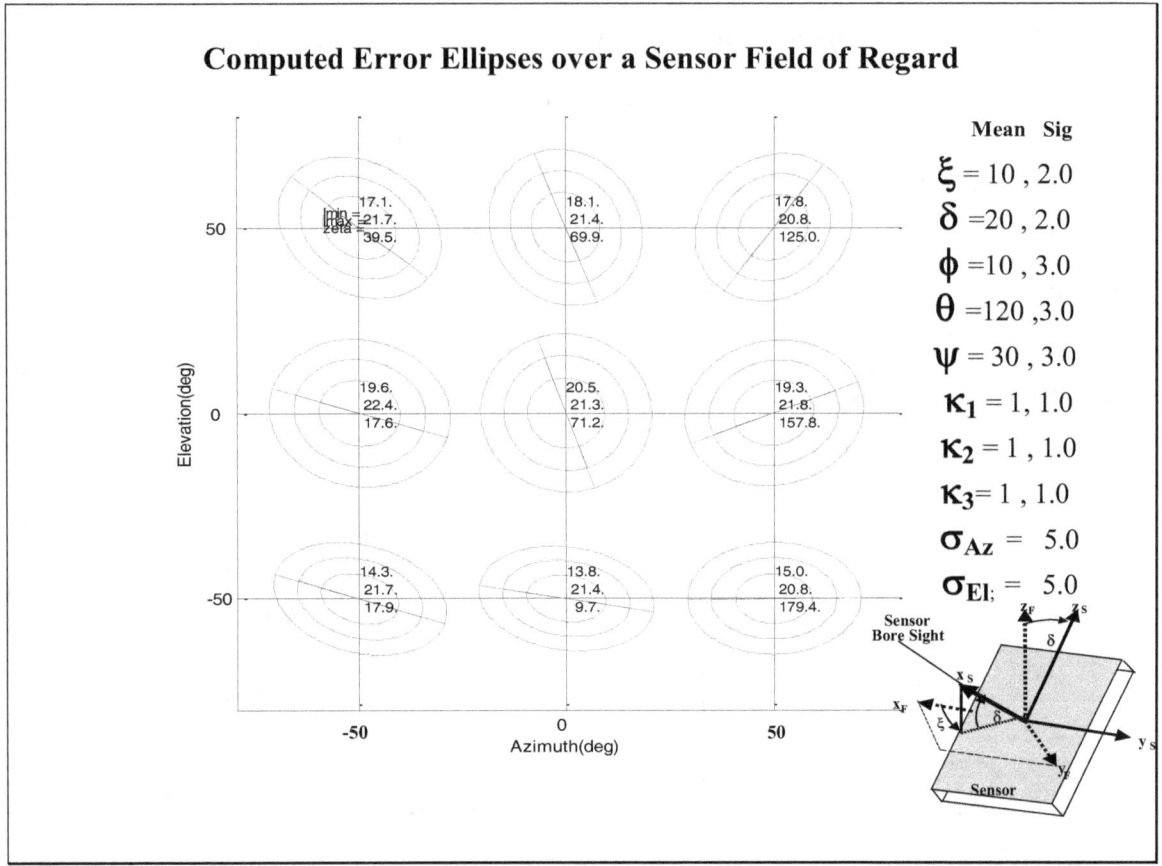

This slide illustrates a "typical" set of probability error ellipses corresponding to 68.3%, 95.4%, and 99.7% concentrations. The view corresponds to grid of azimuths A* and elevations E* relative to an electronic bore sight located at (0, 0). Although we have plotted all these ellipses in the *same plane* for visualization purposes, they actually "reside" in *different planes* perpendicular to the LOS at a given (A*, E*). The values of 10 unspecified system pointing parameters and their variances (sigmas) are listed on the right side of the figure. One can easily visualize the magnitude and directional characteristics of the errors across the $(100^\circ \times 100^\circ)$ *field of regard* of the sensor displayed in the figure. Varying these parameters allows us to visualize the error sensitivity to the pointing direction in a single plot over the entire field of regard.

LOS Error Analysis: A geometrical line-of-sight (LOS) error analysis ignores processing errors and simply sets down a mathematical chain of transformations from the bore sight frame of the pointing device (or sensor) to the desired coordinate frame of the target. The derivatives of the LOS vector with respect to the instrument orientation/alignment parameters provide the sensitivity partials needed to form the total error covariance. The resulting 3^d error ellipsoid degenerates into a 2^d ellipse *via* a projection of the covariance matrix onto a plane perpendicular to the LOS (similarity transformation). This leaves a 2^d eigenvalue problem to be solved numerically for each new set of parameter and variance values over an "*Az-El* grid" of LOS pointing directions. A unique new factorization of this eigenvalue problem yields an *analytic solution* that is formally the same for all LOS problems; the specifics only enter *via* the coordinate transformation chain and the parameter variances. Although this analysis yields the exact analytic solution, the resulting expressions are not easily interpreted. MatLab[®] is used to display a grid of oriented probability concentration ellipses (68.3%, 95.4% and 99.7%) on the sensor Az-El plane, thus providing a useful system analysis tool for sensor pointing errors (see Ref.[7]).

9.3.6 Relation of Bivariate Gaussian to Rayleigh Distribution

---

Relation Bivariate Gaussian to Rayleigh Distribution

Recall: Product of 2 IID Gaussians expressed in polar coordinates yields Rayleigh distribution in variable "r" Uniform distribution in variable θ

$\rho = 0$

Equivalently Bivariate Gaussian Vector with *uncorrelated* components and $\sigma_{X_1} = \sigma_{X_2}$ yields Rayleigh Distribution

$$\mu_x = \begin{bmatrix} 0 \\ 0 \end{bmatrix}; \ K_{xx} = \begin{bmatrix} \sigma_{X_1}^2 & \rho \\ \rho & \sigma_{X_2}^2 \end{bmatrix} = \begin{bmatrix} 1 & 0 \\ 0 & 1 \end{bmatrix}; \ \sigma_{X_1} = \sigma_{X_2} = 1 \ \rho = 0 \ "uncorrelated"$$

$$\Pr[x_1^2 + x_2^2 \le a^2] = \iint\limits_{x_1^2 + x_2^2 \le a^2} \frac{1}{2\pi} e^{-(x_1^2 + x_2^2)/2} dx_1 dx_2 = \int_{\theta=0}^{2\pi} \int_{r=0}^{a} \frac{1}{2\pi} e^{-(x_1^2 + x_2^2)/2} r dr d\theta = \int_{r=0}^{a} e^{-r^2/2} r dr = -e^{-r^2/2} \Big|_{r=0}^{a}$$

Rayleigh Distribution Characterized by *Ellipses of Concentration*

$$\Pr[x_1^2 + x_2^2 \le a^2] = 1 - e^{-a^2/2}$$

Like a "bullseye target" for Bivariate Gaussian Vector

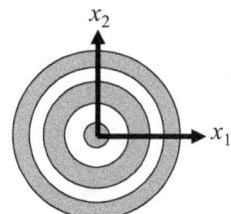

The Rayleigh distribution (Slide#2-18) results from taking the product of two $N(0,1)$ Gaussians to form a joint distribution $f_{X_1X_2}(x_1,x_2)$ and then transforming from Cartesian (x_1,x_2) to polar (r,θ) coordinates. Using the Jacobian determinant $1/r$, we find the new joint density $f_{R,\Theta}(r, \theta)$ is the product of a Uniform density $f_{\Theta}(\theta)=1/2\pi$ and the Rayleigh density $f_R(r)=r \cdot \exp(-r^2/2)$. The probability density surface has rotational symmetry and therefore displays the same Rayleigh $f_R(r)$ for any value of θ; horizontal cuts through the probability surface (not shown) yields level surfaces that are concentric circles ranging over all positive r-values as shown in the figure. A simple integration of the Rayleigh density from $r=0$ to $r = a$ yields the cumulative distribution $F_R(r)=\Pr[R \le a]=1-\exp(-a^2/2)$ which characterizes the Rayleigh distribution in terms of circles of concentration. Thus, the ellipses of concentration function for a Bivariate Gaussian $F_C(c) =1-\exp(-c^2/2)$ is a Rayleigh distribution expressed in terms of the ellipse scaling parameter "c".

This interesting result can be better understood as follows: The polar and Cartesian coordinates are related by the simple quadratic form $R^2 =X_1^2+X_2^2$; accordingly, the cumulative Rayleigh probability $\Pr\{R \le a\}$ can also be expressed in terms of the quadratic form as $\Pr[x_1^2+x_2^2 \le a^2]$. On the other hand, since the Bivariate Gaussian is expressed using a more general quadratic form as $\Pr[\mathbf{x}^T K^{-1}_{xx} \mathbf{x} \le c^2]$, we expect the Rayleigh distribution to be a special case of the Bivariate Gaussian distribution. In fact taking the explicit values for a pair of uncorrelated $N(0,1)$ Gaussians, *i.e.*, $\rho=0$, $\mu_1= \mu_2 =0$, and $\sigma_1 = \sigma_2 =1$ and evaluating $\mathbf{x}^T K^{-1}_{xx} \mathbf{x} \le c^2$ we find $x_1^2+x_2^2=r^2 \le c^2$ which is identical to Rayleigh with "c" replaced by "a". If we were to scale the Rayleigh σ from 1 to "c" , then its distribution is unchanged in terms of a scaled radial variable $u=r/c$ so we have $(r/c)^2 \le a^2$ or $r^2 \le c^2 a^2$ showing the scaling property holds for the Rayleigh distribution as well. Thus even though the Bivariate Gaussian is a product of *two different* and Rayleigh is a product of two identical independent RVs, they share the same scaling properties applied to ellipses and circles respectively.

9.4 Eigenvectors/Eigenvalues for Bivariate Gaussian

Eigenvectors/Eigenvalues for Bivariate Gaussian

Bivariate Gaussian N(0,1)

$$\mu_x = \begin{bmatrix} 0 \\ 0 \end{bmatrix} \; ; \; K_{xx} = \begin{bmatrix} 1 & \rho \\ \rho & 1 \end{bmatrix} \; ; \; \sigma_{X_1} = \sigma_{X_2} = 1 \; \Rightarrow \; K_{xx}^{-1} = \frac{1}{1-\rho^2} \begin{bmatrix} 1 & -\rho \\ -\rho & 1 \end{bmatrix}$$

Ellipses of Concentration:
set $q(x_1, x_2) = c^2 = 1$

$$q(x_1, x_2) = [x_1, x_2] \frac{1}{1-\rho^2} \begin{bmatrix} 1 & -\rho \\ -\rho & 1 \end{bmatrix} \begin{bmatrix} x_1 \\ x_2 \end{bmatrix} = \frac{x_1^2 - 2\rho x_1 x_2 + x_2^2}{1-\rho^2} = c^2 = 1$$

Eigenvalues of K_{XX}^{-1}   *Eigenvalues*

$$\left(K_{XX}^{-1}\right)x = \lambda x = I\lambda x \rightarrow \left(K_{XX}^{-1} - I\lambda\right)x = 0 \quad \text{Solution}: \det\left(K_{XX}^{-1} - I\lambda\right) = 0$$

$$\lambda_+ = 1/(1-\rho)$$
$$\lambda_- = 1/(1+\rho)$$

$$\det \begin{bmatrix} 1/(1-\rho^2) - \lambda & -\rho/(1-\rho^2) \\ -\rho/(1-\rho^2) & 1/(1-\rho^2) - \lambda \end{bmatrix} = \left(1/(1-\rho^2) - \lambda\right)^2 - \left(-\rho/(1-\rho^2)\right)^2 \rightarrow \lambda^2 - 2\lambda/(1-\rho^2) + 1/(1-\rho^2) = 0$$

Eigenvectors of $K_{XX}^{-1} = \begin{bmatrix} \lambda_- & 0 \\ 0 & \lambda_+ \end{bmatrix}$ **Principal Axis Coordinates**   *Eigenvectors*

Case 1:
$\rho = 0$ $\lambda_+ = \lambda_- = 1$: $K_{xx}^{-1} = \begin{bmatrix} 1 & 0 \\ 0 & 1 \end{bmatrix}$ Every Vector $\begin{bmatrix} x_1 \\ x_2 \end{bmatrix}$ is an eigenvector! $e_1 = \begin{bmatrix} x_1 \\ x_2 \end{bmatrix}$

Case 2: $\lambda_+ = 1/(1-\rho)$: $\begin{bmatrix} 1/(1-\rho^2) - 1/(1-\rho) & -\rho/(1-\rho^2) \\ -\rho/(1-\rho^2) & 1/(1-\rho^2) - 1/(1-\rho) \end{bmatrix} \begin{bmatrix} x_1 \\ x_2 \end{bmatrix} = \begin{bmatrix} 0 \\ 0 \end{bmatrix} \rightarrow x_1 = -x_2$ $e_+ = \begin{bmatrix} -1 \\ 1 \end{bmatrix} or \begin{bmatrix} 1 \\ -1 \end{bmatrix}$

$0 < |\rho| \le 1$

$\lambda_- = 1/(1+\rho)$: $\begin{bmatrix} 1/(1-\rho^2) - 1/(1+\rho) & -\rho/(1-\rho^2) \\ -\rho/(1-\rho^2) & 1/(1-\rho^2) - 1/(1+\rho) \end{bmatrix} \begin{bmatrix} x_1 \\ x_2 \end{bmatrix} = \begin{bmatrix} 0 \\ 0 \end{bmatrix} \rightarrow x_1 = +x_2$ $e_- = \begin{bmatrix} 1 \\ 1 \end{bmatrix}$

The eigenvalues and eigenvectors of the inverse covariance matrix K_{XX}^{-1} give the magnitudes and directions of the principal axes of the concentration ellipses and are an important visual tool in understanding the properties of the covariance matrix (see Slide#9-7). Consider the simple case of a Bivariate Gaussian with zero mean vector, unit variance, and arbitrary correlation ρ, displayed in the first equation of this slide; the ellipses of concentration for $c^2 = 1$ are given explicitly in the second equation. Now, in order to determine the eigenvalues and eigenvectors of the *inverse covariance matrix* K_{XX}^{-1}, we only need require that its effect on a vector **x** is to "stretch" it by an amount λ. This leads to a homogeneous matrix equation whose solution exists *if and only if* we set its determinant to zero, viz., $|K_{XX}^{-1} - I\lambda| = 0$. Expansion of this determinant leads to a quadratic equation for the eigenvalues λ with two solutions $\lambda_+ = 1/(1-\rho)$ and $\lambda_- = 1/(1+\rho)$.

For the general case $\rho \ne 0$ the eigenvalues are distinct $\lambda_+ = 1/(1-\rho)$ and $\lambda_- = 1/(1+\rho)$ and they have the distinct eigenvectors $e_+ = [1,-1]^T$ (or $[-1,1]^T$) and $e_- = [1,1]^T$ respectively. For a positive correlation coefficient $\rho > 0$, the eigenvalue $\lambda_+ = 1/(1-\rho) > 1$ corresponds to the larger eigenvalue (but the **smaller** ellipse axis "b") and conversely for the eigenvalue $\lambda_- = 1/(1+\rho) < 1$ corresponds to the smaller eigenvalue (but the **larger** ellipse axis "a"). Thus the *major axis* eigenvector is $e_- = [1,1]^T$ is along the y_1-axis which is rotated $+45°$ from the x_1-axis (1st quadrant), while the *minor axis* eigenvector is chosen to be $e_+ = [-1,1]^T$ so as to form a right-handed y_1-y_2 coordinate system. On the other, hand for negative correlation $\rho < 0$, the roles of the eigenvalues and eigenvectors are reversed; i.e., $\lambda_+ = 1/(1-\rho) < 1$ and the *major axis* eigenvector is $e_+ = [1,-1]^T$ is again along the y_1-axis but now is rotated $-45°$ from the x_1-axis (4th quadrant); the larger eigenvalue $\lambda_- = 1/(1+\rho) > 1$ (corresponding to the *minor ellipse axis*) determines the eigenvector as $e_- = [1,1]^T$ so as to form a right-handed coordinate system y_1-y_2.

On the other hand, for the degenerate $\rho = 0$ case of a double eigenvalue $\lambda_+ = \lambda_- = 1$ there are no unique eigenvectors and in fact every vector is an eigenvector and the ellipse becomes a circle with no preferred direction and this corresponds two uncorrelated Gaussian RVs X_1 and X_2.

9.4.1 Bivariate Gaussian Vector Solution Geometry

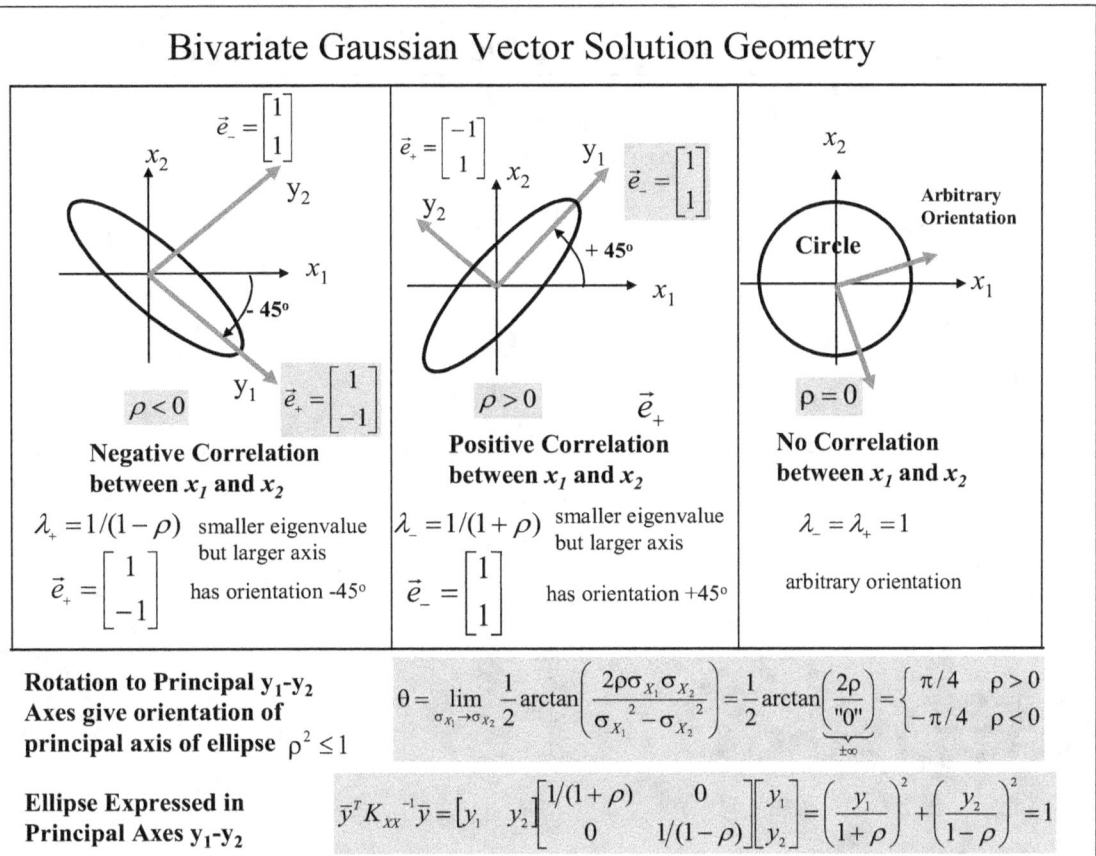

Bivariate Gaussian Vector Solution Geometry

Negative Correlation between x_1 and x_2

$\vec{e}_- = \begin{bmatrix} 1 \\ 1 \end{bmatrix}$

$\rho < 0$ $\vec{e}_+ = \begin{bmatrix} 1 \\ -1 \end{bmatrix}$

$\lambda_+ = 1/(1-\rho)$ smaller eigenvalue but larger axis

$\vec{e}_+ = \begin{bmatrix} 1 \\ -1 \end{bmatrix}$ has orientation -45°

Positive Correlation between x_1 and x_2

$\vec{e}_+ = \begin{bmatrix} -1 \\ 1 \end{bmatrix}$ $\vec{e}_- = \begin{bmatrix} 1 \\ 1 \end{bmatrix}$ +45°

$\rho > 0$

$\lambda_- = 1/(1+\rho)$ smaller eigenvalue but larger axis

$\vec{e}_- = \begin{bmatrix} 1 \\ 1 \end{bmatrix}$ has orientation +45°

No Correlation between x_1 and x_2

Arbitrary Orientation   Circle

$\rho = 0$

$\lambda_- = \lambda_+ = 1$

arbitrary orientation

Rotation to Principal y_1-y_2 Axes give orientation of principal axis of ellipse $\rho^2 \leq 1$

$$\theta = \lim_{\sigma_{x_1} \to \sigma_{x_2}} \frac{1}{2}\arctan\left(\frac{2\rho\sigma_{X_1}\sigma_{X_2}}{\sigma_{X_1}^2 - \sigma_{X_2}^2}\right) = \frac{1}{2}\arctan\left(\frac{2\rho}{"0"}\right) = \begin{cases} \pi/4 & \rho > 0 \\ -\pi/4 & \rho < 0 \end{cases}$$

Ellipse Expressed in Principal Axes y_1-y_2

$$\bar{y}^T K_{XX}^{-1} \bar{y} = \begin{bmatrix} y_1 & y_2 \end{bmatrix}\begin{bmatrix} 1/(1+\rho) & 0 \\ 0 & 1/(1-\rho) \end{bmatrix}\begin{bmatrix} y_1 \\ y_2 \end{bmatrix} = \left(\frac{y_1}{1+\rho}\right)^2 + \left(\frac{y_2}{1-\rho}\right)^2 = 1$$

Here we display the concentration ellipse corresponding to the correlated $\rho \neq 0$ and uncorrelated $\rho=0$ cases we discussed on the previous slide. The figure on the right illustrates the situation for the degenerate case of no correlation ($\rho=0$) in which the ellipse has degenerated into a unit circle with eigenvalues $\lambda_+ = \lambda_- = 1$ and there is no preferred direction for the eigenvectors; we symbolize this by displaying a set of two eigenvectors that are orthogonal, but otherwise pointing in an arbitrary direction.

For the correlated case ($\rho \neq 0$) the figures on the left illustrate the ellipse (again for equal variances $\sigma_{X1}^2 = \sigma_{X2}^2 = 1$). For negative correlation ($\rho<0$) $\lambda_+ = 1/(1-\rho) = 1/(1+|\rho|)<1$ and hence $e_+ = [1,-1]^T$ corresponds to the larger major axis and the orientation is at -45° as shown in the left most figure; on the other hand, for positive correlation ($\rho >0$) $\lambda_- = 1/(1+\rho) < 1$ and hence $e_- = [1,1]^T$ corresponds to the larger major axis and the orientation is at +45° as shown in the center figure. The corresponding minor axes are chosen so as to form a right-handed principal axis coordinate system y_1-y_2 as shown.

The formula giving the rotation angle θ from x_1-x_2 coordinates to the y_1-y_2 principal axis coordinates of the concentration ellipse is displayed again at the bottom of the slide for convenience. In the current case of a Bivariate Gaussian with equal unit variances $\sigma_{X1}^2 = \sigma_{X2}^2 = 1$, the formula yields the rotation angle $\theta = \pm\pi/4$ for positive and negative correlations respectively; inspection of the figures we have drawn for these two cases shows agreement with Slide#9-8. The equation at the bottom of the slide gives the explicit analytic expression for the concentration ellipse expressed in y_1-y_2 principal axis coordinates for the case $\rho>0$; for negative correlation ($\rho<0$), setting $\rho \to -\rho$ exchanges the roles of y_1 and y_2 in the boxed equation so that y_1 remains the major axis. In this regard, it should be emphasized once again that it is the *minimum* (not the maximum) eigenvalue $1/(1+\rho)$ for $\rho>0$ (or $1/(1-\rho)$ for $\rho<0$) that corresponds to the major y_1-axis "a" because the eigenvalues correspond to the inverse covariance $K_{XX}^{-1} = \text{diag}(1/a^2, 1/b^2)$ while the axes of the ellipse are determined by the intercepts of the equation $(y_1/a)^2 + (y_2/b)^2 = 1$.

9.5 Covariance Ellipse "Geometrical Construction Rules"

Covariance Ellipse "Geometrical Construction Rules"

$$K_{XX} = \begin{bmatrix} \sigma_X^2 & \rho\sigma_X\sigma_Y \\ \rho\sigma_X\sigma_Y & \sigma_Y^2 \end{bmatrix} \;;\; K_{XX}^{-1} = \frac{1}{(1-\rho^2)\sigma_X^2\sigma_Y^2} \begin{bmatrix} \sigma_Y^2 & -\rho\sigma_X\sigma_Y \\ -\rho\sigma_X\sigma_Y & \sigma_X^2 \end{bmatrix}$$

Concentration Ellipse with parameter "c"

$$q(x,y) = x^T K_{XX}^{-1} x = [x,y] \cdot \frac{1}{(1-\rho^2)\sigma_X^2\sigma_Y^2} \begin{bmatrix} \sigma_Y^2 & -\rho\sigma_X\sigma_Y \\ -\rho\sigma_X\sigma_Y & \sigma_X^2 \end{bmatrix} \cdot \begin{bmatrix} x \\ y \end{bmatrix} = c^2$$

$$q(x,y) = \frac{\sigma_Y^2 x^2 - 2\rho\sigma_X\sigma_Y xy + \sigma_X^2 y^2}{\sigma_X^2\sigma_Y^2(1-\rho^2)} = \frac{1}{(1-\rho^2)}\left[\left(\frac{x}{\sigma_X}\right)^2 - 2\rho\left(\frac{x}{\sigma_X}\right)\left(\frac{y}{\sigma_Y}\right) + \left(\frac{y}{\sigma_Y}\right)^2\right] = c^2$$

Points on Concentration Ellipse with parameter "c=1"

See Slide#9-13 for relation between the parameter "c" and 1^d parameter σ

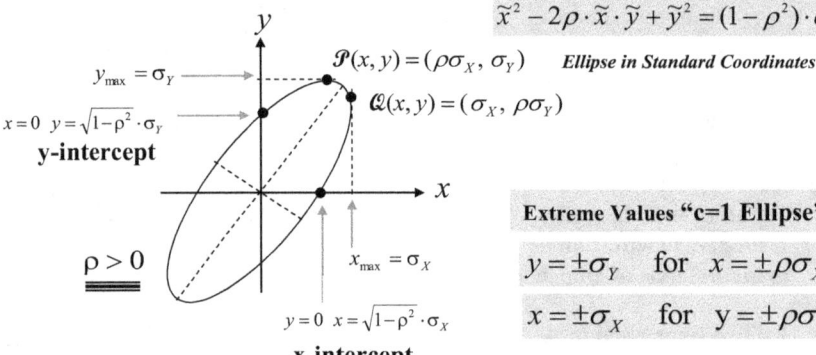

$$\tilde{x}^2 - 2\rho \cdot \tilde{x} \cdot \tilde{y} + \tilde{y}^2 = (1-\rho^2) \cdot c^2$$

$\mathcal{P}(x,y) = (\rho\sigma_X, \sigma_Y)$ *Ellipse in Standard Coordinates*

$\mathcal{Q}(x,y) = (\sigma_X, \rho\sigma_Y)$

$y_{max} = \sigma_Y$

$x=0 \quad y=\sqrt{1-\rho^2}\cdot\sigma_Y$ **y-intercept**

$\rho > 0$

$x_{max} = \sigma_X$

$y=0 \quad x=\sqrt{1-\rho^2}\cdot\sigma_X$ **x-intercept**

Extreme Values "c=1 Ellipse"

$$y = \pm\sigma_Y \quad \text{for} \quad x = \pm\rho\sigma_X$$

$$x = \pm\sigma_X \quad \text{for} \quad y = \pm\rho\sigma_Y$$

For a general covariance with non-zero correlation and different sigmas it would be nice to have a rough sketch of what the correlation ellipse looks like. This can be done very simply by (i) first computing the inverse covariance matrix K_{XX}^{-1} and then (ii) determining values for the following four points: the x and y intercepts, and max (x) and max(y) , (iii) determine the sign of the correlation coefficient from the matrix, and (iv) sketch the elliptical shape through these points making sure the large axis of the ellipse has positive slope for $\rho > 0$ and negative slope for $\rho < 0$.

In order to locate these points, consider the explicit form for the quadratic q(x,y) in the 1st boxed equation for the particular ellipse defined by c=1. Setting x=0 yields the y-intercept $y = (1-\rho^2)^{1/2} \sigma_Y$ and similarly setting y=0 yields the x-intercept $x = (1-\rho^2)^{1/2} \sigma_X$. On the next slide we prove that the extremum x-values are $\pm\sigma_X$; substituting this value $x=\pm\sigma_X$ back into q(x,y) yields a quadratic whose solutions are respectively $y=\pm\rho\cdot\sigma_Y$; similarly, the extremum y-values are $y=\pm\sigma_Y$ and substitution yields $x=\pm\rho\sigma_X$, thus verifying two of the four equations for the extreme values on the bottom right of the slide. Thus we have established the four points on the ellipse as shown by the large black dots and can sketch the symmetric ellipse as shown for positive correlation $\rho > 0$.

If the correlation is negative $\rho < 0$, the two intercept points are unchanged, but $\mathcal{P}(\rho\sigma_X, \sigma_Y) \to \mathcal{P}^*(-\rho\sigma_X, \sigma_Y)$ and $\mathcal{Q}(\sigma_X, \rho\sigma_Y) \to \mathcal{Q}^*(\sigma_X, -\rho\sigma_Y)$; the shape of the ellipse is the same, but it will be oriented perpendicular to the illustration. This suffices to sketch the general shape and orientation of the covariance ellipse. In the next two slides we give a proof of the extremum formulas in the boxed equation at the bottom of this slide and then summarize *all points* that can be used to plot the concentration ellipses.

9.5.1 Geometrical Construction Rules for Covariance Ellipse - Derivation

---

Geometrical Construction Rules for Covariance Ellipse - Derivation

Covariance Ellipse	$$q(x,y)=\frac{1}{(1-\rho^2)}\left[\left(\frac{x-\mu_X}{\sigma_X}\right)^2-2\rho\left(\frac{x-\mu_X}{\sigma_X}\right)\cdot\left(\frac{y-\mu_Y}{\sigma_Y}\right)+\left(\frac{y-\mu_Y}{\sigma_Y}\right)^2\right]=c^2$$
Express in Normalized and Centered Coordinates	$$\tilde{x}=\frac{x-\mu_X}{\sigma_X}\ ;\ \tilde{y}=\frac{y-\mu_Y}{\sigma_Y}\quad\Rightarrow\quad \boxed{\tilde{x}^2-2\rho\tilde{x}\tilde{y}+\tilde{y}^2=\left(1-\rho^2\right)\cdot c^2}$$

Find Extremum for $\tilde{y}=\tilde{y}(\tilde{x})$

$$2\tilde{x}\frac{d\tilde{x}}{d\tilde{x}}-2\rho\left(\frac{d\tilde{x}}{d\tilde{x}}\tilde{y}+\tilde{x}\frac{d\tilde{y}}{d\tilde{x}}\right)+2\tilde{y}\frac{d\tilde{y}}{d\tilde{x}}=0\ \Rightarrow\ \frac{d\tilde{y}}{d\tilde{x}}=\frac{\tilde{x}-\rho\tilde{y}}{\tilde{y}-\rho\tilde{x}}=0\ \Rightarrow\ \tilde{x}=\rho\tilde{y}\quad(1)$$

Alternately, for $\tilde{x}=\tilde{x}(\tilde{y})$

$$2\tilde{x}\frac{d\tilde{x}}{d\tilde{y}}-2\rho\left(\frac{d\tilde{x}}{d\tilde{y}}\tilde{y}+\tilde{x}\frac{d\tilde{y}}{d\tilde{y}}\right)+2\tilde{y}\frac{d\tilde{y}}{d\tilde{y}}=0\ \Rightarrow\ \frac{d\tilde{x}}{d\tilde{y}}=\frac{\tilde{y}-\rho\tilde{x}}{\tilde{x}-\rho\tilde{y}}=0\ \Rightarrow\ \tilde{y}=\rho\tilde{x}\quad(2)$$

Substitute (1) back into boxed eqn to find corresponding value of $\tilde{y}$

$$\left(\rho\tilde{y}\right)^2-2\rho(\rho\tilde{y})\tilde{y}+\tilde{y}^2=\left(1-\rho^2\right)\cdot c^2\ \Rightarrow\ \tilde{y}=\pm c$$

$$\Rightarrow\ \tilde{y}=\pm c\quad \text{for}\quad \tilde{x}=\rho\tilde{y}=\pm\rho c\qquad \textbf{Extreme Values}$$

Set c=1
Use defn of $\tilde{x},\tilde{y}$

$$\tilde{y}=\frac{y-\mu_Y}{\sigma_Y}=\pm 1\ \text{for}\ \tilde{x}=\frac{x-\mu_X}{\sigma_X}=\pm\rho\quad\text{For }\mu_X=\mu_Y=0\qquad y=\pm\sigma_Y\ \text{for}\ x=\pm\rho\sigma_X\quad(1')$$

Similarly for Eq.(2) we find $\qquad x=\pm\sigma_X\ \text{for}\ y=\pm\rho\sigma_Y\quad(2')$

For convenience we repeat the quadratic form q(x,y) from the last slide and put it in simpler form by going to standardized (centered and normalized) coordinates $x_{tilde}=(x-\mu_X)/\sigma_X$ and $y_{tilde}=(y-\mu_Y)/\sigma_Y$; this gives the boxed quadratic expression $x_{tilde}^2 -2\rho\cdot x_{tilde}\cdot y_{tilde}+y_{tilde}^2 =(1-\rho^2)\cdot c^2$ on the top panel of the slide. Considering this formula to express y_{tilde} as a function of x_{tilde}, we differentiate the whole expression implicitly with respect to x_{tilde} and set the result to zero and find Eq.(1) $x_{tilde}=\rho\cdot y_{tilde}$ as the extreme value of x_{tilde}. Upon substitution of this value for x_{tilde} back into the original expression, we find the corresponding value $y_{tilde}=\pm c$; thus, we have the two extreme value coordinate pairs

$$\{x_{tilde}=\pm\rho c,\ y_{tilde}=\pm c\};$$

by symmetry we interchange the roles of x_{tilde} and y_{tilde} to find two more pairs

$$\{x_{tilde}=\pm c,\ y_{tilde}=\pm\rho c\}.$$

Choosing c=1 for a *unit* ellipse and using the definition of the *tilde* coordinates for the zero mean case $\mu_X=\mu_Y=0$ results in Eqs. (1') and (2') which give the extreme value points used on the previous slide, *viz.*,

$$\{x=\pm\rho\cdot\sigma_X,\ y=\pm\sigma_Y\}\ \text{and}\ \{x=\pm\sigma_X,\ y=\pm\rho\cdot\sigma_Y\}.$$

For the case of *non-zero means*, the general results are written as

$$\{x=\mu_X\pm\rho\cdot\sigma_X,\ y=\mu_Y\pm\sigma_Y\}\ \text{and}\ \{x=\mu_X\pm\sigma_X,\ y=\mu_Y\pm\rho\cdot\sigma_Y\}.$$

The next slide summarizes these extrema together with the x- and y- intercepts in all four quadrants thus giving a complete prescription for quickly sketching the ellipses by hand.

9.5.2 Concentration Ellipse Construction - Summary

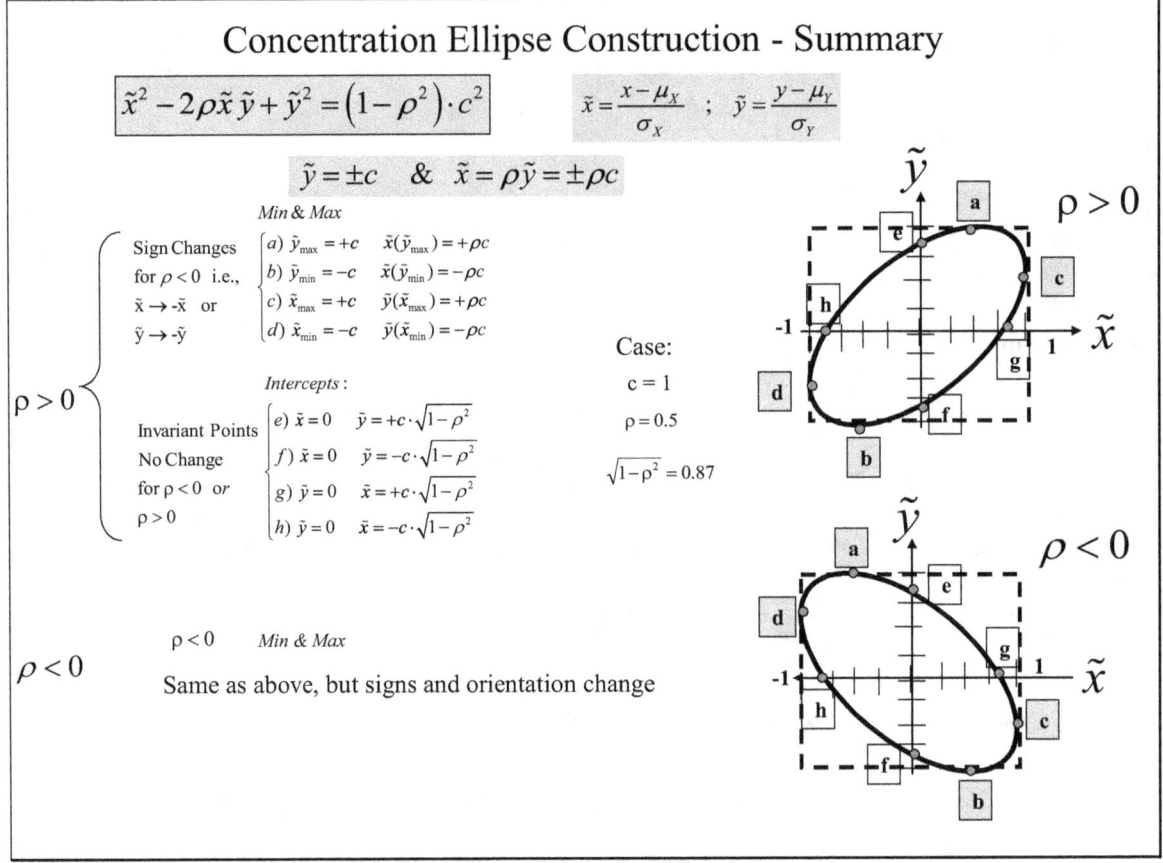

In this slide we show all the extrema labeled {a,b,c,d} and intercepts labeled {e,f,g,h} for positive correlation $\rho > 0$ in the table on the left and show these "labeled points" in the figure on the right. Note that the figure is explicitly for the case c=1 and extends from "scaled-x" value x/σ_X in the interval [-1, 1] and "scaled-y" value y/σ_Y in the interval [-1, 1]; the plot is assumed to be centered on the mean vector (μ_X, μ_Y) as the origin. We see the natural symmetries between the y and x maximum values {a, c} in Quadrant#1 respectively with their counterparts {b,d} in Quadrant#3; similarly for the x- and y- intercepts {g,e} in Quadrant#1 respectively with their counterparts {h,f} in Quadrant#3.

An analogous set of points are displayed for the negative correlation case $\rho < 0$; we see that the sign of the correlation will naturally give the orientation of the ellipse by appropriately shifting the 8 points.

9.6 *Bayes' Rule, Bivariate Gaussian Vectors & Communication Channel*

Bayes' Rule, Bivariate Gaussian Vectors & Communication Channel

Bivariate "Gaussian Vectors" are "closed" i.e., remain Gaussian under two main operations:

1) Linear Transformations: $\boxed{X' = AX + b}$ $\quad X = \begin{bmatrix} x \\ Y \end{bmatrix} \quad X' = \begin{bmatrix} X' \\ Y' \end{bmatrix} \quad A = \begin{pmatrix} a_{11} & a_{12} \\ a_{21} & a_{22} \end{pmatrix}$ **Already Shown**

Note: X vector now has components [X, Y] , not [X_1 , X_2]

---

2) Bayes' Rule

Between Components

$$f_{X|Y}(x \mid y) = \frac{f_{Y|X}(y \mid x) f_X(x)}{f_Y(y)} = \frac{f_{XY}(x,y)}{f_Y(y)}$$

X & Y scalars!! <== **Show this**

Here, we can think in terms of a communications channel in which the scalar **input X** *& scalar* **output Y** *are the two components of the* **Gaussian Vector** $\quad \vec{X} = \begin{bmatrix} X \\ Y \end{bmatrix}$

Communication Channel: Gaussian input "scalar" X & Gaussian output "scalar" Y form a joint gaussian distribution between correlated scalar random variables X & Y , *i.e.*, a **gaussian vector** $\quad \vec{X} = \begin{bmatrix} X \\ Y \end{bmatrix}$

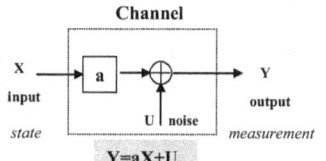

Note "a" is a scalar here!

Assume: zero mean, unit variance Gaussian Vector with correlation coefficient ρ

$$\vec{\mu}_{\vec{x}} \equiv E \begin{bmatrix} X \\ Y \end{bmatrix} = \begin{bmatrix} 0 \\ 0 \end{bmatrix} \qquad K_{XY} = \begin{bmatrix} 1 & \rho \\ \rho & 1 \end{bmatrix} \qquad K_{XY}^{-1} = \frac{1}{1-\rho^2} \begin{bmatrix} 1 & -\rho \\ -\rho & 1 \end{bmatrix}$$

Changing notation slightly, the Bivariate Gaussian is a joint distribution defined by two Gaussian RVs X and Y (instead of X_1 and X_2) which form a 2-vector **X** =[X,Y]^T. This Gaussian Vector **X** is completely characterized by its mean vector and its covariance matrix {μ_X , K_{XY}} and the result of a linear transformation **X'**=A**X**+**b** is a new Gaussian Vector **X'** = [X',Y']^T with a new mean vector and covariance matrix {μ_X', $K_{X'Y'}$}. Gaussian vectors are said to be "closed" under linear vector transformations because they remain Gaussian.

There is another type of closure that exists *relating to the components X and Y* of a Gaussian vector *via* Bayes' rule rather than a linear transformation of coordinates. It simply states that if the joint distribution $f_{XY}(x,y)$ is Gaussian, then the conditional distribution $f_{X|Y}(x|y)$ is also Gaussian; because of the obvious connection to Bayes' rule shown in the boxed equation, this is often called "closure under Bayes' Rule."

The communications channel illustrated in the middle panel shows an input signal X going through a channel consisting of a scalar multiplier "a" and additive noise U to yield an output scalar of the form Y = a·X+U, where it must be emphasized that all quantities in this equation are scalars. Now this equation relating the output Y to the input X can be considered to be the two components of a Bivariate Gaussian vector **X** =[X,Y]^T and in the context of a communication channel we would expect them to be highly correlated.

With such a communication channel in mind, consider the zero mean, unit variance Gaussian vector **X** =[X,Y]^T with correlation ρ. The mean μ_X = **0**, the covariance matrix K_{XY} and its inverse K_{XY}^{-1} are given explicitly in the last panel of the slide. On the next slide we shall use this explicit representation to show closure under Bayes' rule, *i.e.*, that the conditional density $f_{X|Y}(x|y)$ is also Gaussian.

The Bivariate Gaussian Distribution

9.6.1 Computing the Bayes' Update Probability

Computing the Bayes' Update Probability

Joint Distribution	$$f_{XY}(x,y) = \dfrac{1}{2\pi\sqrt{1-\rho^2}}\, e^{-\frac{\left[x^2 - 2\rho xy + y^2\right]}{2(1-\rho^2)}}$$
In order to compute $f_{X\|Y}$ need marginal $f_Y(y)$	$$f_Y(y) = \int_{-\infty}^{\infty} f_{XY}(x,y)\,dx = \dfrac{1}{2\pi\sqrt{1-\rho^2}} \int_{-\infty}^{\infty} e^{-\frac{\left[x^2 - 2\rho xy + y^2\right]}{2(1-\rho^2)}}\, dx$$
"Completing the square" in the exponent & Subst. into marginal $f_Y(y)$ yields	$$(x^2 - 2\rho xy) = (x-\rho y)^2 - \rho^2 y^2$$ $$f_Y(y) = \underbrace{\dfrac{1}{2\pi\sqrt{1-\rho^2}}\, e^{-\frac{\left[y^2 - \rho^2 y^2\right]}{2(1-\rho^2)}}}_{=\,e^{-y^2/2}} \underbrace{\int_{-\infty}^{\infty} e^{-\frac{(x-\rho y)^2}{2(1-\rho^2)}}\, d(x-\rho y)}_{=\sqrt{2\pi(1-\rho^2)}} = \dfrac{e^{-y^2/2}}{\sqrt{2\pi}} \;\Rightarrow\; N(0,1)$$

Marginal $f_Y(y)$ is $N(0,1)$ $\quad f_Y(y) = \dfrac{e^{-y^2/2}}{\sqrt{2\pi}}$	Bayes' Conditional $f_{X\|Y}(x\|y)$ is $\underset{\uparrow\;\;\;\;\uparrow}{N(\rho y,(1-\rho^2))}$ $\quad f_{X\|Y}(x\mid y) = \dfrac{e^{-\frac{\left[(x-\rho y)^2\right]}{2(1-\rho^2)}}}{\sqrt{2\pi(1-\rho^2)}}$
	$\mu_{X\|Y}\;\;\;\sigma_{X\|Y}^2$

Bayes' Conditional Details

$$f_{X\|Y}(x\mid y) = \dfrac{\dfrac{1}{2\pi\sqrt{1-\rho^2}}\, e^{-\frac{\left[x^2 - 2\rho xy + y^2\right]}{2(1-\rho^2)}}}{\dfrac{1}{\sqrt{2\pi}}\, e^{-\frac{y^2}{2}}} = \dfrac{e^{-\frac{\left[x^2 - 2\rho xy + y^2\right]}{2(1-\rho^2)} + \frac{y^2}{2}}}{\sqrt{2\pi(1-\rho^2)}} = \dfrac{e^{-\frac{\left[x^2 - 2\rho xy + y^2 - (1-\rho^2)y^2\right]}{2(1-\rho^2)}}}{\sqrt{2\pi(1-\rho^2)}} = \dfrac{e^{-\frac{\left[(x-\rho y)^2\right]}{2(1-\rho^2)}}}{\sqrt{2\pi(1-\rho^2)}}$$

In order to show that the conditional distribution $f_{X\|Y}(x\|y)$ is also Gaussian, we need to first compute the marginal distribution $f_Y(y)$ and then divide it into the joint distribution $f_{XY}(x,y)$. The marginal distribution is computed by integrating the joint distribution over x; first separating out the y^2 term in the exponent and completing the square of the remaining terms $\{x^2 - 2\rho xy\} + y^2 = \{(x-\rho y)^2 - \rho^2 y^2\} + y^2$ yields a Gaussian normalization integral in the variable $(x-\rho y)$; hence, the marginal is found to be $f_Y(y) = (2\pi)^{-\frac{1}{2}} \cdot \exp(-y^2/2)$ which is in fact an N(0,1) Gaussian.

Subsequently dividing this marginal $f_Y(y)$ into the joint distribution $f_{XY}(x,y)$ yields the Bayes' conditional density $f_{X\|Y}(x\|y)$ which is again a Gaussian Normal distribution $\sim N(\rho y, 1-\rho^2)$. As explicitly shown in the bottom equation, the conditional mean $\mu_{X\|Y} = E[X\|Y] = \rho y$ and the conditional variance $\text{var}(X\|Y) = \sigma_{X\|Y}^2 = 1-\rho^2$ are easily be picked off the exponential term once it is expressed in standard Gaussian form, $\exp\{-(x-\mu_{X\|Y})^2 / 2\cdot\sigma_{X\|Y}^2$.

Thus, in the context of a communication channel in which both the transmitted signal X and the observed signal Y are Gaussian, the Bayes' update density $f_{X\|Y}(x\|y)$ is also a Gaussian with updated mean $\mu_{X\|Y}$ and variance $\sigma_{X\|Y}^2$ that reflect the value of the measurement $Y=y$ and its correlation ρ, with the transmitted signal X.

The Bivariate Gaussian Distribution

9.6.2 Closure Under Bayesian Updates - Summary

Closure Under Bayesian Updates - Summary

Summary:

Started with a pair of N(0,1) RVs X & Y with correlation ρ

1) The joint distribution is a correlated Gaussian in X and Y

$$\vec{X} = \begin{bmatrix} X \\ Y \end{bmatrix} \qquad \vec{\mu}_{\vec{x}} = E\begin{bmatrix} X \\ Y \end{bmatrix} = \begin{bmatrix} 0 \\ 0 \end{bmatrix} \qquad K_{XY} = \begin{bmatrix} 1 & \rho \\ \rho & 1 \end{bmatrix}$$

$$f_{XY}(x,y) = \frac{1}{2\pi\sqrt{1-\rho^2}} e^{-\frac{\left[x^2 - 2\rho xy + y^2\right]}{2(1-\rho^2)}}$$

2) Marginal $f_Y(y)$ is found to be N(0,1): $\quad f_Y(y) = \dfrac{e^{-y^2/2}}{\sqrt{2\pi}}$

3) **Bayes' Update** $f_{X|Y}(x|y)$ is Gaussian $N(\rho y, 1-\rho^2)$

$$f_{X|Y}(x|y) = \frac{1}{\sqrt{2\pi(1-\rho^2)}} e^{-\frac{\left[(x-\rho y)^2\right]}{2(1-\rho^2)}}$$

4) Pick off "conditional" mean & variance from $f_{X|Y}(x|y)$

$$\mu_{X|Y} \equiv E[X|Y] = \rho y \; ; \; Var(X|Y) = 1-\rho^2$$

Conditional Mean represents an "estimate of X given meas. Y" with Var(X|Y) obtained from Bayes' Updated Gaussian

Generalize:

Start with General Gaussian Vector with non-zero mean & Variance

$$\vec{X} = \begin{bmatrix} X \\ Y \end{bmatrix} \qquad \mu = \begin{bmatrix} \mu_X \\ \mu_Y \end{bmatrix} \; ; \; K_{XY} = \begin{bmatrix} \sigma_X^2 & \rho\sigma_X\sigma_Y \\ \rho\sigma_X\sigma_Y & \sigma_Y^2 \end{bmatrix}$$

Conditional Mean and Variance Represents the Bayes' Update Equation

$$\mu_{X|Y} \equiv E[X|Y] = \mu_X + \rho\frac{\sigma_X}{\sigma_Y}(y-\mu_Y)$$

$$Var(X|Y) = \sigma_X^2(1-\rho^2) \; ; \; \sigma_{X|Y} = \sigma_X\sqrt{1-\rho^2}$$

Note 1 "*Gaussian Arena*" we do not need to work with distributions directly since both
1) Linear Xfms & 2)Bayes' Update Equation yield **Gaussian Vector Results** (*surrogates* for the joint and conditional distributions respectively)

Note 2: Y is irrelevant for $\rho=0$
X & Y indep => Conditionals do not depend upon value of y:
$\mu_{X|Y} = \mu_X$ & $\sigma_{X|Y}^2 = Var(X|Y) = \sigma_X^2$

Closure Under Bayesian Updates started with a pair of correlated N(0,1) Gaussian RVs with correlation coefficient ρ and resulted in a Gaussian conditional distribution $f_{X|Y}(x|y)$ with conditional mean is $\mu_{X|Y} = E[X|Y] = \rho y$ and conditional variance is $var(X|Y) = \sigma_{X|Y}^2 = 1-\rho^2$.

If instead, we start with a pair of correlated Gaussian RVs having different means and variances given by the mean vector μ_X and covariance matrix K_{XY} shown in the middle panel of the slide, then the resulting conditional mean and variance are

$$E[X|Y] = \mu_{X|Y} = \mu_X + \rho\sigma_X(y-\mu_Y)/\sigma_Y \quad ; \quad var(X|Y) = \sigma_{X|Y}^2 = \sigma_X^2(1-\rho^2).$$

The lower panel interprets these results in terms of a two dimensional "Gaussian Arena" in which the input and output are related by the underlying joint Gaussian distribution which remains Gaussian for all possible linear coordinate transformations and even maintains its Gaussian character when one of the variables is conditioned on the other. Thus the Gaussian vector remains Gaussian under both linear transformations and Bayes' updates. Also note that if the correlation is zero ($\rho = 0$) then the input and output variables are independent and the update equations reduce to statements that the conditional mean and variance are unchanged, *viz.*, $\mu_{X|Y} = \mu_X$ and $\sigma_{X|Y}^2 = \sigma_X^2$.

We note in passing that because the quadratic form in the joint Gaussian is symmetric in the X and Y variables, we could just as well have computed the output Y conditioned on the input X to find analogous results with X $\Leftrightarrow$ Y corresponding to the forward Bayesian relation. A visual interpretation of this result will be given next in the Slide# 9-25 and further insight into the role of the communication channel and its inverse will be given in the Slide# 9-26 to 9-29.

9.6.3 Visualization of Conditional Mean - Bayes & Geometry

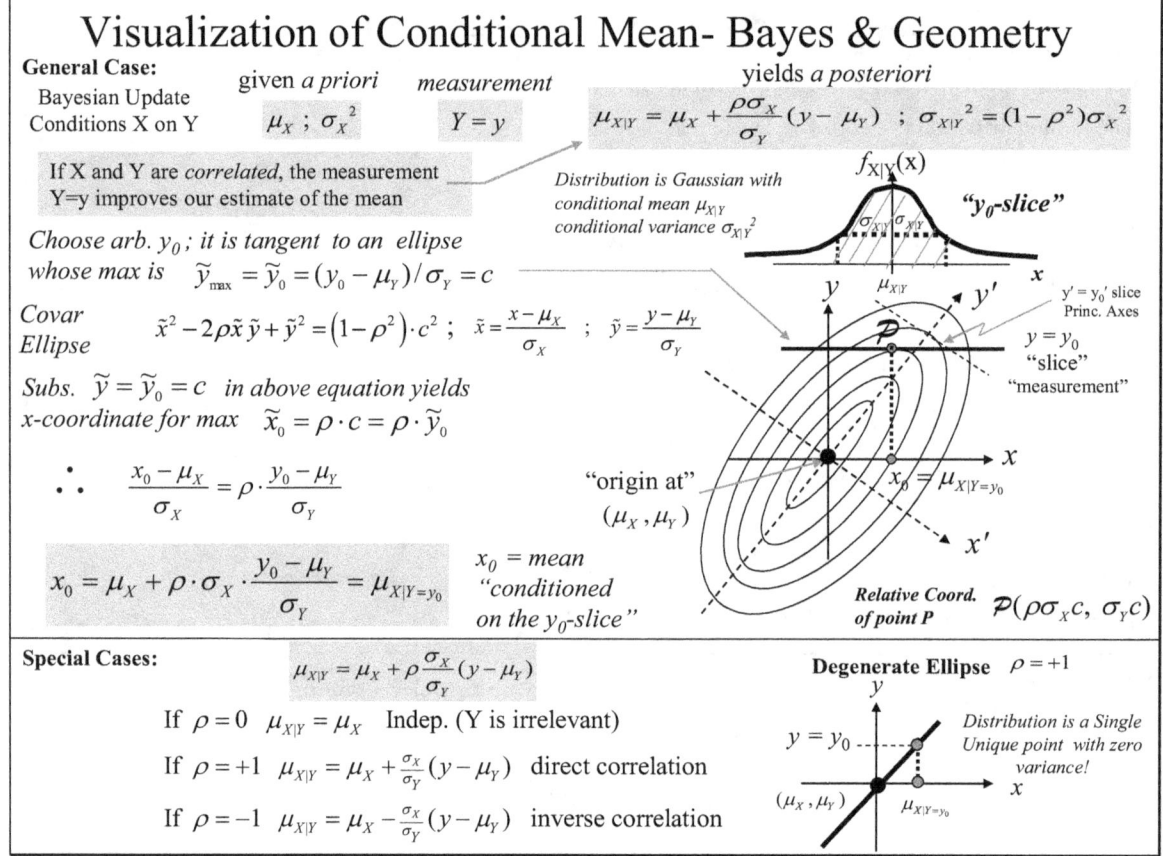

Visualization of Conditional Mean- Bayes & Geometry

General Case:
Bayesian Update
Conditions X on Y

given *a priori* — μ_X ; σ_X^2 — *measurement* — $Y = y$

yields *a posteriori*

$$\mu_{X|Y} = \mu_X + \frac{\rho\sigma_X}{\sigma_Y}(y - \mu_Y) \;\; ; \;\; \sigma_{X|Y}^2 = (1 - \rho^2)\sigma_X^2$$

If X and Y are *correlated*, the measurement Y=y improves our estimate of the mean

Distribution is Gaussian with conditional mean $\mu_{X|Y}$ conditional variance $\sigma_{X|Y}^2$

$f_{X|Y}(x)$

"y_0-slice"

Choose arb. y_0 ; *it is tangent to an ellipse whose max is* $\widetilde{y}_{max} = \widetilde{y}_0 = (y_0 - \mu_Y)/\sigma_Y = c$

Covar Ellipse $\quad \tilde{x}^2 - 2\rho\tilde{x}\tilde{y} + \tilde{y}^2 = (1 - \rho^2)\cdot c^2 \;\; ; \;\; \tilde{x} = \frac{x - \mu_X}{\sigma_X} \;\; ; \;\; \tilde{y} = \frac{y - \mu_Y}{\sigma_Y}$

Subs. $\tilde{y} = \tilde{y}_0 = c$ *in above equation yields* *x-coordinate for max* $\widetilde{x}_0 = \rho \cdot c = \rho \cdot \tilde{y}_0$

$y' = y_0'$ slice Princ. Axes

$y = y_0$ "slice" "measurement"

$$\therefore \quad \frac{x_0 - \mu_X}{\sigma_X} = \rho \cdot \frac{y_0 - \mu_Y}{\sigma_Y}$$

"origin at" (μ_X, μ_Y)

$x_0 = \mu_{X|Y=y_0}$

$$\boxed{x_0 = \mu_X + \rho \cdot \sigma_X \cdot \frac{y_0 - \mu_Y}{\sigma_Y} = \mu_{X|Y=y_0}}$$

$x_0 = mean$ "conditioned on the y_0-slice"

Relative Coord. of point P $\quad \mathcal{P}(\rho\sigma_X c, \; \sigma_Y c)$

Special Cases:

$$\mu_{X|Y} = \mu_X + \rho\frac{\sigma_X}{\sigma_Y}(y - \mu_Y)$$

If $\rho = 0$ $\mu_{X|Y} = \mu_X$ Indep. (Y is irrelevant)

If $\rho = +1$ $\mu_{X|Y} = \mu_X + \frac{\sigma_X}{\sigma_Y}(y - \mu_Y)$ direct correlation

If $\rho = -1$ $\mu_{X|Y} = \mu_X - \frac{\sigma_X}{\sigma_Y}(y - \mu_Y)$ inverse correlation

Degenerate Ellipse $\rho = +1$

$y = y_0$

Distribution is a Single Unique point with zero variance!

(μ_X, μ_Y) $\quad \mu_{X|Y=y_0}$

The results for the conditional mean and variance can be understood graphically as follows. Starting with the Bivariate Gaussian Density we draw the elliptical contours corresponding to the horizontal cuts through the density surface centered at the mean coordinates μ_X and μ_Y indicated by the black dot at the center. If we choose a fixed value of $y=y_0$ the line parallel to the x-axis is tangent to one of the ellipses and hence y_0 represents the maximum y-value for that ellipse as shown by the red dot. This line also results from a vertical plane $y=y_0$ cutting through the distribution and the Gaussian cut through the distribution is shown in a plot just above the contours.

The x-coordinate corresponding to this maximum is found by dropping a perpendicular onto the x-axis at a value $x_0 = \mu_{X|Y=y_0}$ as shown in the figure. Recalling the calculation used for the covariance ellipse construction, the x_0-value corresponding to this maximum at $y=y_0$ is given in standardized coordinates as $x_0 = \rho y_0$. This relationship is converted to the coordinates of the figure by letting $x_0 \rightarrow (x_0 - \mu_X)/\sigma_X$ and $y_0 \rightarrow (y_0 - \mu_Y)/\sigma_Y$ to yield $(x_0 - \mu_X)/\sigma_X = \rho(y_0 - \mu_Y)/\sigma_Y$, or upon solving for x_0, $x_0 = \mu_X + \rho\sigma_X(y_0 - \mu_Y)/\sigma_Y$, which states that this geometrically constructed x_0 is precisely the conditional mean $\mu_{X|Y=y_0}$. Note, however, in (x', y') principal axis coordinates the slice $y' = y_0'$ is perpendicular to the y'-axis (dashed line) and hence cannot geometrically change the center of the conditional distribution. Therefore the conditional mean and covariance are unchanged, *i.e.*, $\mu_{X'|Y'=y'0} = \mu_{X'}$, $\sigma_{X'|Y'=y'0} = \sigma_{X'}$; moreover, the RVs X' and Y' are independent of one another in principal axes coordinates and accordingly the covariance matrix is diagonal and the correlation coefficient $\rho = 0$.

The three special cases $\rho = 0, +1, -1$ shown in the bottom panel are:

(i) $\rho = 0$ no correlation corresponds a coordinate system along the principal axis of the ellipse for which a constant $y=y_0$ cut always leaves the conditional mean unchanged, *i.e.*, $\mu_{X|Y=y0} = \mu_X$ (ellipse $\rightarrow$ circle).

(ii) $\rho = +1$ complete positive correlation corresponds the case where the ellipse collapses to a straight line; the conditional distribution is a single point with zero variance on the line having slope (σ_Y/σ_X) as shown and yields a conditional mean $\mu_{X|Y=y0} = \mu_X + \sigma_X(y_0 - \mu_Y)/\sigma_Y$ (ellipse $\rightarrow$ straight line with positive slope and shift to right).

(iii) $\rho = -1$ complete negative correlation corresponds the case where the ellipse collapses to a straight line; the conditional distribution is a single point with zero variance on the line having slope $(-\sigma_Y/\sigma_X)$ (not shown) and yields a conditional mean $\mu_{X|Y=y0} = \mu_X - \sigma_X(y_0 - \mu_Y)/\sigma_Y$ (ellipse $\rightarrow$ straight line with negative slope and shift to left).

9.6.4 Bayes Update for Joint Gaussian Communication Channel $[X, Y]^T$

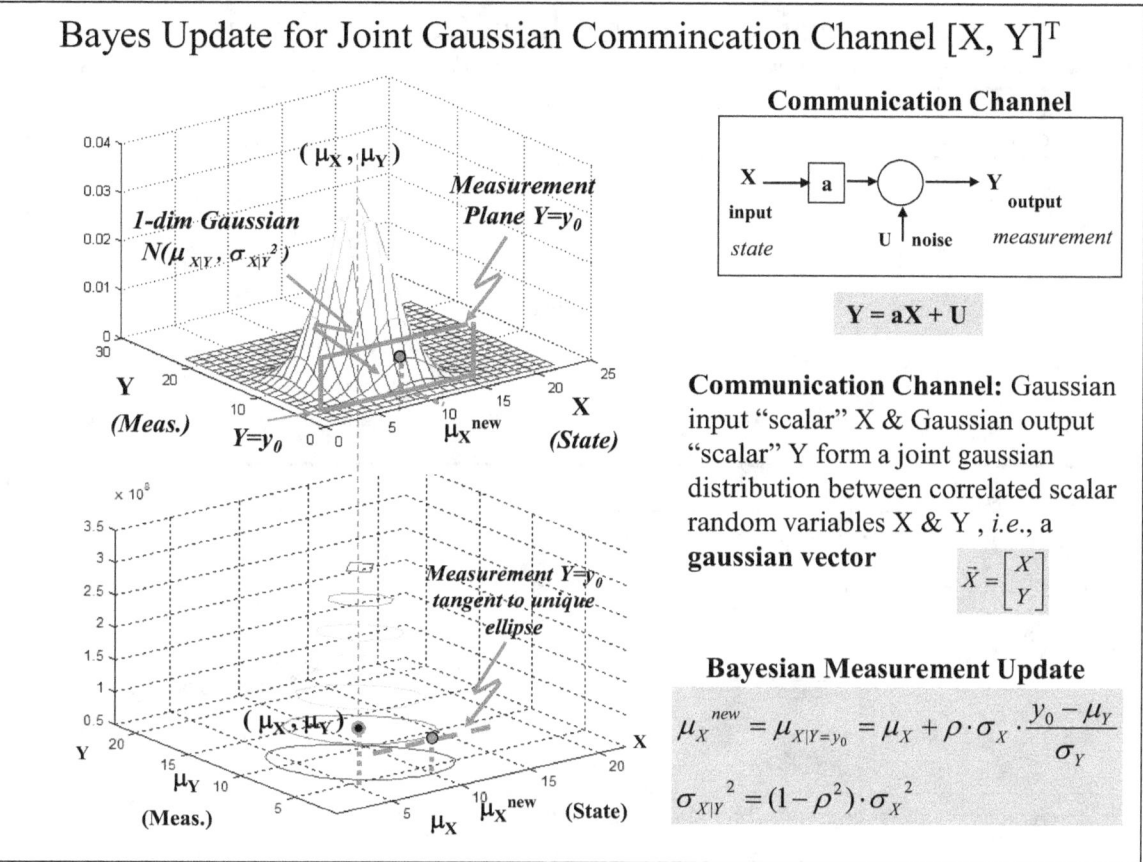

Bayes Update for Joint Gaussian Commincation Channel $[X, Y]^T$

Communication Channel

X input / state ── a ── ○ ── Y output / measurement

U noise

$$Y = aX + U$$

Communication Channel: Gaussian input "scalar" X & Gaussian output "scalar" Y form a joint gaussian distribution between correlated scalar random variables X & Y , *i.e.*, a **gaussian vector**

$$\vec{X} = \begin{bmatrix} X \\ Y \end{bmatrix}$$

Bayesian Measurement Update

$$\mu_X^{new} = \mu_{X|Y=y_0} = \mu_X + \rho \cdot \sigma_X \cdot \frac{y_0 - \mu_Y}{\sigma_Y}$$

$$\sigma_{X|Y}^2 = (1-\rho^2) \cdot \sigma_X^2$$

This slide gives a 3-dimensional visualization of the Bayesian Update for the communication channel with one input (scalar state) X and one output (scalar measurement) Y. The relationship shown in the communication channel box Y=aX+U, where *a* is a scalar multiplier and U is the channel noise, can be reformulated as a vector Gaussian $[X , Y]^T$ consisting of the two correlated random variables X and Y. The resulting Bayesian measurement update yields the conditional mean and variance shown in the lower box on the right.

A measurement at the value $Y=y_0$ is represented by a "plane slice" through the bivariate Gaussian and yields the 1-dimensional Gaussian with mean and variance given by the Bayesian measurement update equations. This "measurement plane" will be tangent to a unique ellipse as shown in the bottom left figure and leads to a new mean value μ_X^{new} which is just the conditional mean $\mu_{X|Y=y_0}$ and a new conditional variance Var(X|Y) $=\sigma_X^2$, again in accordance with the Bayesian measurement update.

Thus measurements are visualized as Y=constant cuts through the bivariate Gaussian and yield new Gaussians corresponding to the Bayesian measurement updates. This 3-dimensional visualization makes the results on the previous slide very intuitive. Note that in this slide, the major axis orientation of the cross-sectional ellipses corresponds to a **negative correlation** and the $Y=y_0$ value is **less than** the mean μ_Y so that according to the update equation the two minus signs yield a positive term and thus the updated value of the mean μ_X^{new} is larger than μ_X as shown in the figures. Clearly, if instead we chose a $Y=y_0$ value **greater than** the mean μ_Y (not shown in figure), the same negative correlation would yield an updated value of the mean μ_X^{new} that is smaller than μ_X.

9.6.5 Communication Channel – Input/Output Symmetry

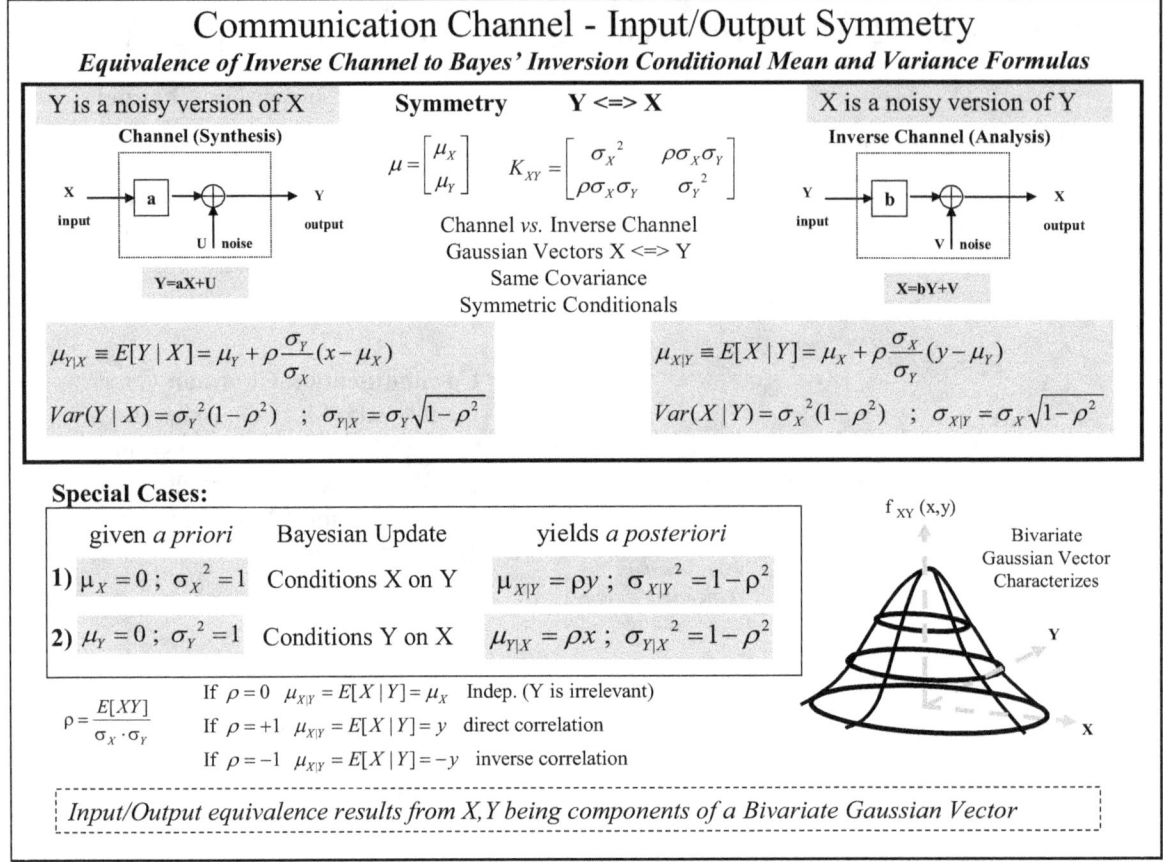

The summary box at the top of this slide shows the X, Y or input/output symmetry for the conditional means and variances that results from their being components of a Bivariate Gaussian Vector. We previously derived the equations for the conditional quantities $\mu_{X|Y}$ and $\sigma_{X|Y}^2$ in the right equation box directly from the joint distribution; clearly, since the joint Gaussian is symmetric in X and Y, we can skip an analogous calculation and simply exchange the roles of X and Y to produce the results shown in the left equation box which now represents the output "Y" conditioned on the input "X."

Above these boxed equations we show diagrams of the Direct Channel which gives the "output Y as a noisy version of the input X" $Y = aX + U$ on the left and the Inverse Channel which gives the "input X as a noisy version of the output Y" $X = bY + V$ on the right. As noted in the center of this summary box both the Direct and Inverse channels have the same mean vector $\mu_X = [\mu_X, \mu_Y]^T$ and covariance matrix K_{XY} and hence the equations on the left and right are obtained from one another by the simple exchange of variables $X \Leftrightarrow Y$.

There is a natural interpretation of these results for the Inverse Channel $X = bY + V$ as a **Bayesian Update (Inversion)** which "looks back" to determine the input signal X given the noisy output Y. The *a priori* mean and variance are Bayesian updated (*via* the boxed formulas) to give the new "conditional" mean and variance; these formulas have been derived directly from the Bayesian relation for the densities $f_{X|Y} = f_{Y|X} f_X / f_Y$ using the explicit form of the distributions. On the other hand, the Direct Channel governed by $Y = aX + U$ can be thought of as "channel synthesis" since it generates the channel output signal Y from the input X. On the next slide we use only the properties of inverse channel to derive the formulas for $\mu_{X|Y}$ and $var(X|Y) = \sigma_{X|Y}^2$ without considering details of the distribution. To reinforce this interpretation, several special cases of Bayesian Updates are shown in the bottom panel of the current slide.

9.7 *Equivalence of Bayes Inversion, Inverse Channel & System Analysis*

Equivalence of Bayes Inversion, Inverse Channel & System Analysis

Inverse channel "b*Y +V" allows direct computation of conditional mean $\mu_{X|Y}$ and variance $Var(X|Y)$ without the need for Bayesian Inversion. Thus inverse channel yields same results as Bayes' "inversion"

Assume *uncorrelated* Gaussian noise $\qquad V = N(\mu_V, \sigma_V^2)$ & $E[VY]=0 \qquad K_{XY} = \begin{bmatrix} 1 & \rho \\ \rho & 1 \end{bmatrix} \qquad X,Y \sim N(0,1)$

Adding this noise V to bY yields "output of Inverse Channel": X=bY+V $\qquad$ *V turns out to be zero-mean noise*

X & Y are N(0,1) Gaussian RVs
Find Channel Parameters:
$b,\ \mu_V,\ and\ \sigma_V^2$

$\mu_V = E[V] = E[X - bY] = \underset{=0}{E[X]} - b\,\underset{=0}{E[Y]} = 0$

$\rho = E[XY] = E[(bY+V)\cdot Y] = b\,\underset{\sigma_Y^2=1}{E[Y^2]} + \underset{=0}{E[V\cdot Y]} = b \qquad$ *Const. 'b"turns out to be the Correlation Coeff. btwn X & Y*

$\sigma_V^2 = E[V^2] - \underset{=0}{E[V]^2} = E[(X-bY)^2] = \underset{=\sigma_X^2=1}{E[X^2]} - 2b\,\underset{=\rho\ \ K(1,2)=\rho\sigma_X\sigma_Y=\rho}{E[XY]} + b^2\,\underset{=\rho^2\ \ \sigma_Y^2=1}{E[Y^2]} = 1-\rho^2$

Channel Parameters: $\qquad b = \rho\ ;\ \mu_V = 0\ ;\ \sigma_V^2 = 1-\rho^2$

Conditional mean and variance computed using inverse channel

Yields *same results as Bayes' inversion* of direct channel.

Thus they are "equivalent"

$\mu_{X|Y} = E[X|Y] = E[bY+V|Y] = b\cdot\underset{=y}{\underline{E[Y|Y=y]}} + \underset{=E[V]=0}{\underline{E[V|Y]}} = \rho y \qquad \boxed{\mu_{X|Y} = \rho y}$

$\sigma_{X|Y}^2 = Var[X|Y] = E[X^2|Y] - E[X|Y]^2 = \underset{b^2Y^2+2bYV+V^2}{E[(bY+V)^2|Y]} - \underset{=(\rho y)^2}{\underline{E[X|Y]^2}}$

$= \underset{\rho^2\ \ =y^2}{\underline{b^2 E[Y^2|Y]}} + \underset{\rho\ \ =E[YV]=0}{\underline{2bE[YV|Y]}} + \underset{=\sigma_V^2=(1-\rho^2)}{\boxed{E[V^2|Y]}} - (\rho y)^2 \qquad \boxed{\sigma_{X|Y}^2 = Var[X|Y] = 1-\rho^2}$

Note: Even though E[Y]=0, the *conditional mean* is not zero but rather E[Y|"Y=y"] has the explicit value "y"

The action of an inverse channel $X = bY+V$ is to multiply a zero-mean, unit-variance Gaussian RV Y~N(0,1) by a scalar "b" and then add uncorrelated Gaussian noise V~ N(μ_V, σ_V^2) to obtain the input RV: X~N(0,1). The underlying assumption is that the two N(0,1) Gaussian RVs X and Y form a Gaussian Vector and thus share a Bivariate Gaussian distribution with correlation ρ; these properties allow us to use the channel transformation $X = bY+V$ directly to compute $\mu_{X|Y}$ and var(X|Y) = $\sigma_{X|Y}^2$ *without considering any other details of the distribution*. The derivation is straightforward, but does involve some care.

The derivation proceeds by first finding the channel parameters b, μ_V, σ_V^2: the expected value of V is found directly to be zero; the scalar b is found by substituting the inverse channel transformation X = bY+V into the off-diagonal covariance term E[XY] and using the properties Y~N(0,1) and E[VY]=0 to give b=ρ ; finally, the channel noise variance is computed by solving the transformation for V = X-bY = X-ρY and computing $\sigma_V^2 =$ E[(X-ρY)²]- E[(X-ρY)] ² = 1- ρ² .

The conditional mean requires a little care because the direct evaluation of E[X|Y] =E[ρY +V|Y] generates the sum of two conditional expectations ρE[Y|Y] and E[V|Y]. The 2nd term is zero because E[VY] =0 , *i.e.,* V and Y are uncorrelated and thus independent. However, the 1st term requires a little thought, because even though E[Y]=0, the conditional expectation "given Y" is explicitly E[Y|Y=y]; this is the expected value of a constant E[y] which must of course be the constant y. The final result is $\mu_{X|Y}$= E[X|Y] = ρy.

The conditional variance requires the computation of the conditional second moment
E[X² |Y] = E[(ρY +V)²|Y] = E[ρ²Y²+ 2ρVY+V²|Y] = ρ²E[Y²|Y=y] + 0 + σ_V^2 = ρ² y² +(1-ρ²)
where again "given Y =y" means Y takes on a fixed value and can no longer be considered a RV, so we have
E[Y²|Y=y] =y² (and **not** σ_Y^2). Finally subtracting E[X|Y]² = (ρy)² from the above expression yields the conditional variance $\sigma_{X|Y}^2$= 1- ρ² .

9.7.1 Rationale for "Inverse Channel" and Generating Correlated RVs

Rationale for "Inverse Channel" and Generating Correlated RVs

Given Y: N(0,1) RV
Generate X: N(0,1) correlated to Y with coeff. ρ

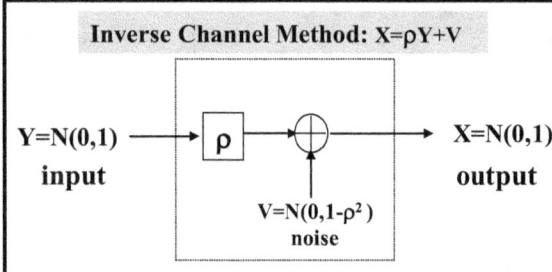

Inverse Channel Method: X=ρY+V

Y=N(0,1) ⟶ ρ ⊕ ⟶ X=N(0,1)
input   **output**

V=N(0,1-ρ^2)
noise

(i) **Generate** samples of RV "Y" using standard method (*e.g.*, sum 12 uniform Variates on [-0.5, 0.5]) to yield N(0,1).

(ii) **Generate** zero mean Gaussian noise "V" with variance (1-ρ^2): $(1-\rho^2)^{\frac{1}{2}} \cdot N(0,1) = N(0, 1-\rho^2)$

(iii) **Multiply** each RV sample "Y" by desired correlation coefficient ρ

(iv) **Add noise** sample "V" to obtain output "X" which is N(0,1) and has the desired correlation coefficient correl(X,Y)= ρ

Rationale: "X=ρY+V" $\quad$ [x,y]^T a Gaussian Vector ~ **N(0,1)**

(i) **If Noise is not added: X=ρY:**
Var(X) =Var(ρY) =ρ^2 Var(Y)= $\rho^2 \neq 1$

(ii) **If uncorrel noise** is added **X=ρY+"V"** with appropriate Var(V)= (1- ρ^2) to cancel correl contrib. to Var(X) then

Var(X) = Var(ρY+V) = ρ^2 Var(Y) +Var(V)+2ρCov(Y,V)

$$= \rho^2 \cdot 1 + (1-\rho^2) + 0 = 1$$

Special Cases: "X=ρY+V" ; $-1 \leq \rho \leq +1$

ρ = 0: No correlation between X & Y.

$$X = 0 \cdot Y + N(0,1-0^2) = N(0,1) \rightarrow X$$

X is simply the uncorrel noise sample N(0,1).

ρ = ±1: Full correlation/anti-correlation (Degenerate Ellipse or St.Line)

$$X = (\pm1) \cdot Y + N(0, 1-(\pm1)^2) = \pm Y \rightarrow X$$

X is simply ±Y – value

-1 < ρ < 1: General correlation

$$X = \rho \cdot Y + N(0,1-\rho^2) \rightarrow X$$

X results from multiplying Y by the correlation ρ and adding noise with variance (1- ρ^2)

The last two slides considered the inverse channel and its relation to a Bayesian update which starts with an *a priori* value of the mean μ_X and variance σ_X^2 and then updates their values as a result of an actual "measurement Y". The conditional mean $\mu_{X|Y}$ and conditional variance $\sigma_{X|Y}^2$ constructed using the inverse channel in this slide is the same as that obtained from the Bayesian update equation applied to the conditional probability densities. This fact strengthens our understanding of these entirely different approaches:

(i) Bayesian update relating conditional probability densities from which $\mu_{X|Y}$ and $\sigma_{X|Y}^2$ are computed using moments of the conditional distribution $f_{X|Y}$

(ii) Inverse Channel model with a Gaussian vector [X , Y]^T [components X (state) and Y (measurement)] and Channel Noise V, from which $\mu_{X|Y}$ and $\sigma_{X|Y}^2$ are computed using statistical constraints on RVs X,Y, and Z.

The box on the left uses the inverse channel model as a computer program flow diagram to actually generate a RV X~N(0,1) from a linear combination of Y ~N(0,1) and noise V~N(0, 1-ρ^2). Note that the input and output RVs are both N(0,1) Gaussians with unit variance yet the noise must have a variance that is less than unity for this to work.

The rationale is simple enough, for consider what might be your first impulse to generate a pair of correlated RVs by setting Y = ρ X (upper right box); taking the expectations E[Y] and E[Y^2] we find $\mu_Y = \rho \cdot \mu_X = \rho \cdot 0 = 0$ and $\sigma_Y^2 = \rho^2 \cdot \sigma_X^2 = \rho^2 \neq 1$ which this does not agree with the assumption that both X and Y are N(0,1). Agreement is possible only if we add zero-mean noise with variance (1-ρ^2) because when added to ρ^2 it yields the desired unit variance for the RV Y.

The special cases of no correlation ($\rho = 0$) and full positive and negative correlation ($\rho = \pm1$) are explicitly shown to be in agreement with this model. For no correlation the model gives X as just N(0,1) random noise; *i.e.*, X takes on random values completely independent Y. On the other hand, for full positive (or negative) correlation the model gives X as N(0,1) which takes on values that are exactly the same as those for Y (or –Y). In the general case -1< ρ <+1 the model gives X as N(0,1) RV which tracks Y more closely for correlations near +1 and tracks the noise more closely for correlations nearer to zero thus giving the expected intermediate behavior. In the next slide we consider reasons why this equivalence is the only reasonable explanation for this important "coincidence".

9.7.2 Bivariate Gaussian Vector Potpourri Discussion

Bivariate Gaussian Vector Potpourri Discussion

Probability Theory	*Geometry*
Cond. Prob. Density of Bivariate Gaussian	**Extremum of Bivariate Ellipsoid y=y(x)**

is also Gaussian with new mean and variance

$$E[X\,|\,Y] = \mu_{X|Y} = \mu_X + \rho\frac{\sigma_X}{\sigma_Y}(y-\mu_Y)$$

$$Var(X\,|\,Y) = \sigma_X^{\,2}(1-\rho^2)$$

a) $\quad \mu_{X|Y} = \rho y$

• Equivalent to Bayesian Update
• Equivalent to Inverse Channel

x-coord for peak value of $y=y_0$

$$x = \mu_X + \rho\frac{\sigma_X}{\sigma_Y}(y-\mu_Y)$$

b) $\quad x = \rho y$

$\tilde{x} = \dfrac{x-\mu_X}{\sigma_X}$; $\tilde{y} = \dfrac{y-\mu_Y}{\sigma_Y}$

> **a)** & **b)** look similar but different origins.
> Is there any relation between them?

---

i) Could Interpret **b)** as a **Coord Xform Btwn X & Y** Then compute probability density of X from that for Y: N(0,1)

Find X is zero-mean Gaussian with new variance

$$f_X(x) = \frac{f_Y(y)}{|\,dx/dy\,|} = \frac{1}{\rho}\cdot\frac{1}{\sqrt{2\pi}}e^{-\frac{y^2}{2}} = \frac{1}{\rho\sqrt{2\pi}}e^{-\frac{x^2}{2\rho^2}}$$

$N(0,\rho^2)$ Gaussian; not a bivariate

---

ii) Or try to form a **"Gaussian Vector"** $[\rho Y, Y]^T$

Find **X & Y not indep RVs!**

$$K_{XY} = E\left[\begin{pmatrix}\rho Y \\ Y\end{pmatrix}\cdot(\rho Y \quad Y)\right] = \begin{bmatrix}\rho^2 & \rho \\ \rho & 1\end{bmatrix}$$

$$\det\begin{bmatrix}\rho^2 & \rho \\ \rho & 1\end{bmatrix} = 0$$

"Pseudo" bivariate; no inverse

---

iii) Only if we **add uncorrelated channel noise V** to ρY do we get the desired Gaussian Vector $[\rho Y+V, Y]^T$

$$K_{XY} = E\left[\begin{pmatrix}\rho Y + V \\ Y\end{pmatrix}\cdot(\rho Y+V \quad Y)\right] = \begin{bmatrix}\rho^2+(1-\rho^2) & \rho \\ \rho & 1\end{bmatrix} = \begin{bmatrix}1 & \rho \\ \rho & 1\end{bmatrix}$$

Again need correct "channel noise"

X & Y are correlated RVs with correct covariance.

$$K_{XY}^{-1} = \frac{1}{(1-\rho^2)}\begin{bmatrix}1 & -\rho \\ -\rho & 1\end{bmatrix}$$

Bivariate Gaussian

In this final slide we try to tie together concepts relating the conditional probability density $f_{X|Y}(x|y)$ constructed from the X, Y components of a Gaussian Vector and Vertical cuts of the Bivariate Gaussian density surface for a constant Y=y which lead to the same conditional density function. On the left side of the top panel we show the conditional mean and variance corresponding to the distribution $f_{X|Y}(x|y)$ and recall that these results are fully equivalent to a Bayesian update and to the Inverse Channel model.

On the right side we show a series of elliptical contours resulting from a number of horizontal cuts through the joint probability density surface $f_{X,Y}(x,y)$. If we now take a vertical cut with a plane at "Y=y", the distribution in the that plane is precisely the conditional distribution $f_{X|Y=y}(x|y)$ given on the left. We can geometrically locate the tangent point by finding the ellipse that is tangent to the Y=y line in the contour plane; the x-value corresponding to this maximum has been previously found to be $x = \rho\cdot y$ which is located at the center of the new conditional Gaussian distribution and must therefore equal the conditional mean $\mu_{X|Y}$, thus establishing the equivalence of these two views.

The three lower panels consider various way one might consider the relation $x = \rho\cdot y$

(i) as a coordinate transformation it simply yields a 1-dimensional Gaussian $N(0, \rho^2)$

(ii) forms bivariate vector $\mathbf{X}=[\rho Y, Y]^T$ which yields a singular covariance matrix $\det(K_{XY})=0$

(iii) forms bivariate vector by adding uncorrelated noise (the inverse channel model) $\mathbf{X}=[\rho Y+V, Y]^T$ which yields the correct covariance matrix K_{XY} for two correlated N(0,1) RVs

Thus, the only correct interpretation is (iii), the inverse channel model for which $X =\rho Y+V$ correctly relates the X and Y Gaussians so that they form a Bivariate Gaussian vector. $\mathbf{X}=[X, Y]^T$.

10 Multivariate Gaussian Distributions

Multivariate Gaussian Distributions

10.1 n-Dimensional Gaussian Distribution

n-Dimensional Gaussian Distribution

n-dimensional Gaussian Vector $X= [X_1, X_2, \ldots X_n]^T$

$$f_X(x) = \frac{1}{(2\pi)^{n/2}\sqrt{\det K_{XX}}} e^{-\frac{1}{2}(x-\mu_X)^T K_{XX}^{-1}(x-\mu_X)}$$

Correlation Matrix

$$\begin{bmatrix} K_{11} & K_{12} & K_{13} & \cdots & K_{1n} \\ K_{21} & K_{22} & K_{23} & \cdots & K_{2n} \\ K_{31} & K_{32} & K_{33} & \cdots & K_{3n} \\ \vdots & \vdots & \vdots & \boxed{K_{rc}} & \vdots \\ K_{n1} & K_{n2} & K_{n3} & \cdots & K_{nn} \end{bmatrix}$$

Matrix components $\quad (K_{XX})_{rc} = E\left[(X_r - \mu_{X_r})(X_c - \mu_{X_c})\right] ; \; r,c = 1,2,\cdots n$

Moment Generating Fcn $\quad \phi_{\vec{X}}(\vec{t}) = E[e^{X \cdot t^T}] = e^{\frac{1}{2}t^T K_{XX} t + \mu_X^T t} \; ; \; t = [t_1, t_2, \cdots t_n]^T$

Still Gaussian After Linear Transformation: $\quad Y = AX + b \quad \mu_Y = A\mu_X + b \quad K_{YY} = AK_{XX}A^T$

(See Next Slide =>)

$$f_Y(y) = \frac{1}{(2\pi)^{n/2}\sqrt{\det K_{YY}}} e^{-\frac{1}{2}(y-\mu_Y)^T K_{YY}^{-1}(y-\mu_Y)}$$

Multivariate Gaussian

1st and 2nd Moments: Vector μ_X & Covariance K_{XX} Uniquely Define Multivariate Gaussian

$$\vec{\mu}_Y = E[\vec{Y}] = E[A\vec{X} + \vec{b}] = A\vec{\mu}_X + \vec{b}$$

$$\vec{Y} - \vec{\mu}_Y = (A\vec{X} + \vec{b}) - (A\vec{\mu}_X + \vec{b}) = A(\vec{X} - \vec{\mu}_X)$$

$$K_{YY} = E\left[(\vec{Y} - \vec{\mu}_Y)(\vec{Y} - \vec{\mu}_Y)^T\right] = E\left[A(\vec{X} - \vec{\mu}_X)(A(\vec{X} - \vec{\mu}_X))^T\right]$$

$$= E\left[A(\vec{X} - \vec{\mu}_X)(\vec{X} - \vec{\mu}_X)^T A^T\right] = A\underbrace{E\left[(\vec{X} - \vec{\mu}_X)(\vec{X} - \vec{\mu}_X)^T\right]}_{=K_{XX}} A^T = AK_{XX}A^T$$

The extension to multivariate Gaussian distributions or vectors is straight forward; taking the product of "n" independent $N(\mu_X, \sigma_X^2)$ Gaussians symbolized by the vector $X = [X_1, X_2, \ldots X_n]^T$ yields an n-dimensional Gaussian characterized by an n-dimensional mean vector μ_X and $n \times n$ covariance matrix K_{XX} whose diagonals equal the variances of the individual RVs and whose off diagonal elements are all zero. The n-dimensional Gaussian written in vector notation appears similar to the Bivariate Gaussian as $f_X(x) = (2\pi)^{-n/2}(\det K_{XX})^{-1/2} \cdot \exp(-q/2)$, where the scalar exponent is $q = [x-\mu_X]^T K_{XX}^{-1}[x-\mu_X]$. Similarly, the moment generating function takes on the form $\phi_X(t) = E[\exp(X t^T)] = \exp(\frac{1}{2} t^T K_{XX} t + \mu_X^T t)$ with $t = [t_1, t_2, \ldots, t_n]^T$.

Even if we start with independent RVs, a linear vector transformation of the form $Y = AX + b$ produces correlations, so the off-diagonal terms of the new covariance matrix are no longer zero. The distribution remains Gaussian taking the form $f_Y(y) = (2\pi)^{-n/2}(\det K_{YY})^{-1/2} \cdot \exp(-q/2)$, where the scalar exponent now is written as $q = [Y-\mu_Y]^T K_{YY}^{-1}[Y-\mu_Y]$ in terms of $Y = [Y_1, Y_2, \ldots Y_n]^T$. The new mean $\mu_Y = A\mu_X + b$ and new covariance $K_{YY} = AK_{XX}A^T$ result directly from the transformation. The latter is found directly from its definition as the expected value of the outer product, viz., $K_{YY} = E[(Y-\mu_Y)(Y-\mu_Y)^T]$; noting that $Y-\mu_Y = (AX + b) - (A\mu_X + b)$ becomes $Y-\mu_Y = A(X -\mu_X)$, direct substitution into the expression for K_{YY} yields the result $K_{YY} = AK_{XX}A^T$.

Here we use a standard notation in which the bolded Gaussian vector **X** has scalar components X_i and the new components resulting from a coordinate transformation are Y_i, $i = 1,2,\ldots,n$. This is temporary, however, because we shall again want to consider the communication channel with multiple inputs and outputs and define a conditional distribution as $f_{X|Y}(x|y)$. Therefore, we will partition the n-dimensional Gaussian vector in the form $X_n = [X_k : Y_{n-k}]$ with k-inputs $X_k = [X_1, X_2, \ldots X_k]^T$ and (n-k)-outputs $Y_{n-k} = [X_{k+1}, X_{k+2}, \ldots X_n]^T = [Y_1, Y_2, \ldots Y_{n-k}]^T$.

10.1.1   Transformation of Multivariate Probability Density - Details

Transformation of Multivariate Probability Density - Details
Jacobian PDF Transformation Method $$f_Y(y) = \frac{f_X(X - \mu_X)}{J\left(\dfrac{Y}{X}\right)} = \frac{f_X\left(A^{-1}(Y - \mu_Y)\right)}{\lvert \det(A) \rvert} = \frac{1}{\lvert \det(A) \rvert} \cdot \frac{e^{-\frac{1}{2}\left(A^{-1}(Y-\mu_Y)\right)^T K_{XX}^{-1}\left(A^{-1}(Y-\mu_Y)\right)}}{(2\pi)^{n/2}\sqrt{\det(K_{XX})}}$$
Finding det(A) $$\sqrt{\det(K_{YY})} = \sqrt{\det(A K_{XX} A^T)} = \sqrt{\det(A)\cdot\det(K_{XX})\cdot\det(A^T)} = \det(A)\cdot\sqrt{\det(K_{XX})}$$ $$\text{or} \qquad \det(A) = \frac{\sqrt{\det(K_{YY})}}{\sqrt{\det(K_{XX})}}$$
Expanding Exponent q $$(Y - \mu_Y)^T \left(A^{-1}\right)^T K_{XX}^{-1} A^{-1} (Y - \mu_Y) = (Y - \mu_Y)^T \underbrace{\left(A^T\right)^{-1} K_{XX}^{-1} A^{-1}}_{=(A K_{XX} A^T)^{-1} = K_{YY}^{-1}} (Y - \mu_Y)$$ $$= (Y - \mu_Y)^T K_{YY}^{-1} (Y - \mu_Y)$$
Transformed Gaussian $$f_Y(y) = \frac{1}{(2\pi)^{n/2}\sqrt{\det K_{YY}}} e^{-\frac{1}{2}(y-\mu_Y)^T K_{YY}^{-1}(y-\mu_Y)}$$

Under the linear vector transformation **Y**= A**X**+**b**, the new probability density is obtained using the PDF (Jacobian) method by expressing the original density $f_X(x)$ in terms of the vector **y** and dividing by the Jacobian determinant of the transformation J(**Y**,**X**) = det(A). Following this procedure, we have $f_Y(y) = f_X(x) / \lvert J(Y,X)\rvert = f_X(A^{-1}(\mathbf{Y}\text{-}\mu_Y)) / \lvert J(Y,X)\rvert$, which requires an evaluation of (i) the quadratic form q = $[A^{-1}(\mathbf{Y}\text{-}\mu_Y)]^T$ K_{XX}^{-1} $[A^{-1}(\mathbf{Y}\text{-}\mu_Y)]$ and (ii) its determinant as indicated by the two circled terms on the slide. Taking the determinant of the defining relation $K_{YY} = A K_{XX} A^T$, we find det(K_{YY})= det(A)^2·det(K_{XX}), so that the Jacobian, which is the det(A), can be expressed as the ratio J(**Y**,**X**)=(det K_{YY})$^{1/2}$ / (det K_{XX})$^{1/2}$. Expanding the matrix product in the exponential quadratic form yields, after a little algebra, q = $[\mathbf{Y}\text{-}\mu_Y]^T$ $K_{YY}^{-1}[\mathbf{Y}\text{-}\mu_Y]$ which is the standard Gaussian exponent in terms of the new variable **Y**. Hence assembling the terms we find that the Jacobian ratio term cancels the term (det K_{XX})$^{1/2}$ leaving (det K_{YY})$^{1/2}$ in the denominator and we arrive at the standard form of a n-dimensional Gaussian in the variable **Y** given in the boxed equation at the bottom of the slide.

10.1.2   Partitioned Multivariate Gaussian – Transform to Block Diagonal

Partitioned Multivariate Gaussian - Transform to Block Diagonal Form

Partition: $[X_{(1)} \mid X_{(2)}]^T$ {**Communication Channel: multiple inputs "X"= $X_{(1)}$ & outputs "Y"= $X_{(2)}$**}

2×1 Partitioned Vectors

2×2 Partitioned Matrix

$$\begin{bmatrix} x_{(1)} \\ \cdots \\ x_{(2)} \end{bmatrix} = \begin{bmatrix} x_1 \\ x_2 \\ \vdots \\ x_k \\ \cdots \\ x_{k+1} \\ \vdots \\ x_n \end{bmatrix} \quad \begin{bmatrix} \mu_{(1)} \\ \cdots \\ \mu_{(2)} \end{bmatrix} = \begin{bmatrix} \mu_1 \\ \mu_2 \\ \vdots \\ \mu_k \\ \cdots \\ \mu_{k+1} \\ \vdots \\ \mu_n \end{bmatrix}$$

$$\begin{bmatrix} K_{11} & \cdots & K_{1k} & K_{1,k+1} & \cdots & K_{1n} \\ \vdots & k \times k & \vdots & \vdots & k \times (n-k) & \vdots \\ K_{k,1} & \cdots & K_{kk} & K_{k,k+1} & \cdots & K_{kn} \\ K_{k+1,1} & \cdots & K_{k+1,k} & K_{k+1,k} & \cdots & K_{k+1,n} \\ \vdots & (n-k) \times k & \vdots & \vdots & (n-k) \times (n-k) & \vdots \\ K_{n1} & \cdots & K_{nk} & K_{n,k+1} & \cdots & K_{n,n} \end{bmatrix} = \begin{bmatrix} K_{(1)(1)} & K_{(1)(2)} \\ K_{(2)(1)} & K_{(2)(2)} \end{bmatrix}$$

Linear Transformation "partitioned form"

$$\begin{bmatrix} y_{(1)} \\ \cdots \\ y_{(2)} \end{bmatrix} = \begin{pmatrix} A_{11} & \vdots & A_{12} \\ \cdots & \vdots & \cdots \\ A_{21} & \vdots & A_{22} \end{pmatrix} \begin{bmatrix} x_{(1)} \\ \cdots \\ x_{(2)} \end{bmatrix}$$

where,

$$A = \begin{bmatrix} I_{k,k} & B_{k,(n-k)} \\ 0_{(n-k),k} & I_{(n-k),(n-k)} \end{bmatrix}$$

Find "B" matrix so that new K_{YY} is block diagonal

$$A K_{XX} A^T = \begin{bmatrix} I_k & B \\ 0 & I_{n-k} \end{bmatrix} \cdot \begin{bmatrix} K_{11} & K_{12} \\ K_{21} & K_{22} \end{bmatrix} \cdot \begin{bmatrix} I_k & B \\ 0 & I_{n-k} \end{bmatrix}^T = \begin{bmatrix} I_k & B \\ 0 & I_{n-k} \end{bmatrix} \cdot \begin{bmatrix} K_{11} & K_{12} \\ K_{21} & K_{22} \end{bmatrix} \cdot \begin{bmatrix} I_k & 0 \\ B^T & I_{n-k} \end{bmatrix}$$

Now drop parentheses notation for partitioned components !!

$$= \begin{bmatrix} K_{11} + BK_{21} & K_{12} + BK_{22} \\ K_{21} & K_{22} \end{bmatrix} \cdot \begin{bmatrix} I_k & 0 \\ B^T & I_{n-k} \end{bmatrix}$$

$$= \begin{bmatrix} K_{11} + BK_{21} & K_{12} + BK_{22} \\ +K_{12}B^T + BK_{22}B^T & \\ K_{21} + K_{22}B^T & K_{22} \end{bmatrix}$$

$$\boxed{K_{21} + K_{22}B^T = 0} \quad (1)$$

$$\boxed{K_{12} + BK_{22} = 0} \quad (2)$$

Consider a multi-dimensional communication channel partitioned into two sets as follows:
"X": k-inputs $X_{(1)} = [X_1, X_2, ..., X_k]^T$ and **"Y": (n-k)-outputs** $X_{(2)} = [X_{k+1}, X_{k+2}, ..., X_n]^T$. The mean vector is partitioned similarly as $\mu_{(1)} = [\mu_1, \mu_2, ..., \mu_k]^T$ and $\mu_{(2)} = [\mu_{k+1}, \mu_{k+2}, ..., \mu_n]^T$ and the corresponding 2×2 partitioned covariance matrix $K_{(I)(J)}$ has the structure shown in the **top panel**:

Row#1 $[K_{11} : K_{12}] = [$ $k \times k$: $(n-k) \times k$ $]$
Row#2 $[K_{21} : K_{22}] = [$ $k \times (n-k)$: $(n-k) \times (n-k)]$

Now let's perform a linear transformation to a new coordinate system according to the equation **Y=AX** where the vectors **Y** and **X** are partitioned as 2×1 column vectors and the matrix A is partitioned as a 2×2 matrix corresponding to the partitioning in the **top panel**. The **second panel** shows the explicit partition of this transformation matrix $A_{(I)(J)}$ as follows: $[A_{(1)(1)} : A_{(1)(2)}]=[I_k : B_{k, (n-k)}]$ and $[A_{(2)(1)} : A_{(2)(2)}]=[0_{n-k, k} : I_{n-k}]$. Note I_k and I_{n-k} are *square identity matrices* labeled by their dimensions, $0_{(n-k),k}$ is a (n-k)×k zero matrix, and $B_{k,(n-k)}$ is a k×(n-k) unknown matrix. The transformed covariance matrix K_{YY} is defined by the following product of three n×n matrices $A K_{XX} A^T$; this simplifies to a product of *three* 2×2 partitioned matrices and yields the 2×2 matrix shown in the **bottom panel** of the slide. The task at hand is to find the unknown matrix B such that the new covariance matrix K_{YY} is **block diagonal**; this is accomplished by forcing the two "off-diagonal" partitions (circled) to be zero. This imposes two independent conditions, one on the matrix B and the other on its transpose B^T as follows: (1) $K_{21} + K_{22} B^T=0$ and (2) $K_{12} +BK_{22} =0$. These equations relate B and B^T to partitioned components of the original matrix K_{XX}, *e.g.*, K_{21} is the 2,1 partition component or $(K_{XX})_{21}$. On the next slide we formally solve for B and write down the explicit form of the block diagonal matrix K_{YY} with just 2 diagonal components, namely, $(K_{YY})_{11}$ and $(K_{YY})_{22}$. This "canonical form" will allow us to factor the multivariate Gaussian in terms of inputs **X** and outputs **Y** and prove a very elegant generalization of a Bayesian update for the conditional mean and conditional covariance known as the Gauss-Markov theorem.

10.1.3 Explicit 2×2 Block Diagonal Form of Multivariate Gaussian Vector

Explicit 2×2 Block Diagonal Form of Multivariate Gaussian Vector

B & B^T   (1) $K_{22}^{-1}\{K_{21} + K_{22}B^T\} = 0 \Rightarrow B^T = -K_{22}^{-1}K_{21}$ (2) $\{K_{12} + BK_{22}\}K_{22}^{-1} = 0 \Rightarrow B = -K_{12}K_{22}^{-1}$

Substitute B & B^T into $(K_{YY})_{11}$

$$(K_{YY})_{11} = K_{11} + BK_{21} + K_{12}B^T + BK_{22}B^T$$
$$(K_{YY})_{11} = K_{11} + (-K_{12}K_{22}^{-1})K_{21} + K_{12}(-K_{22}^{-1}K_{21}) + (-K_{12}K_{22}^{-1})K_{22}(-K_{22}^{-1}K_{21})$$
$$= K_{11} - K_{12}K_{22}^{-1}K_{21} - K_{12}K_{22}^{-1}K_{21} + K_{12}K_{22}^{-1}K_{21}$$
$$= K_{11} - K_{12}K_{22}^{-1}K_{21}$$

Partitioned Forms Matrix A & K_{YY}

$$A = \begin{bmatrix} I_k & -K_{12}K_{22}^{-1} \\ 0_{(n-k),k} & I_{(n-k)} \end{bmatrix} \qquad K_{YY} = \begin{bmatrix} K_{11} - K_{12}K_{22}^{-1}K_{21} & 0 \\ 0 & K_{22} \end{bmatrix}$$

Transformation Y =AX Yields for vectors y and μ_Y

$$\begin{bmatrix} y_1 \\ y_2 \end{bmatrix} = \begin{bmatrix} I_k & -K_{12}K_{22}^{-1} \\ 0 & I_{n-k} \end{bmatrix}\begin{bmatrix} x_1 \\ x_2 \end{bmatrix} = \begin{bmatrix} x_1 - K_{12}K_{22}^{-1}x_2 \\ x_2 \end{bmatrix} \quad \leftarrow \text{Lin Comb.} \quad \text{"Y=AX"}$$
$$\quad \leftarrow \text{Unchanged}$$
$$\begin{bmatrix} \mu_{Y_1} \\ \mu_{Y_2} \end{bmatrix} = \begin{bmatrix} I_k & -K_{12}K_{22}^{-1} \\ 0 & I_{n-k} \end{bmatrix}\begin{bmatrix} \mu_{X_1} \\ \mu_{X_2} \end{bmatrix} = \begin{bmatrix} \mu_{X_1} - K_{12}K_{22}^{-1}\mu_{X_2} \\ \mu_{X_2} \end{bmatrix} = \begin{bmatrix} \mu_{X_1}^* \\ \mu_{X_2}^* \end{bmatrix}$$

$\begin{bmatrix} \mu_{X_1} \\ \mu_{X_2} \end{bmatrix} \Rightarrow \begin{bmatrix} \mu_{X_1}^* \\ \mu_{X_2}^* \end{bmatrix}$ "E[Y]=AE[X]"

Subtracting to Form Partitioned Vector y-μ_Y

$$\begin{bmatrix} y_1 - \mu_{Y_1} \\ y_2 - \mu_{Y_2} \end{bmatrix} = \begin{bmatrix} x_1 - \mu_{X_1} - K_{12}K_{22}^{-1}(x_2 - \mu_{X_2}) \\ x_2 - \mu_{X_2} \end{bmatrix} = \begin{bmatrix} x_1 - \mu_{X_1}^* \\ x_2 - \mu_{X_2}^* \end{bmatrix}$$

$\mu_{X_1}^* \equiv \mu_{X_1} + K_{12}K_{22}^{-1}(x_2 - \mu_{X_2})$

← **2nd component unchanged**

Block Diagonal Covariance Inverse K_{YY}^{-1}

$$K_{YY}^{-1} = \begin{bmatrix} \left[K_{11} - K_{12}K_{22}^{-1}K_{21}\right]^{-1} & 0 \\ 0 & K_{22}^{-1} \end{bmatrix}$$

Y_1 & Y_2 are independent
$\Longrightarrow \text{Cov}(Y_1\ Y_2) = K_{YY}(1,2) = 0$

← **2-2 component unchanged**

$$(K_{YY}^{-1})_{22} = (K_{XX}^{-1})_{22}$$

Left-multiplying the 1st off-diagonal equation by K_{22}^{-1} yields B = - $K_{12}K_{22}^{-1}$ and right-multiplying the 2nd off-diagonal equation by K_{22}^{-1} yields B^T = - $K_{22}^{-1}K_{21}$. Substituting these results for B and B^T into the 1-1 element of K_{YY} yields $K_{(1)(1)} = K_{11} - K_{12}K_{22}^{-1}K_{21}$; the 2-2 element $(K_{YY})_{22} = (K_{XX})_{22}$ is unchanged and given the explicit expression for B, the transformation matrix is also known. The explicit form of the partitioned 2×2 matrices K_{YY} and A are displayed in the boxed equations of the **panel #3**.

In order to write down the transformed Gaussian we need the partitioned vector **Y- μ_Y**; this is found by transforming both the RV **X** and its mean μ_X using the explicit partitioned form of the 2×2 transformation matrix A. Subtracting the results given in **panel #4**, we find the desired form for the partitioned vector **Y- μ_Y**, which together K_{YY}^{-1} is shown in the **bottom panel** of the slide.

Note that because the partitioned matrix K_{YY} is block diagonal, its inverse matrix K_{YY}^{-1} is found by simply taking the matrix inverses of its two block diagonal elements $(K_{YY})_{11}^{-1}$ and $(K_{YY})_{22}^{-1}$; this is expressed in terms of the indexed elements of the "original matrix" using the compact notation "K_{12}", where it is understood that this means $(K_{XX})_{12}$, etc..

Note that from the partitioned forms in the last panel we see that two quantities are **unchanged by the transformation A**, namely

(i) the 2-2 element of the inverse block diagonal covariance matrix : $(K_{YY}^{-1})_{22} = (K_{XX}^{-1})_{22}$, and

(ii) the 2-component of the partitioned vector: $(\mathbf{Y}- \mu_Y)_2 = (\mathbf{X}- \mu_X)_2$.

These two facts will be the key in factoring the partitioned Multivariate Gaussian and subsequently proving the generalization of the Bayesian update result found for the Bivariate Gaussian.

10.1.4 Gaussian in Canonical Coordinates and Conditional Density

---

Gaussian in Canonical Coordinates and Conditional Density

After Linear Xform the Joint PDF is

$$f_{Y_1 Y_2}(y_1, y_2) = \frac{1}{(2\pi)^{n/2} \sqrt{\det K_{YY}}} e^{-\frac{1}{2}(y-\mu_Y)^T K_{YY}^{-1}(y-\mu_Y)}$$

Note: Block indices are surrounded by circles to distinguish them from the original K_{ij} matrix components in the the $x_1, ..., x_n$ coordinates

with Inverse in Block Diagonal Form

$$K_{YY}^{-1} = \begin{bmatrix} (K_{11} - K_{12}K_{22}^{-1}K_{21})^{-1} & 0 \\ 0 & K_{22}^{-1} \end{bmatrix}$$

$$\det K_{YY} = \det(K_{YY})_{\textcircled{11}} \cdot \det(K_{YY})_{\textcircled{22}}$$

Therefore **Joint PDF factors** into *two Independent Gaussians corresponding to* $(K_{YY}^{-1})_{\textcircled{11}}$ and $(K_{YY}^{-1})_{\textcircled{22}}$

$$f_{Y_1 Y_2}(y_1, y_2) = \underbrace{\frac{e^{-\frac{1}{2}(y_1 - \mu_{Y_1})^T (K_{YY}^{-1})_{\textcircled{11}} (y_1 - \mu_{Y_1})}}{(2\pi)^{k/2} \sqrt{\det(K_{YY})_{\textcircled{11}}}}}_{f_{Y_1}(y_1) \text{ is k-dim Gaussian}} \cdot \underbrace{\frac{e^{-\frac{1}{2}(y_2 - \mu_{Y_2})^T (K_{YY}^{-1})_{\textcircled{22}} (y_2 - \mu_{Y_2})}}{(2\pi)^{(n-k)/2} \sqrt{\det(K_{YY})_{\textcircled{22}}}}}_{f_{Y_2}(y_2) \text{ is (n-k)-dim Gaussian}}$$

Conditional Density depends only on y_1

$$f_{Y_1|Y_2}(y_1 | y_2) = \frac{f_{Y_1 Y_2}(y_1, y_2)}{f_{Y_2}(y_2)} = \frac{f_{Y_1}(y_1) \cdot f_{Y_2}(y_2)}{f_{Y_2}(y_2)} = f_{Y_1}(y_1) = \frac{e^{-\frac{1}{2}(y_1 - \mu_{Y_1})^T \overset{(K_{YY}^{-1})_{\textcircled{11}}}{\left((K_{11} - K_{12}K_{22}^{-1}K_{21})^{-1}\right)}(y_1 - \mu_{Y_1})}}{(2\pi)^{k/2} \sqrt{\det(K_{11} - K_{12}K_{22}^{-1}K_{21})}}$$

Factoring occurs only in the ***canonical y_1, y_2 coordinates*** where Covariance matrix K_{YY} is ***Block Diagonal***
The last step is to re-express $f_{Y_1|Y_2}(y_1 | y_2)$ as a function of x_1, x_2 and identify it with $f_{X_1|X_2}(x_1 | x_2)$
This is done by noting the transformation A leaves $y_2 = x_2$, so that "conditioning on y_2" is the same as "conditioning on x_2"; also, we can replace $(y_1 - \mu_{Y_1}) = (x_1 - \mu_{x_1}*)$ leaving everything in terms of x_1, x_2
It is then easy to "pick off" the conditional mean and variance $\mu_{X_1|X_2}$ & $K_{X_1|X_2}$ from the quadratic form $q(x_1, x_2)$. (see Next Slide)

The linear transformation $\mathbf{Y} = A\mathbf{X}$ yields a block diagonal covariance matrix K_{YY} and also a block diagonal inverse K_{YY}^{-1}; this allows us to write down the transformed Gaussian as shown in the top panel. The block diagonal form of the inverse allows us to factor the quadratic form in the exponent into two distinct quadratic terms, while the block diagonal form of K_{YY} itself allows us to write its determinant as the product of its two constituent block determinants, *viz.*, $\det(K_{YY}) = \det(K_{YY})_{11}$ $\det(K_{YY})_{22}$. Since both the quadratic form and the determinant factor into products, the distribution $f_{Y1Y2}(y_1, y_2)$ itself factors into the product of two Gaussians $f_{Y1}(y_1)$ $f_{Y2}(y_2)$ as shown in **panel #2**.

The **canonical coordinate system** Y in which the *covariance is block diagonal* and the partitioned Gaussian *factors into the product of two Gaussians* is the n-dimensional equivalent of the **principal axis coordinate system** for the 2-dimensional Bivariate Gaussian. In the latter case we used the principal axis coordinates to factor the joint distribution $f_{XY}(x,y)$ into the product of two Gaussians and then divide it by the $N(0,1)$ marginal distribution $f_Y(y)$ to find a Gaussian conditional distribution $f_{X|Y}(x|y)$. The conditional mean $\mu_{X|Y}$ and conditional variance $\sigma_{X|Y}^2$ were then picked off this Gaussian to give a Bayesian update in terms of the original means μ_X, μ_Y, variances σ_X^2, σ_Y^2, and the correlation coefficient ρ.

Similarly, here the conditional density $f_{Y1|Y2}(y_1|y_2) = f_{Y1Y2}(y_1,y_2)/f_{Y2}(y_2) = f_{Y1}(y_1)$ is a Gaussian depending only on y_1 for the exact same reasons. All that needs to be done is to return to the original non-conical coordinates replacing $\mathbf{Y}$ by $A\mathbf{X}$ and then group terms so that we can "pick off" the corresponding conditional mean $\mu_{X1|X2}$ and conditional covariance $K_{X1|X2}$ from the quadratic form in the exponent of the Gaussian; the details are given on the next slide.

10.1.5   Conditional Density Re-expressed in Non-Canonical Coordinates

---

Conditional Density Re-expressed in Non-Canonical Coordinates

Using Partitioned Vector $(y-\mu_Y)$

$$\begin{bmatrix} y_1 - \mu_{Y_1} \\ \hline y_2 - \mu_{Y_2} \end{bmatrix} = \begin{bmatrix} x_1 - [\mu_{X_1} + K_{12}K_{22}^{-1}(x_2 - \mu_{X_2})] \\ \hline x_2 - \mu_{X_2} \end{bmatrix} = \begin{bmatrix} x_1 - \mu_{X_1}^{*} \\ \hline x_2 - \mu_{X_2} \end{bmatrix} \longrightarrow (x_1 - \mu_{X_1}^{*}) = (y_1 - \mu_{Y_1})$$

Subs. $\mu_{X_1}^{*} \equiv \mu_{X_1} + K_{12}K_{22}^{-1}(x_2 - \mu_{X_2})$

& $K_{X_1|X_2} = (K_{YY})_{(1)} = (K_{11} - K_{12}K_{22}^{-1}K_{21})$

into $f_{Y_1|Y_2}(y_1 \mid y_2)$ equation yields

$$f_{Y_1|Y_2}(y_1 \mid y_2) = f_{Y_1}(y_1) = \frac{e^{-\frac{1}{2}[(x_1 - \mu_{X_1}^{*})^T (K_{11} - K_{12}K_{22}^{-1}K_{21})^{-1}(x_1 - \mu_{X_1}^{*})]}}{(2\pi)^{n/2}\sqrt{\det(K_{11} - K_{12}K_{22}^{-1}K_{21})}} = f_{X_1|X_2}(x_1 \mid x_2)$$

Fcn of x_1, x_2 only

Finally, identifying: $f_{Y_1|Y_2}(y_1 \mid y_2) \rightarrow f_{X_1|X_2}(x_1 \mid x_2)$

we obtain the conditional PDF of Gaussian Vector X_1 given Gaussian Vector X_2
Everything is expressed in terms of x_1, x_2

$$f_{X_1|X_2}(x_1 \mid x_2) = \frac{e^{-\frac{1}{2}[(x_1 - \mu_{X_1|X_2})^T K_{X_1|X_2}^{-1}(x_1 - \mu_{X_1|X_2})]}}{(2\pi)^{k/2}\sqrt{\det(K_{X_1|X_2})}}$$

"Pick off" the conditional mean and covariance

$$\mu_{X_1|X_2} \equiv \mu_{X_1}^{*} = \mu_{X_1} + K_{12}K_{22}^{-1}(x_2 - \mu_{X_2}) \qquad\qquad K_{X_1|X_2} \equiv K_{11} - K_{12}K_{22}^{-1}K_{21}$$

Reduces to Bivariate Gaussian

"Block Diagonal Components" are just scalars

$$\mu_{X_1|X_2} = \mu_{X_1} + (\rho\sigma_{X_1}\sigma_{X_2})(\sigma_{X_2}^2)^{-1}(x_2 - \mu_{X_2}) \qquad Var(X_1 \mid X_2) = \sigma_{X_1}^2 - (\rho\sigma_{X_1}\sigma_{X_2})(\sigma_{X_2}^2)^{-1}(\rho\sigma_{X_1}\sigma_{X_2})$$

$$\mu_{X_1|X_2} = \mu_{X_1} + \rho\frac{\sigma_{X_1}}{\sigma_{X_2}}(x_2 - \mu_{X_2}) \qquad\qquad \sigma_{X_1|X_2}^2 = \sigma_{X_1}^2(1 - \rho^2)$$

$$\vec{X} = \begin{bmatrix} X_1 \\ X_2 \end{bmatrix} = \begin{bmatrix} "X" \\ "Y" \end{bmatrix} \qquad\qquad \vec{\mu} = \begin{bmatrix} \mu_{X_1} \\ \mu_{X_2} \end{bmatrix} \quad ; \quad K_{X_1 X_2} = \begin{bmatrix} \sigma_{X_1}^2 & \rho\sigma_{X_1}\sigma_{X_2} \\ \rho\sigma_{X_1}\sigma_{X_2} & \sigma_{X_2}^2 \end{bmatrix}$$

The results of the last slide were in **canonical coordinates Y** where everything separated nicely to allow us to obtain an explicit expression for the conditional density $f_{Y1|Y2}(y_1|y_2)$. Now in order to re-express the quadratic form in terms of the original **non-canonical coordinates** X, we use the results previously derived from the coordinate transformation **Y = AX** shown here again in the **top panel**. This transformation equation allows us to identify the 2nd partition component **Y₂** with the original partition component **X₂** which means that the conditioning with respect to **Y₂** is equivalent to (and may be replaced by) conditioning with respect to **X₂** ,*i.e.*, we can write "$f_{Y1|X2}(y_1|x_2)$" in place of $f_{Y1|Y2}(y_1|y_2)$; it also allows us to replace the mean μ_{Y2} with μ_{X2}. The 1st partition component is the one appearing in the quadratic form and the transformation equation shows that the term $y_1 - \mu_{Y1}$ may be replaced by $x_1 - \mu_{X1}^{*}$, where the effective mean is $\mu_{X1}^{*} = \mu_{X1} - K_{12}K_{21}^{-1}(x_2 - \mu_{X2})$.

Thus, making these replacements in $f_{Y1|X2}(y_1|x_2)$ we re-express it as a conditional Gaussian which is now a function of the original **X** partitioned coordinates $[X_{(1)}: X_{(2)}]$ and the associated partitioned mean vector μ_X and covariance K_{XX}. The boxed equation in the **middle panel** gives this conditional density in terms of the conditional mean defined by $\mu_{X1|X2} = \mu_{X1}^{*} = \mu_{X1} - K_{12}K_{21}^{-1}(x_2 - \mu_{X2})$ and the conditional covariance defined by $K_{X1|X2} = K_{11} - K_{12}K_{22}^{-1}K_{21}$; it is important to point out that all the matrix elements are those of the original partitioned covariance matrix K_{XX}.

In the **bottom panel** of the slide we show that in the case n=2 these boxed expressions for conditional mean $\mu_{X1|X2}$ and conditional covariance $K_{X1|X2}$ reduce to the results previously found for the Bivariate Gaussian. In this case the "block diagonal" sub-matrices are just scalars. The general result for n-dimensional Multivariate Gaussians is known as the **Gauss-Markov Theorem** which is summarized on the next slide in a revised notation which is applicable to a multiple input **X** multiple output **Y** communication channel.

10.2 Gauss-Markov Theorem

Gauss-Markov Theorem

Updating Gaussian Vectors under Bayes'Rule

Given X and Y are jointly Gaussian Random input and output vectors with **dim k** and **n-k** respectively
Combine to form **n-dim** vector with partitioned mean and covariance as follows :

$$\underset{n\times1}{\vec{X}} \equiv \begin{bmatrix} X_{(k)} \\ \hline Y_{(n-k)} \end{bmatrix} \begin{array}{l} \textbf{State} \\ \textbf{Meas.} \end{array} \qquad \underset{n\times1}{\vec{\mu}} \equiv \begin{bmatrix} \mu_{X(k)} \\ \hline \mu_{Y(n-k)} \end{bmatrix} \qquad \underset{n\times n}{K} \equiv \left[\begin{array}{c|c} \underset{k\times k}{K_{XX}} & \underset{k\times(n-k)}{K_{XY}} \\ \hline \underset{(n-k)\times k}{K_{YX}} & \underset{(n-k)\times(n-k)}{K_{YY}} \end{array} \right] \qquad \begin{array}{l} \textbf{Block Correlations} \\ \textbf{for State \& Meas.} \end{array}$$

Gauss-Markov Theorem states that the conditional PDF of "X given Y" is also Gaussian
with conditional mean & covariance given by

$$\underset{k\times1}{\mu_{X|Y}} = \underset{k\times1}{\mu_X} + \underset{k\times(n-k)}{K_{XY}} \underset{(n-k)\times(n-k)}{K_{YY}^{-1}} \underset{(n-k)\times1}{(y-\mu_Y)} \qquad \underset{k\times k}{K_{X|Y}} = \underset{k\times k}{K_{XX}} - \underset{k\times(n-k)}{K_{XY}} \underset{(n-k)\times(n-k)}{K_{YY}^{-1}} \underset{(n-k)\times k}{K_{YX}}$$

Note: Although Covariance K is symmetric, the **blocks themselves are not** (they are in fact different shapes), *i.e.*,

$$\underset{k\times(n-k)}{K_{XY}} \neq \underset{(n-k)\times k}{K_{YX}}$$

Different shapes

Symmetry of K requires the following relationship for the off diagonal **blocks**

$$\left(\underset{k\times(n-k)}{K_{XY}} \right)^{T}_{(n-k)\times k} = \underset{(n-k)\times k}{K_{YX}}$$

The 2×2 block partition expressing the Gauss-Markov theorem can be cast into a more convenient form by replacing the block indices (1) → (**X**) and (2) → (**Y**). A direct transcription of results from the previous slide yields the Gaussian vector **X**, mean vector μ, covariance matrix K as well as the Gauss-Markov results for the conditional mean $\mu_{X|Y}$ and conditional covariance $K_{X|Y}$ (boxed equations). This (**X**), (**Y**) index notation makes explicit the input/output (or state/measurement) interpretation, where the block label (**X**) represents the k inputs $X_{(1)} = [X_1, ...,X_k]^T$, and the block label (**Y**) represents the (n-k) outputs $X_{(2)} = [X_{k+1}, ...,X_n]^T = [Y_1 ...Y_{n-k}]^T$.

The Gauss-Markov theorem is interpreted as a Bayesian-type update of the *a priori* mean vector $\mu_X = E[X]$ (corresponding to a k-dimensional state vector **X**) to the *a posteriori* conditional mean vector (estimate) $\mu_{X|Y}$ by virtue of an (n-k)-dimensional measurement vector **Y**. Moreover, the state covariance K_{XX} is also updated to give $K_{X|Y}$ by virtue of the inherent accuracy of the measurements K_{YY} and the correlations between measurements and state parameters expressed by the off-diagonal covariance terms K_{XY} and K_{YX}. For this to be true, the state and measurement vectors must be part of the same multivariate Gaussian distribution, *i.e.*, components of a partitioned Gaussian vector $[X : Y]^T$ whose partitioned mean $\mu = [\mu_X : \mu_Y]$ and covariance K, are shown at the top of the slide. They indeed form a Gaussian "Arena".

Note1: Although the full n×n covariance matrix is symmetric $K_{jk} = K_{kj}$ with respect to all n indices (i.e., $K = K^T$), this is no longer true for the 2×2 partitioned components $K_{(X)(Y)} \neq K_{(Y)(X)}$ as evidenced by the fact that K_{XY} and K_{YX} have transposed dimensions. The symmetry of the full matrix only requires that blocks with transposed partition indices in fact be transposes of one another, *i.e.*, $K_{XY}^T = K_{YX}$ which is possible now because these "transposed partitioned" matrices have the *same dimensions*.

Note 2: The state/measurement (**X**)(**Y**) block indexing notation usually drops the parentheses and labels with **X,Y**; this abbreviated notation should not be conflated with the transformation to canonical coordinates **Y**=A**X** which takes place in n-dimensions; clearly, the abbreviated block component labels $X = X_{(1)}$ and $Y = X_{(2)}$ do not satisfy such a coordinate transformation since they have different dimensions k and (n-k) respectively.

10.2.1   Gauss-Markov Estimator

Gauss-Markov Estimator

New RVs:

$$\mu_{X|Y} \rightarrow \hat{X} = \mu_X + K_{XY}K_{YY}^{-1}(\hat{Y} - \mu_Y) \qquad \textbf{Estimator RV}$$

$$e = X - \hat{X} = (X) - [\mu_X + K_{XY}K_{YY}^{-1}(\hat{Y}) - \mu_Y)] \quad \textbf{Error RV}$$

Note: The "Estimator" and the "Error" depend upon the specific values of X="x" and Y="y" and hence generate samples of two new random variables $\hat{X}$ & e whose statistics can be inferred from those of X and Y.

Following remarkable properties can be shown for these RVs

Error **e** and Conditional Mean Estimator $\hat{X}$ satisfy the following:

1) $E[e\hat{X}] = 0$ & $E[eY] = 0$ $e \perp \hat{X}$ & $e \perp Y$ *i.e.*, **e** is uncorrelated with the
   "orthogonal"   estimator $\hat{X}$ and the data Y

2) $K_{\hat{X}Y} = K_{XY}$ Estimator $\hat{X}$ and RV X   have **same correlation with measurements** Y

3) **Distributions** for $\hat{X}$ and e satisfy "Pythagorean Right Triangle Relationship" as shown

$$\hat{X} = N(\mu_X, \underbrace{K_{XY}K_{YY}^{-1}K_{YX}}_{\equiv Q}) = N(\mu_X, Q)$$

$$e = N(0, \underbrace{K_{XX} - K_{XY}K_{YY}^{-1}K_{YX}}_{\equiv P}) = N(0, P)$$

$$X = \hat{X} + e$$

X : $N(\mu_X, K_{XX})$ *Random Variable*

e : $N(0, P)$ Error

$\hat{\textbf{X}}$: $N(\mu_X, Q)$ Gauss-Markov Estimator

Gaussian Means & Variances Add

$$N(\mu_X, K_{XX}) = N(\mu_X, Q) + N(0, P)$$

Recall Inverse channel with scalars X & Y: $\boxed{X = \rho Y + V \quad N(0,1) = N(0, \rho^2) + N(0, 1 - \rho^2)}$

The conditional mean is evaluated for a specific "realization" of a Gaussian RV X="x" and Y="y" and hence looking at many realizations allows us to consider the conditional mean $\mu_{X|Y}$ as a random variable itself. Thus we replace the specific realizations $\mu_{X|Y}$ and "y" in the update equation by RVs denoted respectively as X_{hat} and Y as shown in the first equation. Now the difference between the true state X and the conditional mean estimate of that state X_{hat} is a RV that represents the Estimation Error **e** $= X - X_{hat}$ as shown in the second equation.

These two equations can be shown to have the following remarkable properties: 1) the error is uncorrelated with either the estimator X_{hat} or the data Y, 2) the X_{hat} estimator and the true state X correlate with the measurements in the same way, and 3) the distributions for the RVs X_{hat} and **e** satisfy a "Pythagorean Right Triangle Relationship between their Gaussian $N(\mu_X, K_{XX})$ designations. Looking at the figure, we see the true state X $\sim N(\mu_X, K_{XX})$ is on the hypotenuse, while the estimator $X_{hat} \sim N(\mu_X, Q)$, with variance Q= $K_{XY}K_{YY}^{-1}K_{YX}$ is in the plane, and the error **e** $\sim N(0, P)$ with variance P= $K_{XX} - K_{XY}K_{YY}^{-1}K_{YX}$ is perpendicular to the plane. The vector relation is **X** = **X**$_{hat}$ + **e** which in fact forms a vector right triangle and the means and variances add as follows

$$\mu_X = \mu_X + 0 \quad \text{and} \quad K_{XX} = P + Q = (K_{XX} - K_{XY}K_{YY}^{-1}K_{YX}) + (K_{XY}K_{YY}^{-1}K_{YX}).$$

For the normal distributions this may be written in the suggestive form as a statistical right triangle relationship

$$N(\mu_X, K_{XX}) = N(\mu_X, Q) + N(0, P).$$

Recall results for the scalar communication ***inverse channel*** with a single input Y $\sim$ N(0,1), a single output X $\sim$ N(0, 1), and zero-mean noise V $\sim$ N(0, $1-\rho^2$) which satisfied the scalar relation X=ρY+V and produces an analogous statistical relation N(0,1) = N(0,ρ^2) + N(0,1-ρ^2) which sums variances and means to yield

$$N(0+0, \rho^2 + 1 - \rho^2) = N(0,1).$$

10.2.2   Representation Models of Gaussian Processes

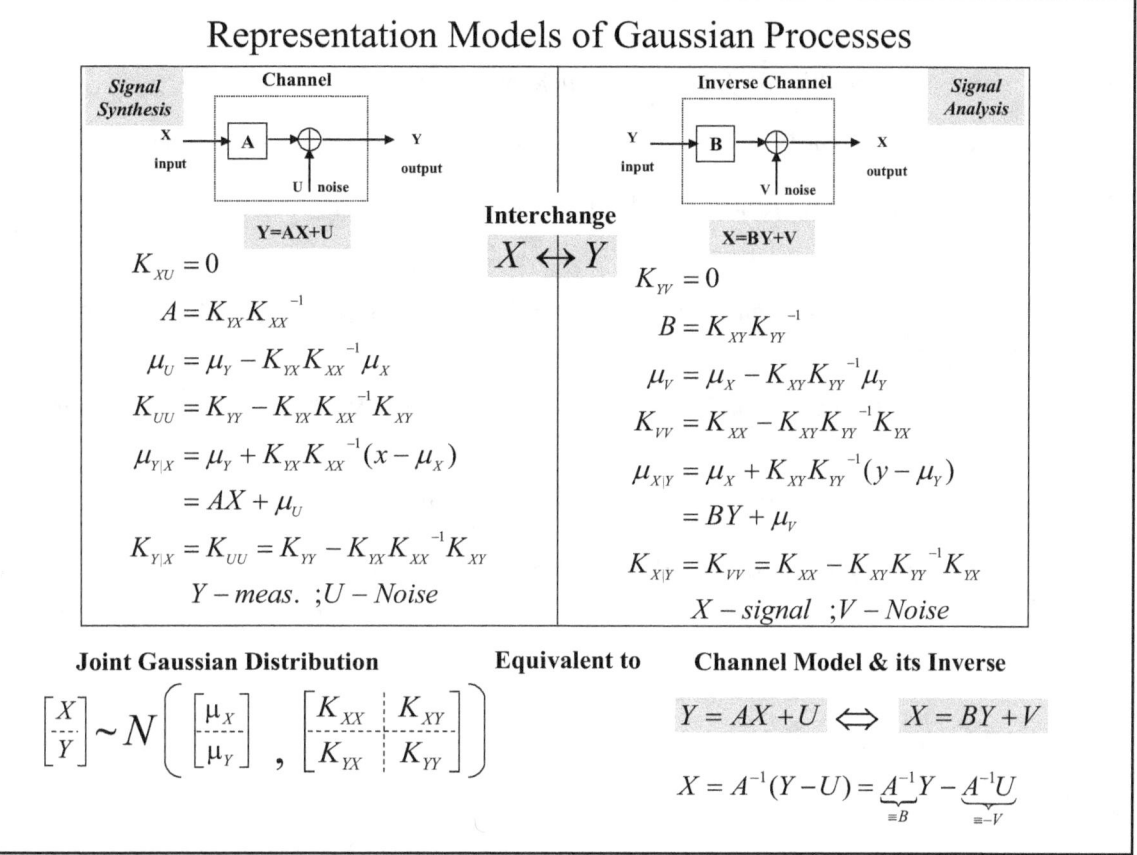

Representation Models of Gaussian Processes

This summary slide is an extension of the single input, single output communication channel we previously discussed (Slide#9-26). The summary box at the top of this slide shows the X, Y or input/output symmetry for the conditional means and variances that result from their being components of a Multivariate Gaussian Vector.

The Gauss-Markov Theorem corresponds to the Inverse Analysis Channel in the right box obtained by direct calculation using the partitioned matrices in the canonical coordinates. This allows a factoring of the two partitions and the subsequent expressions for the conditional distribution $f_{X|Y}(x|y)$ which is again a multivariate Gaussian, and yields the conditional mean vector $\mu_{X|Y} = E[X|Y]$, and conditional covariance cov($X|Y$) representing *Bayesian updates* of the quantities as detailed in the right box.

Clearly, we could just as well have conditioned on the **X** vector and solved for the other conditional $f_{Y|X}(y|x)$ to find the analogous results of the left box. Above these boxed equations we show the Direct (Synthesis) Channel which gives the "output vector **Y** as a noisy version of the input vector **X**" *via* the equation **Y** = A**X** +**U** on the left and the Inverse Channel which gives the "input **X** as a noisy version of the output **Y**" *via* the equation **X**= B**Y** +**V** on the right.

As noted in the center of this summary box both the Direct and Inverse channels have the same partitioned covariance matrix K and have conditionals related by the simple exchange of variables X↔Y.

Below the table we show the elements of the two equivalent multivariate Gaussian representations corresponding to the *Synthesis* or Direct Channel Model on the left and and the *Analysis* or Inverse Channel Model on the right.

11 References

1. "*A First Course in Probability*, 7th Ed., " Ross, Sheldon, Prentice-Hall, 2006.

2. "*An Introduction to Applied Probability*," Roberts, Robert A., Addison-Wesley, 1992.

3. "*Probability Models and Applications*, 2nd Ed.," Olkin, I., Gleser, L.J., Derman, C., Prentice-Hall, 1994.

4. "*Probability, Statistics, and Random Processes for Electrical Engineering*, 3rd Ed.," Leon-Garcia, Leon, Prentice-Hall, 2008.

5. "*Introduction to Mathematical Statistics*, 5th Ed.," Hogg, Robert V., Craig, Allen T., Prentice-Hall, 1995.

6. "*Discrete Probability: Lecture Slide Notes*," Morganstern, Ralph E., CreateSpace Publications, 2013.

7. "*Geometrical Line-of-Sight Error Analysis*," Morganstern, Ralph E., Lockheed-Martin Technical Memorandum EM-01493, Sept., 1997.

Index

12 Index

Index

Index

Index

Index